D1330820

Edexcel
GCSE MATHEMATICS
FOUNDATION COURSE

1
2
3
4
5
6
7
8
9
10
11
12A
12B
13

Success through qualifications

About this book

This book is designed to provide you with the best possible preparation for your Edexcel GCSE Mathematics Examination. The authors are examiners and coursework moderators themselves and have a good understanding of Edexcel's requirements.

Finding your way around

To help you find your way around when you are studying and revising use the:

- **edge marks** (shown on the front pages) – these help you get to the right unit quickly;
- **contents list** – this lists the headings that identify key syllabus ideas covered in the book so you can turn straight to them. (Codes are included to show which part of the programmes of study and left-hand column of the syllabus these relate to. For example **S4d** means the content relates to Shape, space and measures section **4** understanding and using measures subsection **d**.)

Remembering key ideas

We have provided clear explanations of the key ideas and techniques you need throughout the book. **Key points** you need to remember are listed in a **summary** at the end of each unit and marked like this where they appear in the units themselves:

■ **Each digit in a number has a face value and a place value.**

Exercises and examination questions

In this book questions are carefully graded so they increase in difficulty and gradually bring you up to examination standard.

- **past examination questions** are marked with an [E];
- **worked examples** and **worked examination questions** show you how to answer questions;
- **examination practice paper** – this is included to help you prepare for the examination itself;
- **answers** are included at the end of the book.

Investigations and information technology

Three units focus on particular skills required for your course and examination:

- **using and applying mathematics** (unit 12A) – shows how investigative work is assessed, and the skills required to carry out such work;
- **data handling** (unit 12B) – shows you how to approach your data handling coursework project, and how the project will be assessed;
- **calculators and computers** (unit 24) – shows you how to use a variety of methods of solving problems using calculators and computers.

14
15
16
17
18
19
20
2
22
23
24
25

Contents

1 Number

2 Algebra 1

3 Angles and turning

4 Fractions

5 Two-dimensional shapes

6 Decimals

7 Measure 1

8 Collecting and recording data

12A Using and applying mathematics

12B Data handling

13 Measure 2

14 Percentages

15 Algebra 3

16 Pie charts

17 Ratio and proportion

18 Symmetry

19 Measure 3

20 Averages

21 Algebra 4

22 Transformations

23 Probability

24 Calculators and computers

25 Scatter diagrams

Examination practice paper (non-calculator)

Examination practice paper (calculator)

Formulae sheet

Answers

Index

Heinemann Educational Publishers
Halley Court, Jordan Hill, Oxford OX2 8EJ
a division of Reed Educational & Professional Publishing Ltd
Heinemann is a registered trademark of Reed Educational & Professional Publishing Ltd

OXFORD   MELBOURNE   AUCKLAND
JOHANNESBURG   BLANTYRE   GABORONE
IBADAN   PORTSMOUTH NH (USA)   CHICAGO

First published 2001

ISBN 0 435 53269 3

02   01
10 9 8 7 6 5 4 3 2

Designed and typeset by Tech-Set Ltd, Gateshead, Tyne and Wear
Original edition produced by Gecko Limited, Bicester, Oxon
Cover design by Miller, Craig and Cocking
Printed in Great Britain by Bath Press Colourbooks, Glasgow

Acknowledgements

The publisher's and authors' thanks are due to Edexcel for permission to reproduce questions from past examination papers. These are marked with an [E]. The answers have been provided by the authors and are not the responsibility of Edexcel.

p2: Powerstock Zefa; p4: Robert Harding/Simon Harris; p24 Robert Harding/G. Renner; p24 Robert Harding/Robert Francis; p40: Tony Stone Worldwide; p49: Tony Stone Worldwide; p95: Rex Features; p97: Tony Stone Images; p99: Empics/Steve Mitchell; p100: Associated Sports Photography; p191: Richard Greenhill; p200: PA News Photo Library; p221: Topham Picture Point; p315: Robert Harding/N. Francis; p317: Richard Greenhill; p356: Associated Sports Photography.

Publishing team	Design	Author team	
Editorial	Phil Richards	John Casson	David Kent
Sue Bennett	Colette Jacquelin	Tony Clough	Andrew Killick
Philip Ellaway	Mags Robertson	Gareth Cole	Christine Medlow
Susanna Geoghegan		Ray Fraser	Graham Newman
Maggie Rumble	**Production**	Barry Grantham	Keith Pledger
Nick Sample	David Lawrence	John Hackney	Sally Russell
Harry Smith	Jason Wyatt	Karen Hughes	Rob Summerson
Juliet Smith		Trevor Johnson	John Sylvester
		Peter Jolly	Roy Woodward

Tel: 01865 888058   email: info.he@heinemann.co.uk

1 Number

1.1 Face value and place value

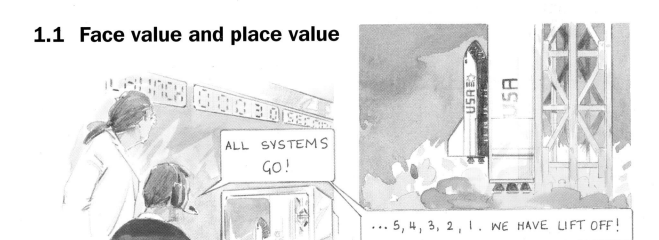

ALL SYSTEMS GO!

...5, 4, 3, 2, 1. WE HAVE LIFT OFF!

■ **Each digit in a number has a face value and a place value.**

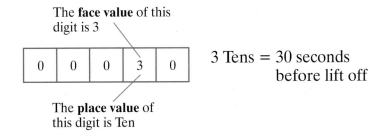

The **face value** of this digit is 3

| 0 | 0 | 0 | 3 | 0 |

The **place value** of this digit is Ten

3 Tens = 30 seconds before lift off

Look at this place value diagram:

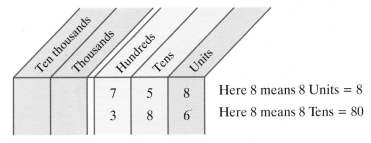

Ten thousands	Thousands	Hundreds	Tens	Units
		7	5	8
		3	8	6

Here 8 means 8 Units = 8
Here 8 means 8 Tens = 80

758 has three digits. It is also called a three figure number.

Exercise 1A

1   Draw a place value diagram and write in:
   (a) a four figure number with a 4 in the Thousands column
   (b) a two figure number with a 3 in the Tens column
   (c) a five figure number with a 1 in the Hundreds column

(d) a three figure number with a 9 in the Units column
(e) a four figure number with a 0 in the Tens column
(f) a five figure number with a 4 in the Hundreds column
(g) a three figure number with a 7 in every column
(h) a four figure number with a 6 in the first and last
columns

1.2 Reading, writing and ordering numbers

Sometimes you will need to write in words a number that has
been written in figures.

Ten thousands	Thousands	Hundreds	Tens	Units	
		9	8	7	Nine hundred and eighty seven
1	6	4	1	2	Sixteen thousand four hundred and twelve

16 412
↑
The thin space separates
Thousands from
Hundreds and makes it
easier to read the number.

Sometimes you will need to write in figures a number that has
been written in words:

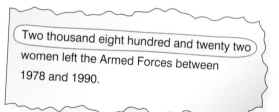

Two thousand eight hundred and twenty two
women left the Armed Forces between
1978 and 1990.

Part of the thirty one thousand
two hundred and fifty eight people
at an outdoor rock concert.

The word numbers from the newspaper can be written in
figures like this:

	Ten thousands	Thousands	Hundreds	Tens	Units	
Two thousand eight hundred and twenty two		2	8	2	2	or 2822
Thirty one thousand two hundred and fifty eight	3	1	2	5	8	or 31 258

Remember: zeros have a face value too.

Thousands	Hundreds	Tens	Units	
5	0	4	2	Five thousand and forty two
	4	0	6	Four hundred and six

Sometimes you will need to rewrite a set of numbers in order of size:

    15   8400   6991   2406   2410   84 000

The size of a number depends on how many digits it has.
The more digits, the bigger the number.
So 84 000 is bigger than 8400 because it has more digits.

If two numbers have an equal amount of digits, the face value of the digit in the highest place value column tells which is the bigger number.
So 8400 is bigger than 6991 because 8 is bigger than 6.

If two numbers have an equal amount of digits and the face value of the digits in the highest place value column is the same, then the face value of the digit in the next place value column tells which is the bigger and so on.
So 2410 is bigger than 2406 because 1 is bigger than 0.

So starting with the biggest number, the list above written in order of size is:

    84 000   8400   6991   2410   2406   15

Exercise 1B

1 Write these numbers in words:
   (a) 36   (b) 95   (c) 598
   (d) 246   (e) 5623
   (f) There are 1251 people in James St. School.

2 Write these numbers in words:
   (a) 709   (b) 890   (c) 6054
   (d) 9201   (e) 26 007   (f) 40 200
   (g) 32 000 (h) 70 090
   (i) The number of school leavers in Axeshire last year
      was 13 406.

3 Write these numbers in figures:

  (a) Sixty three

  (b) Seven hundred and eight

  (c) Seven thousand

  (d) Eighteen thousand six hundred

  (e) Seventy five thousand

  (f) Eight hundred and nine thousand

  (g) Four million

  (h) One million one thousand

  (i) Nine thousand and twenty

  (j) Forty thousand six hundred

4

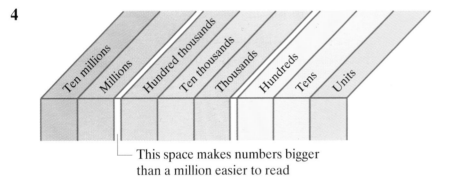

            └── This space makes numbers bigger
                than a million easier to read

The table below gives the populations of five member states of the European Union. Use the place value diagram above to help you write the numbers in words.

	Country	Population
(a)	Belgium	9 910 000
(b)	Luxembourg	403 000
(c)	Spain	37 900 000
(d)	Portugal	9 913 000
(e)	France	54 200 000

5 Write the numbers below in figures:

  (a) The numbers of people employed by a local police authority were:

    •   Traffic wardens: sixty nine

    •   Civilian support staff: one thousand and ten

    •   Police officers: two thousand three hundred and six

(b) The tonnages of three cruise liners are:

- Canberra: forty five thousand two hundred and seventy
- QE 2: eighty three thousand six hundred and sixty three
- Queen Mary: eighty thousand seven hundred and seventy four

(c) The average daily readership of four newspapers was:

- Financial Times: two hundred and ninety four thousand
- Times: six hundred and eighty two thousand
- Daily Mirror: two million six hundred thousand
- Sun: three million nine hundred and ninety thousand

6 Rearrange these lists of numbers into order of size, starting with the largest number:

(a) 86   104   79   88   114   200

(b) 3000   3003   30 300   330 000   3033

(c) 6 000 006   660 000   600 006   990 000   6 102 000

7

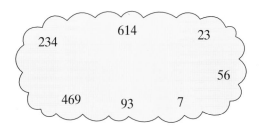

Put the numbers in the cloud in order. Start with the *smallest* number.   [E]

8 This table gives the prices of several popular motor cars:

Car	Price
Vauxhall Astra	£10 530
Volkswagen Polo	£9 110
Ford Granada	£20 270
Rover 623	£19 435
Citroen Xantia	£17 065
Jaguar Sport	£30 479
Mercedes Benz	£31 394
Marcos Mantera	£31 007
Volvo 940	£17 815
Ford Fiesta	£9 010

Rewrite the list in price order, starting with the least expensive.

9 This table shows the numbers of people who died in road accidents in Britain:

Year	1961	1971	1976	1982	1984
Deaths	20 121	19 246	17 099	15 113	14 259

In which year did:
(a) the smallest number of people die in road accidents
(b) more than twenty thousand people die in road accidents
(c) between fifteen and sixteen thousand people die in road accidents
(d) less than fifteen thousand die in road accidents?

10 The table below gives the areas, in km², of five member states of the European Union.
(a) Write the area of the countries in words.
(b) Write the countries given in order of size.

Country	Area
Belgium	30 513
Luxembourg	2576
Spain	504 782
Portugal	92 082
France	547 026

1.3 Combining numbers

Using a number line

Here is a number line. It goes up from 0 to 20.

It goes down from 20 to 0.

You can use the number line to increase or decrease numbers.

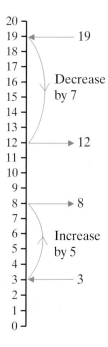

Example 1

(a) Increase 3 by 5.

(b) Decrease 19 by 7.

(a) Start at 3 on the line.
From 3 count 5 upwards.
The answer is 8.

(b) Start at 19 on the line.
From 19 count 7 downwards.
The answer is 12.

Exercise 1C

Draw number lines going from 0 to 20 to help you with these questions:

1 Increase:

    (a) 1 by 6   **(b)** 5 by 4   **(c)** 9 by 10

    (d) 8 by 5   **(e)** 12 by 3

2 Decrease:

    (a) 13 by 4   **(b)** 7 by 6   **(c)** 18 by 7

    (d) 9 by 6   **(e)** 15 by 8

3 What increase moves:

    (a) 6 to 10   **(b)** 3 to 11   **(c)** 12 to 14

    (d) 18 to 20   **(e)** 16 to 19

4 What decrease moves:

    (a) 7 to 4   **(b)** 8 to 1   **(c)** 12 to 3

    (d) 16 to 1   **(e)** 19 to 8

5 What change moves:

    (a) 6 to 11   **(b)** 10 to 16   **(c)** 19 to 14

    (d) 13 to 4   **(e)** 20 to 0

Remember to say whether your change is an increase or a decrease.

Adding

Here are some of the different ways of writing 6 add 9:

Find the **sum** of 6 and 9.
Work out 6 **plus** 9.
Find the **total** of 6 and 9.
Add 6 and 9.
Work out 6 **+** 9.

Only use a calculator to check your answers when you have *finished* each exercise.

Exercise 1D

Work out the answer.

1 Find the total of 26 and 17.

2 Work out 58 plus 22.

3 Work out 236 + 95.

4 In four maths tests, Anna scored 61 marks, 46 marks, 87 marks and 76 marks.
   How many marks did she score altogether?

5 Find the sum of all the single figure numbers.

6 In a fishing competition, five competitors caught 16 fish, 31 fish, 8 fish, 19 fish and 22 fish.
   Find the total number of fish caught.

7 The number of passengers on a bus was 36 downstairs and 48 upstairs.
   How many passengers were on the bus altogether?

8 For her birthday party, Anna provided 86 bottles of cola, 58 bottles of orangeade and 72 bottles of mineral water.
   How many bottles did she provide in total?

9 Work out 38 + 96 + 127 + 92 + 48.

10 The numbers of caravans sold by the six branches of a nationwide caravan dealer were 86, 43, 75, 104, 38 and 70.
    What was the total number of caravans sold?

Subtracting

Here are some different ways of writing 38 subtract 16:

38 **minus** 16.
Take 16 **from** 38.
38 **–** 16
the **difference** between 38 and 16.
How many is 16 **less** than 38?
38 **take away** 16.

Exercise 1E

Work out the answer.

1   Take 1007 from 2010.

2   How much is 7260 minus 4094?

3   Work out 611 – 306.

4   Take 17 from 29.

5   When playing darts, James scored 111 with his first three darts, Iris scored 94 with her first three darts and Nadine scored 75 with her first three darts.
   (a) How many more than Nadine did Iris score?
   (b) What is the difference between James' and Iris' scores?
   (c) How many less did Nadine score than James?

6   The winner of a darts match is the first one to score 601 points.
   James has now scored 513, Iris 542 and Nadine 468.
   (a) How many is James short of 601?
   (b) How many more does Iris need to score to win?
   (c) How many is Nadine's total less than 601?

Name	Score
James	513
Iris	542
Nadine	468

7   The 'thermometer' shows the yearly progress of a Cathedral Appeal towards a Target of £100 000.
   (a) How much did the appeal raise between the end of 1988 and the end of 1989?
   (b) How much money was raised between the end of 1989 and the end of 1990?
   (c) At the start of 1991 how much was still needed to be raised to reach the target at the end of 1992?

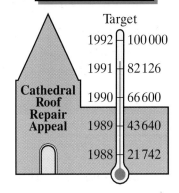

	Target
1992	100 000
1991	82 126
1990	66 600
1989	43 640
1988	21 742

Cathedral Roof Repair Appeal

Multiplying

Here are some different ways of writing 80 multiplied by 16.

Find the **product** of 80 and 16.
Times 16 by 80
Multiply 80 by 16
Work out 80 × 16.

Exercise 1F

Work out the answer.

1 Find the product of 36 and 30.

2 How many is 321 multiplied by 14?

3 Championship Music bought 19 boxes of assorted CDs.
Each box contained 24 CDs. How many CDs did they
buy in total?

4 Work out 106 times 43.

5 What is the product of 63 and 36?

6 Wasim and Jan are training for the school sports day.
During each session Wasim runs 50 metres 30 times and
Jan runs 7 times round the 400 metre track.

(a) How far does Wasim run each session?

(b) How far does Jan run each session?

By sports day, Wasim has trained for 12 sessions and Jan
for 20 sessions.

(c) How far has each run in training by sports day?

7 Packets of Cheese Flips come in three sizes.
There are 36 Cheese Flips in the Economy size,
3 times as many in the Large size and 6 times as
many in the Family size.
How many Cheese Flips are in each size packet?

8 The distance from Mr. Singh's home to work is 17 miles.
He makes the journey 12 times a week.
How far does he travel:

(a) in one week

(b) in ten weeks

(c) in a year? (Hint: 52 weeks make 1 year.)

Dividing

Here are some of the different ways of writing 140 divided by 20.

Divide 140 by 20.
Share 140 by 20. ◄
How many times does 20 go into 140?
Work out 140 ÷ 20.
Work out $\frac{140}{20}$.
How many 20s are there in 140?

Note: this is written
140 ÷ 20 not 20 ÷ 140.

Exercise 1G

Work out the answer.

1 (a) Work out 315 ÷ 15.
   (b) How many 50s make 750?
   (c) Work out $\frac{680}{17}$.
   (d) Work out 600 divided by 30.
   (e) Divide 8 into 112.

2 Five people shared a prize draw win of £2400 equally.
   How much did each person receive?

3 Edward packs books into boxes.
   Each box will hold 24 books.
   How many boxes will he need to pack:
   (a) 240 books
   (b) 720 books
   (c) 864 books?

4 A delivery van is used to carry crates of books
   from a printer to Heinemann's warehouse.
   The van can carry 18 crates at once.
   How many deliveries are needed to transfer:
   (a) 126 crates
   (b) 234 crates
   (c) 648 crates?

5 A packing case will hold 72 economy size boxes,
   24 large size boxes or 12 family size boxes.
   How many packing cases would be needed to pack:
   (a) 864 economy size boxes
   (b) 984 large size boxes
   (c) 960 family size boxes?

1.4 Solving number problems

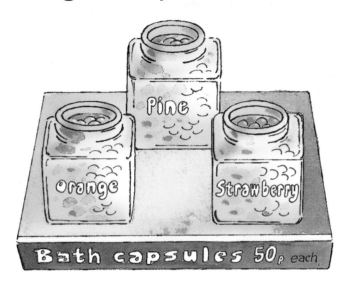

Louise chooses 7 bath capsules.

She could find the cost by adding:

50p + 50p + 50p + 50p + 50p + 50p + 50p

or by multiplying:

50p × 7

Here multiplying is quicker than adding.

When you are solving number problems like this you need to choose whether to add, subtract, divide or multiply.

Example 2

The key words here are **equal shares**. They tell you to **divide**:

£2000 ÷ 5 = £400

Exercise 1H

This exercise will help you choose whether to add, subtract, multiply or divide.

Look at the size of each answer when you have worked it out. This may help you check whether you have chosen + − × or ÷ correctly.

1 Owen keeps 12 pet hamsters and 21 pet rabbits.
How many pets has he altogether?

2 Ranjit and Jane both collect stamps.
Ranjit has 1310 stamps and Jane has 942 stamps.
How many more stamps has Ranjit than Jane?

3 42 packing cases of tins of beans are delivered to Simpsons Superstore. The packing cases each hold 48 tins of beans.
How many tins of beans are delivered altogether?

4 A librarian has 343 cassettes to display in the music library. They fill exactly 7 shelves.
How many cassettes are on each shelf?

5 Robina was given 20 face paints for her birthday. 4 of the paints were blue, 6 were green and 7 were red. The others were brown.

   (a) How many of Robina's paints were brown?

   (b) How many more green paints than blue paints did she have?

   Robina gave 2 paints of each colour to her sister Amanda.

   (c) How many paints did Robina give to Amanda altogether?

   (d) How many paints did Robina have left?

6 Jason bought three boxes of toffees. There were 30 toffees in each box. Jason ate 14 toffees himself then shared all the rest equally between himself and his three sisters.

   (a) How many toffees did Jason have to start with?

   (b) When they were shared out, how many toffees did each person get?

7 This table shows the numbers of pupils in each year group at Gordon School:

Year 7	Year 8	Year 9	Year 10	Year 11
112	121	104	98	126

(a) How many pupils are at the school altogether?

(b) How many less pupils are in Year 10 than in Year 8?

Each class in Year 11 has 21 pupils.

(c) How many classes are there in Year 11?

(d) There are 3 times as many pupils in Bennett School as in Gordon School.
How many pupils are in Bennett School?

8 A bus company owns twelve 48–seater coaches.
Every Saturday all the coaches are used to take people to the seaside.

(a) How many people can be carried at the same time in the 12 coaches?

One Saturday seven of the coaches each carried 39 people. The other coaches each had 11 empty seats.

(b) How many people were carried altogether that Saturday?

1.5 Rounding numbers

Sometimes an exact answer is not needed because:

- an approximate answer is good enough;
- *or* an approximate answer is easier to understand than an exact answer;
- *or* there is no exact answer.

To give an approximate answer you can **round** to the nearest 10, 100, 1000 and so on.

(These numbers are called 'powers of ten' and are explained in Section 1.9.)

■ **To round to the nearest 10:**
look at the digit in the Units column.
- If it is less than 5 round down.
- If it is 5 or more round up.

Example 3

Round 687 to the nearest ten.

687 to the nearest ten is 690.

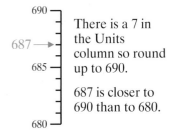

There is a 7 in the Units column so round up to 690.

687 is closer to 690 than to 680.

■ **To round to the nearest 100:**
look at the digit in the Tens column.
- If it is less than 5 round down.
- If it is 5 or more round up.

Example 4

Round 7650 to the nearest hundred.

7650 to the nearest hundred is 7700.

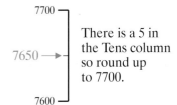

There is a 5 in the Tens column so round up to 7700.

■ **To round to the nearest 1000:**
look at the digit in the Hundreds column.
- If it is less than 5 round down.
- If it is 5 or more round up.

Example 5

Round 614 652 to the nearest thousand.

614 652 to the nearest thousand is 615 000.

Exercise 1I

1 Round to the nearest ten:
   (a) 57   **(b)** 63   **(c)** 185
   (d) 194   **(e)** 991   **(f)** 2407

2 Round to the nearest hundred:
   (a) 314   **(b)** 691   **(c)** 2406
   (d) 3094   **(e)** 8777 **(f)** 29 456

3 Round to the nearest thousand:
   (a) 2116   **(b)** 36 161   **(c)** 28 505
   (d) 321 604 **(e)** 717 171 **(f)** 2 246 810

4

HOW IT COMPARES		Launched	Length	Tonnage	Speed
	Queen Elizabeth	1938	1,029 ft	83,673	32 knots
	Queen Mary	1934	1,019 ft	80,774	32 knots
	New P&O cruise liner	(1994-5)	850 ft	67,000	24 knots
	QE2	1967	963 ft	65,863	32.5 knots
	United States	1951	990 ft	53,329	40 knots
	Canberra	1960	820 ft	45,270	29.3 knots

Source: Great Passenger Ships of the world, A. Kludes

For each of these liners, round:

(a) the lengths to the nearest ten feet

(b) the tonnages to the nearest hundred tonnes

(c) the speeds to the nearest ten knots.

5 The table gives the areas and populations of five European Union states.

Country	Area (km^2)	Population
Greece	131 944	9 804 266
Italy	301 225	57 436 280
Netherlands	33 812	14 292 416
Germany	356 733	80 180 660
Ireland	70 283	3 521 000

Round:

(a) the areas to the nearest thousand km^2

(b) the populations to the nearest hundred thousand.

6 Give an approximate answer for each number below. Explain why you chose your answer.

(a) James has a flock of 142 chickens.

(b) Mrs Wilson sold 306 portions of fish and chips.

(c) Asif needed 6318 bricks to build his new bungalow.

(d) A crowd of 35 157 spectators watched Chelsea last night.

(e) A pop group earned £45 376 290 in a year.

(f) Newhouse General Hospital treated 13 296 patients last year.

Rounding to 1 significant figure

Sometimes you will be asked to round a number to *1 significant figure*.

- To write a number to 1 significant figure, look at the place value of the first digit and round the number to this place value.

Example 6

Write these numbers to 1 significant figure.

(a) 32   **(b)** 452   **(c)** 8780

(a) ③ 2

    The first digit is in the tens column.

    32 to the nearest ten is 30.
    32 to 1 significant figure is 30.

(b) ④ 52

    The first digit is in the hundreds column, so round to the nearest hundred.

    452 to 1 significant figure is 500.

(c) ⑧ 780

    The first digit is in the thousands column, so round to the nearest thousand.

    8780 to 1 significant figure is 9000.

Is your answer reasonable?

When you do a calculation it helps to have a rough estimate of what answer to expect.

■   **Rounding is used to help you estimate an answer.**
    To estimate an answer round each number in the question to 1 significant figure.

Example 7

Estimate the answer to $\dfrac{289 \times 96}{184}$

Rounding each of the numbers to 1 significant figure gives:

$$\frac{300 \times 100}{200}$$

Working out the simplified calculation gives 150.

Example 8

Work out the exact value of $\dfrac{18 \times 104}{48}$ by calculator.

Rough check your answer by estimating

$\dfrac{18 \times 104}{48}$ is about $\dfrac{20 \times 100}{50} \approx 40$

By calculator:

$\dfrac{18 \times 104}{48} = \dfrac{1872}{48} = 39$

Exercise 1J

1   Write down these numbers to 1 significant figure.
   (a) 12   (b) 49   (c) 4   (d) 203
   (e) 4960   (f) 501   (g) 3497   (h) 65
   (i) 6034   (j) 8921·   (k) 78 321   (l) 81 476

2   There are 165 pupils in Year 11 in Russell High School.
   Write this number to 1 significant figure.

3   The population of Clifton is 2437. What is the population
   of Clifton to 1 significant figure?

4   Showing all your rounding, make an estimate of the
   answer to:

   (a) $\dfrac{63 \times 57}{31}$ (b) $\dfrac{206 \times 311}{154}$ (c) $\dfrac{9 \times 31 \times 97}{304}$

   (d) $\dfrac{2006}{12 \times 99}$ (e) $\dfrac{498}{11 \times 51}$ (f) $\dfrac{103 \times 87}{21 \times 32}$

5   For each of the following calculations:
      (i) work out the exact value by calculator
      (ii) rough check your answer by estimating.

   (a) $\dfrac{201 \times 96}{51}$ (b) $\dfrac{11 \times 999}{496}$ (c) $\dfrac{146 \times 51}{69}$

   (d) $\dfrac{1000}{9 \times 12}$ (e) $\dfrac{5206}{131 \times 7}$ (f) $\dfrac{913 \times 81}{39 \times 298}$

If you use your calculator
to find an answer you can
check it by estimating.

6 A football grandstand has 49 rows of seats. Each row seats 98 people.

(a) Estimate the capacity of the grandstand.

(b) Is your estimate bigger or smaller than the actual number? Explain your answer.

7 For Inglefield's school outing 1335 pupils went to the seaside by coach. Each coach carried 47 people. Estimate the number of coaches needed.

8 Thelma worked out the answer to the sum
$916 \times 402 \div 1010$ on her calculator.
Her answer was 36.458
By estimating show whether her answer was right.

9 The Headteacher lives 16 miles from Inglefield School. He went to school on 195 days last year. Estimate how many miles he travelled to and from school.

10 A machine operator produces 121 memory chips every hour. He works 39 hours each week for 47 weeks in the year. Estimate the operator's annual chip production.

1.6 Even, odd and prime numbers

You need to be able to recognize these types of numbers:

- **Even** numbers – all divide exactly by 2: 2, 4, 6, 8, 10, . . .
- **Odd** numbers – none of these divide exactly by 2: 1, 3, 5, 7, . . .
- **Prime** numbers – only divide exactly by two numbers: themselves and 1: 2, 3, 5, 7, 11, . . .

Note: the number 1 is *not* a prime number because it can only be divided exactly by *one* number – itself.

Exercise 1K

1 Write down any even numbers from the list:
2, 18, 37, 955, 1110, 73 536, 500 000

2 Write down any odd numbers from the list:
108, 537, 9216, 811, 36 225, 300 000

3 The first six prime numbers are:
2, 3, 5, 7, 11, 13.
Write down the next seven prime numbers.

4 Here is a list of numbers.
15, 20, 25, 30, 37, 39, 49, 69, 70, 71, 400, 450
Write down:
(a) the largest even number
(b) the largest odd number
(c) the largest prime number.

1.7 Factors and multiples

• A **factor** is a whole number which will divide into another
whole number without a remainder.
For example:
1, 3, 5, 15 are all factors of 15 because they divide into 15
without a remainder.

• **Multiples** of 5 are made by multiplying the whole numbers
1, 2, 3 . . . by 5 like this:

$$1 \times 5 = \ 5$$
$$2 \times 5 = 10$$
$$3 \times 5 = 15$$

These are multiples of 5.

Multiples of 7 are 7, 14, 21 . . .

• A common factor is a whole number which will divide into
more than one other whole number without a remainder.
For example: Factors of 12 are 3 and 4.
Factors of 33 are 3 and 11.
3 is a common factor of 12 and 33.

Exercise 1L

1 Find one factor other than 1 of:
(a) 9   (b) 24   (c) 32   (d) 55   (e) 108   (f) 625

2 Find all six factors of 12.

3 Find all the factors of:
   (a) 32   **(b)** 200   **(c)** 340 **(d)** 1000

4 Find common factors of:
   (a) 4 and 6   **(b)** 9 and 12   **(c)** 15 and 25
   (d) 6 and 14   **(e)** 14 and 35

5 Find common factors of:
   (a) 20 and 30   **(b)** 90 and 100   **(c)** 12 and 28
   (d) 6 and 18   **(e)** 18 and 45

6 **(a)** Find two common factors of 20 and 50.
   (b) Find two common factors of 12 and 40.
   (c) Find two common factors of 70 and 105.
   (d) Find two common factors of 30 and 42.
   (e) Find two common factors of 18 and 42.

7 Find three multiples of 9.

8 Find the nearest number to 100 that is a multiple of 9.

9 Find:
   (a) three multiples of 6 that are bigger than 50
   (b) three multiples of 20 that are between 1000 and 2000
   (c) three multiples of 15 that are bigger than 100.

10 Write down:
   (a) three numbers that have a factor of 8
   (b) three numbers that have a factor of 150
   (c) three numbers that only have odd number factors.

11 **(a)** Copy this number square:

10	20	30	40	50
60	70	80	90	100
110	120	130	140	150
160	170	180	190	200
210	220	230	240	250

   (b) Draw a line — through any number with a factor of 3.
   (c) Draw a line | through any number with a factor of 20.
   (d) Draw a line \ through any multiples of 30.
   (e) Draw a line / through any multiples of 90.
   (f) Which number do all four lines pass through?

1.8 Squares, cubes and square roots

- A **square number** is the result of multiplying one number by itself.
  49 is a square number because $7 \times 7 = 49$
  Sometimes 7×7 is written as 7^2 (seven squared).

- A **cube number** is the result of multiplying one number by itself, then multiplying by the number again.
  343 is a cube number because $7 \times 7 \times 7 = 343$
  Sometimes $7 \times 7 \times 7$ is written as 7^3 (seven cubed).

- A **square root** is a number that has been multiplied by itself to make another number.

  $7 \times 7 = 49$

  7 is the positive square root of 49
  The sign $\sqrt{}$ is usually used for square root.
  $\sqrt{49} = 7$ says the square root of 49 is 7.

This is a 3 by 3 square:

○ ○ ○
○ ○ ○
○ ○ ○

$3 \times 3 = 3^2 = 9$
is a square number.

Exercise 1M

1 Find:
   (a) the next three square numbers after 1, 4, 9, —, —, —.
   (b) the square numbers represented by:
       (i) 10^2 (ii) 20^2 (iii) 30^2
   (c) the square number nearest to:
       (i) 80   (ii) 150   (iii) 200
   (d) the next three cube numbers after 1, 8, 27, —, —, —.
   (e) the cube numbers represented by:
       (i) 8^3 (ii) 10^3 (iii) 20^3
   (f) the cube number nearest to:
       (i) 80   (ii) 150   (iii) 200

2 (a) Draw a number square from 1 to 100 on squared paper.
   (b) Draw a ○ round all the square numbers in the square.
   (c) Draw a — through all the cube numbers in the square.
   (d) Which two numbers are both square and cube numbers?

3 State the positive square root of:
   (a) 4   (b) 100   (c) 36   (d) 25   (e) 400

4 State the value of:
   (a) $\sqrt{36}$ (b) $\sqrt{81}$ (c) $\sqrt{225}$ (d) $\sqrt{9}$ (e) $\sqrt{64}$

1.9 Indices and powers

In Section 1.8, you learned that 7×7 can be written as
7 squared or 7^2.

- **The ^{2} is called an *index* or a *power*.
  It tells you how many times the given number must be
  multiplied by itself.**

7^2 says seven squared (or seven to the power two).
7^3 says seven cubed (or seven to the power three).
7^4 says seven to the power four.

Example 9
Find the value of 10^5.

$$10^5 = 10 \times 10 \times 10 \times 10 \times 10 = 100\,000$$

Exercise 1N

1 Find the value of:

    (a) 5 squared   **(b)** 2 cubed

    (c) 10 to the power 4   **(d)** 10 to the power 6

    (e) 3 cubed

2 Find the value of:

    (a) 4^2 **(b)** 10^2 **(c)** 5^3 **(d)** 10^4 **(e)** 10^7

    (f) 1^2 **(g)** 1^3 **(h)** 3^2

3 Complete the table below for powers of 10.

Power of 10	Index	Value	Value in words
10^1		10	
10^2			A hundred
10^3	3		
10^4			
10^5			
10^6		1 000 000	
10^7	7		
10^8			A hundred million

1.10 Negative numbers

The temperature here is 31 degrees below zero Celsius.

- We use negative numbers to represent quantities that are less than zero.

  –31 °C is 31 degrees below zero.

+15 °C
+10 °C
+5 °C
0 °C — This thermometer is showing a temperature of –5 °C
–5 °C
–10 °C
–15 °C

–396 m is 396 m below sea level.

The Dead Sea is the lowest place on Earth. It is 396 m below sea level.

Exercise 1O

1 Write down the largest and the smallest number in each list.
   (a) 5, –10, –3, 0, 4
   (b) –7, –2, –9, –13, 0
   (c) –3, 6, 13, –15, –6
   (d) –13, –2, –20, –21, –5

2 Write the missing two numbers in each sequence.
   (a) 4, 3, 2, 1, —, —, –2
   (b) 10, 7, 4, 1, —, —, –8
   (c) –13, –9, –5, –1, —, —, –11
   (d) 13, 8, 3, –2, —, —, –17
   (e) 21, 12, 3, –6, —, —, –33
   (f) –13, –10, –7, –4, —, —, 5

3 Use this number line to find the number that is:
   (a) 5 more than 2   **(b)** 4 more than –7
   (c) 7 less than 6   **(d)** 2 less than –3
   (e) 6 less than 0   **(f)** 10 more than –7
   (g) 6 more than –6   **(h)** 4 less than –3
   (i) 10 less than 5   **(j)** 1 more than –1

```
 9
 8
 7
 6
 5
 4
 3
 2
 1
 0
-1
-2
-3
-4
-5
-6
-7
```

4 What number is:
   (a) 30 more than –70   **(b)** 50 less than –20
   (c) 80 greater than –50   **(d)** 90 smaller than 60
   (e) 130 smaller than –30 **(f)** 70 bigger than 200
   (g) 170 bigger than –200 **(h)** 100 bigger than –100
   (i) 140 more than –20   **(j)** 200 less than –200?

5 The table gives the highest and lowest temperatures in several cities during one year.

	New York	Brussels	Tripoli	Minsk	Canberra
Highest temperature	27 °C	32 °C	34 °C	28 °C	34 °C
Lowest temperature	–9 °C	–6 °C	8 °C	–21 °C	7 °C

   (a) Which city recorded the lowest temperature?
   (b) Which city recorded the biggest difference between its highest and lowest temperatures?
   (c) Which city recorded the smallest difference between its highest and lowest temperatures?

6 The temperature of the fridge compartment of a fridge–freezer is set at 8 °C. The freezer compartment is set at –10 °C.
   What is the difference between these temperature settings?

7 The deep freeze of a freezer shop should be set at –12 °C. It is set to –18 °C by mistake.
   What is the difference between these temperature settings?

1.11 Adding, subtracting and multiplying negative numbers

You need to be able to add negative numbers.
$-3 + -4$ is the same as $-3 - 4 = -7$

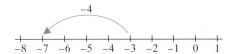

- **When you add two negative numbers the result is a negative number.**

  You also need to be able to subtract negative numbers.

  $2 - -3 = 2 + 3 = 5$
  $-4 - -3 = -4 + 3 = -1$

- **Subtracting a negative number has the same effect as adding a positive number.**

  You also need to be able to multiply negative numbers.

  $2 \times -3 = -6$
  $-4 \times 2 = -8$

- **Multiplying a positive number and a negative number together always gives a negative answer.**

  You also need to be able to multiply two negative numbers together.

  $-3 \times -4 = 12$

- **Multiplying two negative numbers together always gives a positive answer.**

  Here are the rules for multiplying positive and negative numbers:

First number		Second number		Answer
+	×	+	=	+
+	×	−	=	−
−	×	+	=	−
−	×	−	=	+

Exercise 1P

1 Work out:
   (a) $4 + -2$ (b) $3 + -1$ (c) $5 + -6$ (d) $3 + -3$
   (e) $2 + -5$ (f) $7 + -4$ (g) $-4 + -2$ (h) $-3 + -3$
   (i) $-2 + -5$ (j) $-6 + -3$ (k) $-4 + -1$ (l) $-7 + -2$

2 Find the value of:
   (a) $-3 - -5$ **(b)** $-3 - -1$ **(c)** $-1 - -3$ **(d)** $-5 - -2$
   (e) $-7 - -4$ **(f)** $-4 - -2$ **(g)** $-2 - -3$ **(h)** $-2 - -9$
   (i) $-5 - -4$ **(j)** $-4 - -7$ **(k)** $-1 - -3$ **(l)** $-3 - -6$

3 Work out the value of:
   (a) -3×-2 **(b)** -4×5 **(c)** -2×-5 **(d)** 4×-3
   (e) -2×-2 **(f)** -6×-3 **(g)** -6×-5 **(h)** -7×-3
   (i) 5×-6 **(j)** -7×2 **(k)** 6×-7 **(l)** -5×-3

Summary of key points

1 Each digit in a number has a face value and a place value.

The face value
of this digit is 3.

325

The place value of this
digit is Hundred.

2 To round to the nearest 10 look at the digit in the Units column:
   - If it is less than 5 round down.
   - If it is 5 or more round up.

3 To round to the nearest 100 look at the digit in the Tens column:
   - If it is less than 5 round down.
   - If it is 5 or more round up.

4 To round to the nearest 1000 look at the digit in the Hundreds column:
   - If it is less than 5 round down.
   - If it is 5 or more round up.

5 Rounding is used to help you estimate an answer.

6 The ^{2} in 7^2 is called an index or a power.
   It tells you how many times the given number must be multiplied by itself.

7 When you add two negative numbers the result is a negative number.

8 Subtracting a negative number has the same effect as adding a positive number.

9 Mutiplying a positive number and a negative number together always gives a negative answer.

10 Multiplying two negative numbers together always gives a positive answer.

2 Algebra 1

2.1 Using letters to represent numbers

Algebra is the branch of mathematics in which letters are used to represent numbers.

This can help solve some mathematical problems.

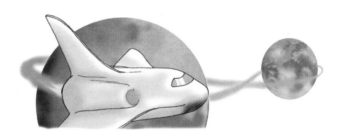

If you know how fast a cyclist is travelling you can use this algebra to work out how far he will get in a given time:

$$s = v \times t$$
(distance = velocity × time)

Very complicated algebra is used to calculate the right orbit for a spacecraft to visit the moon. Luckily you don't have to do this in your GCSE course!

You can use letters even when you do not know the number itself.

Example 1

Peter has some CDs. How many CDs has he got?

You do not know how many CDs Peter has, but using algebra you could say:

'Peter has x CDs.'

Example 2

If Peter buys 3 more CDs, how many will he have altogether?

x CDs and 3 CDs is $x + 3$ CDs.

Example 3

Ann used 6 thank-you cards to thank friends for her Birthday presents. How many thank-you cards has she got left?

You do not know how many thank-you cards she had to start with, but you can say she had y. After using 6 she will have $y - 6$ left.

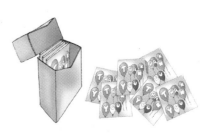

Example 4

Yasmin has some sweets. So has Ali. How many sweets do they have altogether?

Use two different letters: x for Yasmin's sweets and y for Ali's.

Altogether they have $x + y$ sweets.

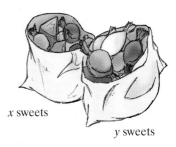

x sweets

y sweets

Exercise 2A

Use algebra to write:

1 3 more than a	**2** x with 4 added
3 x more than 7	**4** 2 less than b
5 c with 3 taken away	**6** p less than q
7 x more than y	**8** 4 together with a
9 $3b$ with 6 subtracted	

2.2 Adding with letters

In algebra you can add letters that are the same. For example:

$a + a$ can be written as $2a$ a means $1a$
$a + a + a$ can be written as $3a$ so $1a + 1a = 2a$

Exercise 2B

Write these in a shorter form. The first one is done for you.

1 $a + a + a + a + a + a = 6a$

2 $p + p + p + p$

3 $b + b + b + b + b$

4 $q + q + q + q + q + q$

5 $c + c$

6 $n + n + n$

7 $w + w + w + w + w$

8 $y + y + y + y + y$

9 $z + z + z + z + z + z + z$

10 $a + a + a + a + a + a + a$

11 $b + b + b + b + b$

12 $p + p + p + p + p + p + p + p$

Write these out in a longer form. The first one is done for you.

13 $4a = a + a + a + a$ **14** $2p$

15 $5p$ **16** $7a$ **17** $2y$ **18** $4q$ **19** $3c$

20 $5d$ **21** $10a$ **22** $5h$ **23** $6g$ **24** $4z$

2.3 Expressions and terms

■ **An algebraic expression is a collection of letters and symbols.** For example:

$$a + 3b - 2c \text{ is an \textbf{algebraic expression}}$$

These are each called **terms**.

Terms which use the same letter or arrangement of letters are called **like terms**.

a and $3a$ are like terms

$2g$ and $8g$ are like terms

Sometimes you can make algebraic expressions simpler by adding or subtracting like terms:

■ **You can combine like terms by adding them:**

$2a + 3a = 5a$

$3b + 4b + b = 8b$

■ **You can combine like terms by subtracting them:**

$5a - 3a = 2a$

$7a - a = 6a$ (Remember: this is $7a - 1a$)

Exercise 2C

Make these expressions simpler by adding or subtracting like terms:

1 $2a + 4a$ **2** $3b + 4b$ **3** $5c + 2c$

4 $5d - 3d$ **5** $7e - 3e$ **6** $5f - f$

7 $3a + 2a + 2a$ **8** $2a + 5a + a$ **9** $5c + 3c + 4c$

10 $6g + 7g + g$ **11** $7g - 3g$ **12** $9s - 6s$

13 $15q - q$ **14** $3p + 7p + 8p$ **15** $12p - 6p$

2.4 Collecting like terms

■ **You can simplify algebraic expressions by collecting like terms together.** For example:

$$2a - 4b + 3a + 5b$$

collect the
a terms $2a + 3a - 4b + 5b$ collect the
                                                            b terms

combine the   combine the
a terms $5a + b$ b terms

Notice that the + or – sign is part of each term.
The minus sign is part of the term $- 4b$

Here are some more examples:

$3a + 5b + 3b + a$	simplifies to:	$4a + 8b$
$5p + 3q - 2p + q$	simplifies to:	$3p + 4q$
$5a + 7 - 3a - 4$	simplifies to:	$2a + 3$

Exercise 2D

Simplify these expressions by collecting like terms:

1 $3a + 4b + 4a + 2b$ **2** $6m + 5n + 3m + 2n$

3 $2p + 3q - p + 2q$ **4** $8e + 6c + 8e$

5 $5y + 7p - 3y + 5p$ **6** $2a + 8g + 3a + 5g - a$

7 $4k + 3q + 5k - 2q$ **8** $9d + 7f - 8d + 3f$

9 $5h + 8 - 2h + 2$ **10** $3f + 2f + 4e - 2e$

11 $7g + 8n - 3g - n$ **12** $5 + g + 2 + 3g - 3 - 2g$

13 $2a + 3a + 4a + 5a$ **14** $3b + b + 5b - 4b$

15 $7c + 8c - 5c - 6c$ **16** $5d + 4d - 3d + 6d - 7d$

17 $2a + 7b + 5a - 6b$ **18** $2c + 4b + 6c - 3b + 7b - 2c$

19 $3p + 2p - 4p + 3 - 2$ **20** $7y + 4z + 2y - 3z + 5y + 3z$

21 $12a - 7a + 9a - 2a + 5a$ **22** $3p + 4p + 5p - 8p - 4p$

23 $6s + 4s - 3s + 5s - 5s$ **24** $3a + 2 + 5a - 1 - 7a - 1$

2.5 Multiplying with letters and numbers

Remember $2a$ is $a + a$ and $3a$ is $a + a + a$, but:

> $2a$ also means   2 lots of a or 2 multiplied by a or $2 \times a$
>
> $3a$ means   3 lots of a or 3 multiplied by a or $3 \times a$

In algebra when you want to multiply two items you just write them next to each other like this:

> $2 \times a$ is written $2a$ $c \times d$ is written cd
>
> $a \times b$ is written ab $3 \times e \times f$ is written $3ef$

Don't forget that $12ab$ is $12 \times a \times b$ *not* $1 \times 2 \times a \times b$!

Exercise 2E

Use multiplication signs to write these expressions out in a longer form. The first one is done for you.

1 $pq = p \times q$	**2** rst	**3** $2ef$	**4** $5abc$
5 $7klm$	**6** $9ab$	**7** $15abc$	**8** $3pqrs$
9 $16st$	**10** $6yz$	**11** $8defg$	**12** $20abcd$

Write these expressions in a simpler form. The first one is done for you.

13 $p \times q = pq$	**14** $e \times f \times g$	**15** $r \times s \times t$
16 $2 \times e \times f$	**17** $2 \times c \times d$	**18** $h \times d \times s$
19 $2 \times s \times f$	**20** $3 \times d \times e \times f$	**21** $4 \times p \times q$
22 $3 \times h \times j$	**23** $5 \times k \times v$	**24** $12 \times r \times s \times t$

2.6 Multiplying algebraic expressions

Sometimes you can simplify algebraic expressions, such as $2a \times 3b$, by multiplying them by each other:

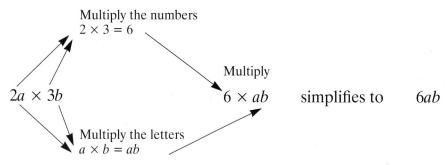

Multiply the numbers
$2 \times 3 = 6$

$2a \times 3b$

Multiply
$6 \times ab$ simplifies to $6ab$

Multiply the letters
$a \times b = ab$

Here are some more examples:

$5a \times 4b$ so $5 \times 4 = 20$
$a \times b = ab$ and $20 \times ab = 20ab$

$3p \times 4q$ so $3 \times 4 = 12$
$p \times q = pq$ and $12 \times pq = 12pq$

Exercise 2F

Simplify these expressions by multiplying them by each other.
The first one is done for you.

1	$2a \times 4b = 8ab$	**2**	$3c \times 5d$
3	$3p \times 4q$	**4**	$5s \times 4t$
5	$6f \times 5g$	**6**	$7p \times 4q$
7	$9m \times 4n$	**8**	$3a \times 4b \times 2c$
9	$3r \times 4s \times 2t$	**10**	$5p \times 4q$
11	$2a \times 6b$	**12**	$5p \times 4q \times 2r$
13	$3t \times 4s$	**14**	$4p \times 7t$
15	$9e \times 5c$	**16**	$8g \times 4q \times 2r$
17	$4d \times 12r$	**18**	$12s \times 5t$
19	$5y \times 6t$	**20**	$2s \times 4t \times 5r$

2.7 Using powers to multiply letters

When you want to multiply two letters that are the same
together you can write them as powers:

	How you write it:	How you say it:	
$a \times a$	a^2	a to the power 2	usually called a squared
$a \times a \times a$	a^3	a to the power 3	usually called a cubed
$a \times a \times a \times a$	a^4	a to the power 4	often called a to the fourth
$a \times a \times a \times a \times a$	a^5	a to the power 5	often called a to the fifth

You can use powers for numbers too:

	How you write it:	How you say it:	
3×3	3^2	3 to the power 2	usually called 3 squared
$3 \times 3 \times 3$	3^3	3 to the power 3	usually called 3 cubed
$3 \times 3 \times 3 \times 3$	3^4	3 to the power 4	often called 3 to the fourth

Exercise 2G

1 Write these expressions in a simpler way using powers.
For example: $d \times d = d^2$

(a) $b \times b \times b$ (b) $p \times p$

(c) $r \times r \times r \times r \times r \times r \times r$ (d) $s \times s \times s \times s \times s$

(e) $q \times q \times q \times q$ (f) $c \times c \times c \times c \times c$

2 Simplify these expressions using powers:

(a) $a \times a \times a \times a$ (b) $s \times s \times s$

(c) $t \times t \times t \times t \times t \times t$ (d) $v \times v \times v$

(e) $f \times f \times f \times f \times f$ (f) $y \times y \times y \times y \times y$

3 Write out these expressions in full.
For example: $c^2 = c \times c$

(a) a^3 (b) a^4 (c) d^2 (d) e^5

(e) f^4 (f) p^5 (g) a^7 (h) s^2

(i) k^6 (j) n^3 (k) n^7 (l) a^{12}

4 Write in a simpler form:

(a) a to the power 5   (b) c to the power 6

(c) d squared   (d) e cubed

(e) f to the power 7

5 Find the values of these powers.
For example: $6^2 = 6 \times 6 = 36$

(a) 3^2 (b) 2^3 (c) 4^2 (d) 5^4

(e) 2^5 (f) 4^3 (g) 2^4 (h) 2^7

(i) 6^3 (j) 5^3 (k) 5^2 (l) 10^5

2.8 Putting in the punctuation

In maths brackets help show the order in which you should
carry out the operations $\div \times + -$ For example:

$$2 + (3 \times 4) \quad \text{is different from} \quad (2 + 3) \times 4$$

even though they both use the same numbers and the $+ \times$
symbols in the same order.

Always deal with the operations in brackets first.
Then $\div$ and $\times$. Then $+$ and $-$.

■ **BIDMAS is a made-up word to help you remember the order of operations:**

BIDMAS

Brackets   Indices   Divide   Multiply   Add   Subtract

Here are some examples:

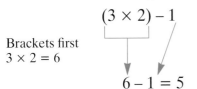

Brackets first
$3 \times 2 = 6$

$6 - 1 = 5$

Brackets first
$2 - 1 = 1$

$3 \times 1 = 3$

This one has no Brackets or Divide, so start with Multiply, then Add, then Subtract.

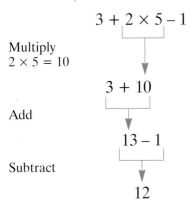

$3 + 2 \times 5 - 1$

Multiply
$2 \times 5 = 10$

$3 + 10$

Add

$13 - 1$

Subtract

12

Work out:

Brackets first $(2 + 3)^2$ $2^2 + 3^2$ Indices first

Indices next $(5)^2$ $4 + 9$ Add next

       25 13

You get different answers. $(2 + 3)^2$ *is not* the same as $2^2 + 3^2$.

■ **When the signs are the same you do them in the order they appear.**

$24 \div 4 \div 2$
$= 6 \div 2$
$= \quad 3$

Exercise 2H

1 Use BIDMAS to help you find the value of these expressions:

  (a) $5 + (3 + 1)$ **(b)** $5 - (3 + 1)$

  (c) $5 \times (2 + 3)$ **(d)** $5 \times 2 + 3$

  (e) $3 \times (4 + 3)$ **(f)** $3 \times 4 + 3$

  (g) $20 \div 4 + 1$ **(h)** $20 \div (4 + 1)$

  (i) $6 + 4 \div 2$ **(j)** $(6 + 4) \div 2$

  (k) $24 \div (6 - 2)$ **(l)** $24 \div 6 - 2$

  (m) $7 - (4 + 2)$ **(n)** $7 - 4 + 2$

  (o) $(15 - 5) \times 4 \div (2 + 3) \times 2$

2 Make these expressions correct by replacing the • with
+ or – or × or ÷ and using brackets if you need to.
The first one is done for you:

(a) $4 • 5 = 9$ becomes $4 + 5 = 9$ (b) $4 • 5 = 20$

(c) $2 • 3 • 4 = 20$ (d) $3 • 2 • 5 = 5$

(e) $5 • 2 • 3 = 9$ (f) $5 • 2 • 3 = 30$

(g) $5 • 4 • 5 • 2 = 27$ (h) $5 • 4 • 5 • 2 = 23$

3 Work out:

(a) $(3 + 4)^2$ (b) $3^2 + 4^2$

(c) $3 × (4 + 5)^2$ (d) $3 × 4^2 + 3 × 5^2$

(e) $2 × (4 + 2)^2$ (f) $2^3 + 3^2$

(g) $2 × (3^2 + 2)$ (h) $\dfrac{(2 + 5)^2}{3^2 - 2}$

(i) $\dfrac{5^2 - 2^2}{3}$ (j) $4^2 - 2^4$

(k) $2^5 - 5^2$ (l) $4^3 - 8^2$

2.9 Using brackets in algebra

Brackets are often used in algebra. For example:

$2 × (a + b)$ means add a to b *before* multiplying by 2

Usually this is written: $2 (a + b)$ without the $×$. This avoids
confusion with the letter x which is used a lot in algebra.

$2(a + b)$ means $2 × a$ + $2 × b$ = $2a + 2b$

Working this out is called **expanding the brackets**. Actually the
brackets disappear!

Here are some more examples of expanding the brackets:

$$3(b + c) = 3 × b + 3 × c = 3b + 3c$$
$$3(2a - b) = 3 × 2a + 3 × -b = 6a - 3b$$

Exercise 2I

Expand the brackets in these expressions.

 1 $2(p + q)$ **2** $3(c + d)$ **3** $5(y - n)$

 4 $3(t + u)$ **5** $7(2p + q)$ **6** $2(3a - 2b)$

 7 $4(2a + b)$ **8** $3(a - 2b)$ **9** $3(4r - 5s)$

10 $10(a - 7b)$ **11** $4(6s + 4t)$ **12** $5(6p + 4q - 2r)$

13 $12(3a + 4b)$ **14** $7(4s - 5t)$ **15** $3(5a - 4b + 2c)$

Adding expressions with brackets

Example 5

Expand the brackets in the expression $2(a + 3b) + 5(2a - b)$.
Then collect like terms.

$$2\,(a + 3b) \quad + \quad 5\,(2a - b)$$

$$2 \times a + 2 \times 3b \quad + \quad 5 \times 2a + 5 \times -b$$

$$2a + 6b \quad + \quad 10a - 5b$$

Collect the a terms
$2a + 10a = 12a$

Collect the b terms
$6b - 5b = 1b = b$

$$12a \quad + \quad b$$

Exercise 2J

Expand the brackets in these expressions and then collect like terms.

Remember:
$4a$ means $a + a + a + a$
and $4 \times a$
$3ab$ means $3 \times a \times b$
p^4 means $p \times p \times p \times p$
$3(c + d)$ means $3c + 3d$

1 $2(a + b) + 3(a + b)$

2 $3(a - 2b) + 4(2a + 3b)$

3 $5(2a - b) + 4(2a + b)$

4 $3(p + q) + 4(p + q)$

5 $4(5a + c) + 2(3a - c)$

6 $3(p + 2q) + 3(5p - 2q)$

7 $5(4t - 3s) + 8(3t + 2s)$

8 $7(2d + 3e) + 6(2e - 2d)$

9 $5(3z + b) + 4(b - 2z)$

10 $2(a + b + 2c) + 3(2a + 3b - c)$

11 $3(4a - 2b) + 5(a + b)$

12 $6(p + 2q + 3r) + 2(3p - 4q - 9r)$

13 $4(3a + 5b) + 5(2a - 4b)$

14 $5(5g + 4h) + 4(h - 5g)$

15 $2(a + 2b - 3c) + 3(5a - b + 4c) + 4(a + b + c)$

Subtracting expressions with brackets

Example 6

Expand the brackets in the expression $15 - (6 + 2)$

Remember: the minus sign
belongs with the term in brackets so:

$15 - (6 + 2)$ means $\quad 15 \quad + \quad -1 \times (6 + 2)$

$$= 15 - 6 - 2$$

$$= 7$$

Example 7

Expand $-(a + b)$

$-(a + b)$ means $-1 \times (a + b)$

$$= -a - b$$

$$\boxed{-1 \times b = -b}$$

Example 8

Expand $-(a - b)$

$-(a - b)$ means $-1 \times (a - b)$

$$= -a - b$$

$$\boxed{-1 \times -b = +b}$$

Remember:
When you multiply:

like signs give +
$$2 \times 3 = 6$$
$$-2 \times -3 = 6$$

unlike signs give –
$$2 \times -3 = -6$$
$$-2 \times 3 = -6$$

Exercise 2K

1 Work out:

(a) $10 - (3 + 2)$ (b) $20 - (5 - 2)$ (c) $12 - (6 - 4)$

(d) $10 - (3 \times 2)$ (e) $12 - (5 + 3)$ (f) $6 - (7 - 5)$

(g) $14 - (8 + 6)$ (h) $15 - (3 - 2)$

2 Remove the bracket in these expressions:

(a) $-(p + q)$ (b) $-(p - q)$ (c) $-(a + b + c)$

(d) $-(a + b - c)$ (e) $-(r + s)$ (f) $-(r - s)$

(g) $-(p + q - r)$ (h) $-(p - q + r)$

Example 9

Write $3a + 2b - (a + b)$ as simply as possible.

$$3a + 2b - (a + b) = 3a + 2b - a - b$$
$$= 2a + b$$

Exercise 2L

Write these expressions as simply as possible:

1 $4a + 3b - (a + b)$ **2** $5p + 2q - (p + q)$

3 $3(2a + 4) - (3a + 5)$ **4** $2y - 3z - (y + z)$

5 $3(3r + 4s) - 2(r + s)$ **6** $3(2a + 5) - 2(3a + 4)$

7 $5(2s + 3t) - 4(s + 2t)$ **8** $4(3a + b) - 3(2a + 5b)$

9 $2(m + 3n) - (2m + n)$ **10** $5(8h - 3k) - 4(7h + 2k)$

11 $3(c + 2d) - 2(c + 3d)$ **12** $4a - (3a + 5b)$

Example 10

Write $3a + 2b - 2(a - b)$ as simply as possible. Watch out for the $-$ sign *inside* the bracket.

Multiplying gives
$-2 \times -b = +2b$

$$3a + 2b - 2(a - b) \quad = 3a + 2b - 2a + 2b$$
$$= a + 4b$$

Exercise 2M

Simplify these expressions. They *all* have $-$ signs inside brackets.

1 $4a + 3b - (a - b)$

2 $5p + 2q - (p - q)$

3 $3(2a + 4) - (3a - 5)$

4 $2y - 3z - (y - z)$

5 $3(3r + 4s) - 2(r - s)$

6 $3(2a + 5) - 2(3a - 4)$

7 $5(2s + 3t) - 4(s - 2t)$

8 $4(3a + b) - 3(2a - 5b)$

9 $2(m + 3n) - (2m - n)$

10 $5(8h - 3k) - 4(7h - 2k)$

11 $3(c + 2d) - 2(c - 3d)$

12 $4a - (3a - 5b)$

Exercise 2N Mixed questions

Simplify these expressions as much as you can:

1 $a + a + a + a + a$

2 $a \times a \times a \times a \times a \times a$

3 $p + p + p + p + p + p$

4 $p \times p \times p \times p \times p$

5 $2p + 4p + 5p$

6 $8k - 4k + 3k$

7 $9s - 3s + 5s$

8 $6t + 8t - t$

9 $3d - 5d$

10 $4j - 7j$

11 $2(a + b)$

12 $5(3p - q)$

13 $4(s + 2t)$

14 $4(9p - 2q + 5r)$

15 $2(4m + 2n) + 3(5m + 3n)$

16 $3(4s + 2t) + 2(3s - 2t)$

17 $3(2a - 5b) + 2(3a - 4b)$

18 $5(2p + 3d) - 2(3p + 2d)$

19 $5(3a - 2b) - 2(3a - 4b)$

20 $2(3a - 4b) + 2(a + 4b) - 4(2a - b)$

21 $3(2t - 50) + 2(3t + 50) - t + 20$

22 $5(3p - 2q) - 4(3p - q) + 2(p - 3q)$

23 $2(5m + 3n) - 3(2m - n) - 4(m + n)$

24 $3(3a - 4b) + 2(4a + b) + 4(2a - 3b)$

25 $5(2b - 3a) - 3(2a + 3b) - 4(2b - 5a)$

Factorising algebraic expressions

The opposite process of multiplying out brackets is called factorising. Factorising means splitting an expression into parts.

You can factorise numbers:

$12 = 4 \times 3$ 4 and 3 are both factors of 12.

$12 = 2 \times 6$ 2 and 6 are factors of 12 as well.

You can also factorise algebraic expressions:

$ab = a \times b$ a and b are factors of ab

$2pq = 2 \times p \times q$ $2, p$ and q are factors of $2pq$

$x^2 = x \times x$ x and x are factors of x^2

Exercise 2O

Factorise:

1 20	**2** 6	**3** 4	**4** pq
5 pqr	**6** xy	**7** $3ab$	**8** $4a$
9 a^2	**10** b^3	**11** $6rs$	**12** $8ef$

Usually, however, you will be asked to take out the common factors from an expression.

Example 11

Factorise $12a + 4b$

You can split 12 into 4×3:

$$4 \times 3 \times a + 4 \times b$$

4 is now common to both expressions.

You can take a 4 outside a bracket:

$4(3a + b)$

This process is called factorising.

$12a + 4b = 4(3a + b)$

■ **Factorising means splitting up an expression using brackets.**

Example 12

Factorise $3a^2 - 4ab$

This can be written as:

$3 \times a \times a - 4 \times a \times b$

a is common in both expressions, so you can take a and put it outside a bracket:

$a(\qquad)$

You then have to make the expression the same as the original. The new expression is:

$a(3a - 4b)$

Exercise 2P

1   Factorise:

(a) $x^2 + 3x$ (b) $a^2 - ab$ (c) $p^2 + pq$

(d) $3a + 12b$ (e) $5a + 10$ (f) $2b - 4c$

(g) $4 + 8a$ (h) $2a - 2$ (i) $3a + 9$

(j) $5p + 25$ (k) $4a + 16$ (l) $4p - 8$

(m) $7x - 14$ (n) $7y + 7$ (o) $7y^2 + y$

(p) $5q - 15$ (q) $x^2 + 2x$ (r) $y^2 + 3y$

(s) $3a - 3$ (t) $2a^4 + 3a$ (u) $3xy - 4xz$

(v) $4a^2 - 5a$ (w) $5a^5 - 4a$ (x) $5x^2 + 4x$

Exercise 2Q

Simplify the following expressions:

1    $2(x + 3) + 4(x + 9)$ 2 $3(y + 2) + 2(y + 7)$

3    $4(p + 6) + 2(p - 3)$ 4 $5(x + 3) - 2(x - 6)$

5    $3(6 - q) + 4(2q + 3)$ 6 $3(4t + 5) - 2(5 - 4t)$

7 $7(2t - 3) - 3(2t + 5)$ **8** $x(x + 1) + x(x + 2)$

9 $p(p + 4) + p(p - 3)$ **10** $t(2t - 2) - t(t + 1)$

11 $y(2y - 3) + y(3y - 4)$ **12** $c(2c - 3) - c(5c - 7)$

2.10 Number patterns

Sometimes you will need to find the missing numbers in a number pattern like this one:

2, 4, 6, 8, 10, —, —, 16, 18

Algebra can help you do this. But first let's explore some number patterns.

2, 4, 6, 8, 10, —, —, 16, 18

The two missing numbers are 12 and 14.

The rule for this pattern is: **add 2 each time**. This pattern is also the two times table, and all the members are multiples of 2.

The rule for this pattern is: **add 4 each time**.

0, 4, 8, 12, 16, —, —, 28, 32

The two missing numbers are 20 and 24.

Exercise 2R

Find the two missing numbers in these number patterns. Write down the rule for each pattern too.

1 3, 6, 9, —, —, 18, 21

2 5, 10, 15, 20, —, —, 35, 40

3 1, 2, 3, 4, —, —, 7, 8

4 7, 14, 21, 28, —, —, 49, 56

5 0, 6, 12, —, —, 30, 36

6 10, 20, 30, —, —, 60, 70

7 5, 7, 9, 11, —, —, 17, 19

8 4, 7, 10, 13, —, —, 22, 25

9 3, 8, 13, 18, —, —, 33, 38

10 1, 5, 9, 13, —, —, 25, 29

Smaller and smaller

Sometimes the numbers in a pattern get smaller each time. The rule for this pattern is:
take away 2 each time.

18, 16, 14, 12, —, —, 6, 4, 2, 0

The missing numbers are 10 and 8.

42, 36, 30, __, __, 12

The rule for this pattern is I eat six of them each time!

Exercise 2S

Find the two missing numbers in these number patterns. Write down the rule for each pattern too.

1 21, 18, 15, 12, —, —, 3   **2** 24, 20, 16, —, —, 4, 0

3 30, 25, 20, —, —, 5, 0   **4** 49, 42, 35, 28, —, —, 7

5 28, 25, 22, 19, 16, —, —, 7   **6** 37, 32, 27, 22, —, —, 7

7 19, 17, 15, —, —, 9, 7   **8** 25, 21, 17, —, —, 5, 1

9 33, 28, 23, —, —, 8, 3   **10** 45, 38, 31, 24, —, —, 3

Larger and larger

Example 13

Find the missing numbers in this number pattern:

1, 2, 4, 8, —, —, 64, 128

The missing numbers are 16 and 32.

The rule for this pattern is: **multiply by 2 each time.**

The numbers in the pattern are also all powers of 2:

1, 2, 4, 8, 16, 32, 64, 128

$$2 = 2^1$$
$$2 \times 2 = 2^2$$
$$2 \times 2 \times 2 = 2^3$$

Exercise 2T

Find the missing numbers in these number patterns. Write down the rule for each pattern too.

1 1, 3, 9, —, —, 243   **2** 1, 4, 16, —, 256

3 1, —, 25, 125, —, 3125   **4** 1, 10, 100, —, —, 100 000

5 3, 6, 12, —, —, 96   **6** 2, 6, 18, —, —, 486

7 2, 8, 32, —, —, 2048   **8** 2, 20, 200, —, —, 200 000

9 2, 10, 50, —, 1250   **10** 3, 15, 75, —, 1875

Example 14

Find the missing number in this number pattern:

243, 81, 27, —, 3, 1

The missing number is 9.

The rule for this pattern is: **divide by 3 each time**.

The numbers in the pattern are also all powers of 3:

$$243 = 3 \times 3 \times 3 \times 3 \times 3 = 3^5$$
$$81 = 3 \times 3 \times 3 \times 3 = 3^4$$
$$27 = 3 \times 3 \times 3 = 3^3$$
$$9 = 3 \times 3 = 3^2$$
$$3 = 3 = 3^1$$

Exercise 2U

Find the missing numbers in these number patterns and write down the rule for each pattern.

1 128, 64, 32, —, 8, 4, —, 1

2 256, 64, —, 4, 1

3 100 000, 10 000, —, —, 10

4 625, 125, —, 5, 1

5 96, 48, 24, —, —, 3

6 486, 162, 54, —, —, 2

7 2048, 512, 128, —, —, 2

8 200 000, 20 000, 2000, —, —, 2

9 1250, 250, 50, —, 2

10 1875, 375, 75, —, 3

2.11 Finding the rule for a number pattern

Here are some more difficult number patterns with some hints on how to find their rules:

Example 15

Pattern: 1, 4, 7, 10, 13, ...

Differences between each pair of numbers: + 3 + 3 + 3 + 3 +3

> The rule is: **add 3 each time**.
> So the next number is 16.

Example 16

Pattern: 1, 4, 9, 16, 25, ...

Difference: + 3 + 5 + 7 + 9

> The rule is: **add the next odd number each time**.
> The *difference* goes up by 2 each time.

Example 17

Pattern: 1, 1, 2, 3, 5, 8, 13, ...

Difference: + 0 + 1 + 1 + 2 + 3 + 5

> The rule is: **add the previous two numbers each time**.
> The *difference has the same pattern* as the pattern itself.

This is called the Fibonacci sequence. It is named after a famous Italian mathematician.

Exercise 2V

Write out each pattern in the same way as Examples 16, 17 and 18. Find the differences and rule for each one, and the next number.

1 1, 3, 5, 7, 9, ...

2 1, 5, 9, 13, 17, ...

3 1, 8, 27, 64, 125, ...

4 2, 4, 6, 8, 10, ...

5 2, 5, 8, 11, 14, ...

6 3, 7, 11, 15, 19, ...

7 2, 2, 4, 6, 10, 16, ...

8 3, 3, 6, 9, 15, 24, ...

9 3, 5, 7, 9, 11, 13, ...

10 4, 7, 10, 13, 16, ...

11 2, 7, 12, 17, 22, ...

12 3, 8, 13, 18, 23, ...

2.12 Using algebra to write the rule for a number pattern

You can use algebra to write a rule to find any number in a pattern.

Each number in a pattern is called a **term**:

$$2, \quad 4, \quad 6, \quad 8, \quad 10, \quad \ldots \qquad \ldots$$

This is the **first term**, or term number 1.

The **fourth term**, or term number 4.

The nth number in the pattern is called the **nth term** or term number n.

> The word **term** is also used for the parts of an algebraic expression. There is more about this on page 30.

To find the 20th number in a pattern you need a rule to find the **20th term**.

Start by writing the pattern next to the term numbers in a table:

					+2	
Pattern	2	4	6	8	10	...
Term number	1	2	3	4	5	

> This pattern is easy to spot. It is the two times table. The rule for finding the *next* number is: add 2 each time.

The rule to find any term is: **multiply 2 by the term number**.
So the **third term** should be:

term number × 2

$$3 \times 2 = 6$$

The **nth term** will be:

term number × 2

$$n \times 2 = 2n$$

The general rule for the nth term of this pattern is:
nth term $= 2n$

Example 18

Find the general rule for the nth term in this pattern:
1, 4, 7, 10, 13, . . .

Use your rule to find the 20th term.

Write the pattern next to the term numbers in a table:

					+ 3	
Pattern	1	4	7	10	13	...
Term number	1	2	3	4	5	

> The rule for finding the *next* number is: add 3 each time.

The rule to find any term is: **multiply 3 by the term number then take away 2**.

So the fourth term should be:

(term number × 3) – 2

$(4 \times 3) - 2 = 10$

The nth term will be:

$(n \times 3) - 2$ which simplifies to: $3n - 2$

So the general rule for the nth term is: nth term $= 3n - 2$

The 20th term will be: $(3 \times 20) - 2 = 58$

Exercise 2W

Write each pattern in a table in the same way as Example 18. Find the general rule for the nth term. Then use your rule to find the 20th term.

1 3, 6, 9, 12, 15, 18, 21, . . .   **2** 5, 10, 15, 20, 25, 30, 35, 40, . . .

3 1, 2, 3, 4, 5, 6, 7, 8, . . .   **4** 7, 14, 21, 28, 35, 42, 49, 56, . . .

5 0, 6, 12, 18, 24, 30, 36, . . .   **6** 10, 20, 30, 40, 50, 60, 70, . . .

7 5, 7, 9, 11, 13, 15, 17, 19, . . .   **8** 4, 7, 10, 13, 16, 19, 22, 25, . . .

9 3, 8, 13, 18, 23, 28, 33, 38, . . .   **10** 1, 5, 9, 13, 17, 21, 25, 29, . . .

11

    4 matches   7 matches   10 matches

(a) Draw the next 2 patterns.

(b) Complete this table.

Term number	1	2	3	4	5
Matches used	4	7	10		

(c) Write down the rule to find the 6th term.

(d) Find the general rule for the nth term.

12

Find the general rule for the number of matches needed to make the nth pattern in this sequence.

Summary of key points

1   An algebraic expression is a collection of letters and symbols:

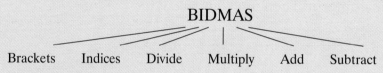

$a + 3b - 2c$ is an algebraic expression

These are each called terms.

2   You can combine like terms by adding or subtracting them:

$2a + 3a = 5a$ and $3b + 4b - b = 6b$

3   You can simplify algebraic expressions by collecting like terms together:

$2a - 4b + 3a + 5b$ simplifies to: $5a + b$

4   BIDMAS is a made-up word to help you remember the order of operations:

BIDMAS

Brackets   Indices   Divide   Multiply   Add   Subtract

5   When the signs are the same you can do them in the order they appear.

6   Factorising means splitting up an expression using brackets:

$12a + 4b = 4(3a + b)$

3 Angles and turning

3.1 Turning

The big wheel is turning . . .

You can show clockwise ↻ and anticlockwise ↺ turns like this:

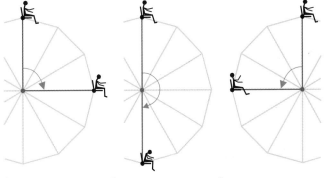

$\frac{1}{4}$ turn clockwise $\frac{1}{2}$ turn clockwise $\frac{1}{4}$ turn anticlockwise

Example 1

An aeroplane is flying North. In which direction will it be flying after:
(a) a $\frac{1}{4}$ turn clockwise
(b) a $\frac{1}{2}$ turn clockwise
(c) a $\frac{1}{4}$ turn anticlockwise?

(a) a $\frac{1}{4}$ turn clockwise   (b) a $\frac{1}{2}$ turn clockwise   (c) a $\frac{1}{4}$ turn anticlockwise

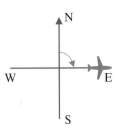   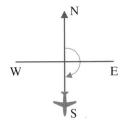   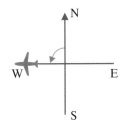

It will be flying East.   It will be flying South.   It will be flying West.

Exercise 3A

1   Write down which of these are turning movements:
     (a) a door opening   **(b)** a ship changing direction
     (c) opening a book   **(d)** a person crossing the road
     (e) a skier skiing down a mountain
     (f) a weather vane as the wind changes.

2 Lisa is facing East. Which way will she be facing after:

   (a) a quarter turn clockwise **(b)** a $\frac{1}{4}$ turn anticlockwise

   (c) a $\frac{1}{2}$ turn clockwise **(d)** a $\frac{1}{2}$ turn anticlockwise?

3 Ajay is walking South-West. Which direction is he walking in after:

   (a) a $\frac{1}{4}$ turn clockwise **(b)** a half turn

   (c) a $\frac{1}{4}$ turn anticlockwise?

4 How much does the hour hand of a clock turn between:

   (a) 3pm and 6pm **(b)** 1pm and 7pm **(c)** 11am and 2pm?

3.2 Measuring angles

■ **An angle is a measure of *turn*. It is a change of direction. There is no change of position.**

■ **An angle which is a $\frac{1}{4}$ turn is called a right angle. Lines that meet at right angles are called perpendicular lines.**

■ **An angle which is less than a $\frac{1}{4}$ turn is called an acute angle.**

■ **An angle which is more than a $\frac{1}{4}$ turn is called an obtuse angle.**

Example 2

Name the different types of angles in this diagram:

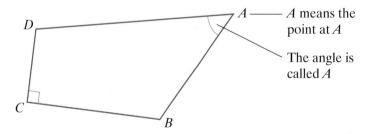

Angle A is less than a $\frac{1}{4}$ turn. It is an **acute angle**.

Angle B is more than a $\frac{1}{4}$ turn. It is an **obtuse angle**.

Angle C is a quarter turn. It is a **right angle**. Lines BC and CD are **perpendicular**.

Angle D is more than a $\frac{1}{4}$ turn. It is an **obtuse angle**.

Measuring angles in degrees

So far angles have been described as $\frac{1}{4}$ and $\frac{1}{2}$ turns. You will need to measure smaller and larger turns than these. The unit for measuring angles is called a **degree**.

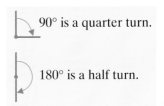

90° is a quarter turn.

180° is a half turn.

■ **There are 360 degrees in a full turn. The sign for a degree is °.**

Exercise 3B

In questions **1–6** write down whether the marked angles are acute, obtuse or right angles.

1

2

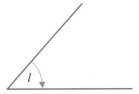

3

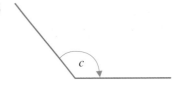

4

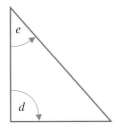

5

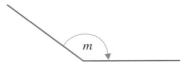

6

In questions **7–12** estimate the size of the marked angle in degrees.

7

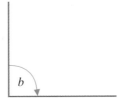

8

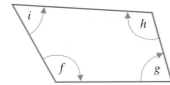

9

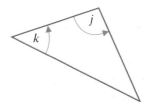

10

11

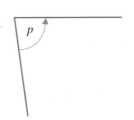

12

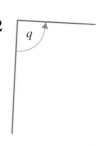

3.3 Naming angles

You can use letters to name the sides and angles of shapes.
This shape is named *ABCD* using the letters for the corners
and going round clockwise:

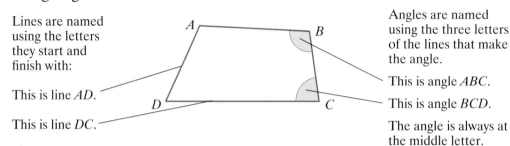

Lines are named
using the letters
they start and
finish with:

This is line *AD*.

This is line *DC*.

Angles are named
using the three letters
of the lines that make
the angle.

This is angle *ABC*.

This is angle *BCD*.

The angle is always at
the middle letter.

Exercise 3C

Use letters to identify the lines and shaded angles in each
diagram:

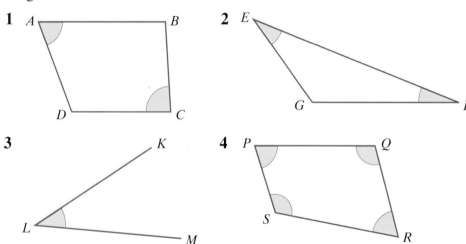

1

2

3

4

3.4 Measuring angles with a protractor

You can use a **protractor** to measure angles accurately.

Protractor

Use the inside
scale to measure
anticlockwise
turns.

Place the cross
at the point of
the angle you
are measuring.

Use the outside
scale to measure
clockwise turns

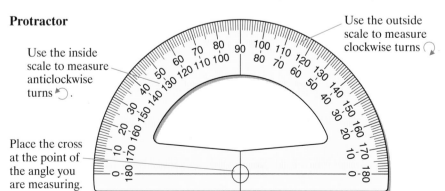

Angle measurer

You can use an angle measurer
instead of a protractor.

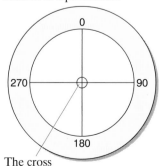

The cross
is at the centre.

Example 3

Use a protractor to measure the angle *CBA*.

Here the lines of
angle *CBA* are
long enough to
reach the outer
edge of the
protractor.

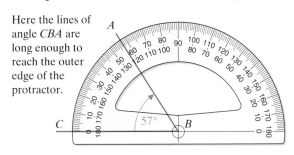

The angle is 57°.

Example 4

Use a protractor to measure the angle *BCD*.

Use the inside scale to
measure Angle *BCD*.

When the line is
too short to reach
the scale, extend
it with a straight
edge like this
piece of paper.

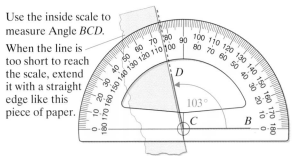

The angle is 103°.

Exercise 3D

Measure and name these angles:

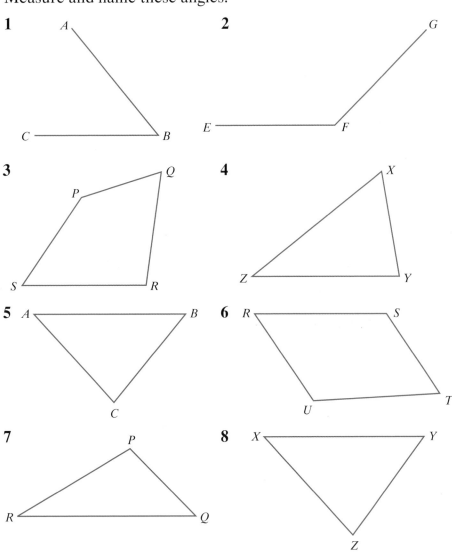

1

A

C ——————— B

2

G

E ——————— F

3

Q
P
S
R

4

X
Z
Y

5 A ——————— B

C

6 R ——————— S

U
T

7

P

R ——————— Q

8 X ——————— Y

Z

3.5 Drawing angles

You need to be able to draw angles which are accurate to within two degrees.

Example 5

Draw these two angles on a line DE which is 8 cm long:

(a) a clockwise angle
$DEF = 79°$

(b) an anticlockwise angle
$EDC = 123°$

(a) Drawing angle *DEF*:

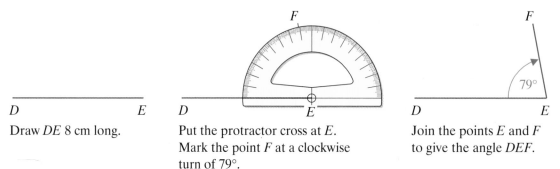

Draw *DE* 8 cm long.

Put the protractor cross at *E*.
Mark the point *F* at a clockwise turn of 79°.

Join the points *E* and *F* to give the angle *DEF*.

(b) Drawing angle *EDC*:

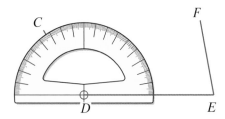
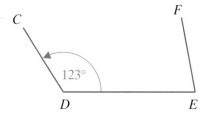

Put the protractor cross at *D*.
Mark the point *C* at an anticlockwise turn of 123°.

Join the points *D* and *C* to give the angle *EDC*.

Exercise 3E

You need a protractor, ruler and pencil.

1   Draw and label these angles:

    (a) $ABC = 40°$ **(b)** $DEF = 65°$ **(c)** $GHK = 125°$

    (d) $LMN = 34°$ **(e)** $OPQ = 136°$ **(f)** $RST = 162°$

    (g) $UVW = 78°$ **(h)** $XYZ = 97°$

2   Make accurate drawings of these diagrams.

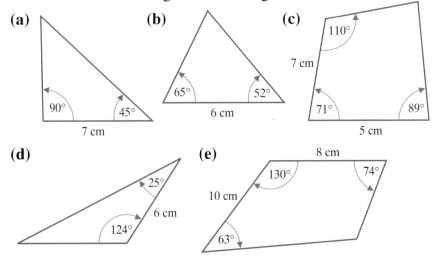

(a) **(b)** **(c)**

(d) **(e)**

3.6 Angles on a straight line

Here are two angles *ABC* and *CBD*.

Joined together they make the angle *ABD* which is a straight line.

These two angles add up to 180°:

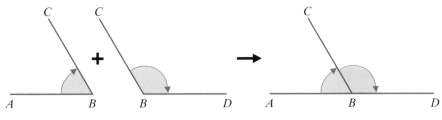

angle *ABC* + angle *CBD* = 180°

■   **The angles on a straight line add up to 180°.**

Example 6

(a) What size is angle *a*?   (b) What size is angle *b*?

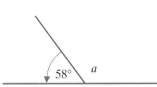 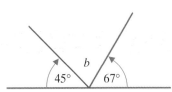

The angles make a straight line so:

$58 + a = 180$

$a = 180 - 58 = 122°$

The 3 angles make a straight line so:

$45 + b + 67 = 180$

$b = 180 - 45 - 67 = 68°$

This can also be written
$b = 180 - (45 + 67) = 68°$

Exercise 3F

In each question find the value of the letter.

1

2
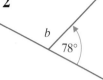

3
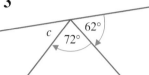

4
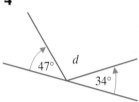

5
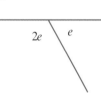

6
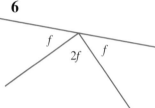

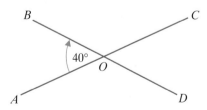

■ **Where two straight lines cross, the opposite angles are equal. They are called vertically opposite angles.**

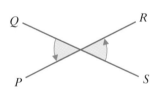

The shaded angles opposite each other are the same.

To see why, imagine that line *QS* has turned anticlockwise to give line *PR*. Both 'ends' of the line have moved through the same angle.

Example 7

Find all the angles in this diagram. Give reasons for your answers.

Angle *AOB* = 40°

So angle *COD* = 40° (vertically opposite angle *AOB*)

Angle *AOD* = 180 − 40 = 140° (the angles make a straight line)

So angle *BOC* = 140° (vertically opposite angle *AOD*)

Give reasons for your answers when you can.

Exercise 3G

Find the angles represented by letters in these questions:

1

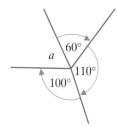

2

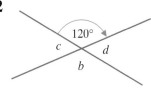

3

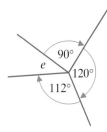

4

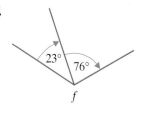

5

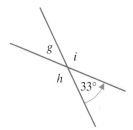

6

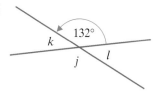

7

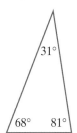

8

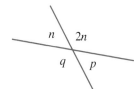

9

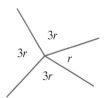

3.8 Sums of angles for triangles and quadrilaterals

■ **The interior angles of a triangle always add up to 180°.**

You can see this by checking that the angles in these triangles add up to 180°.

Another way to see this is to cut out a triangle and tear the corners off like this:

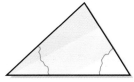

Tear these corners off.

Put three corners together. They make a straight line which is an angle of 180°.

In your examination you might be asked to prove this.

A triangle has 3 angles a, b, c. You need to prove that $a + b + c = 180°$. You will see how to do this in section 3.10.

To prove something in maths you have to explain **why it is true**.

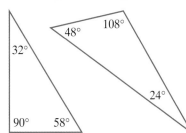

■ **The interior angles of a quadrilateral (a four-sided shape) always add up to 360°.**

You can see this by measuring the angles . . .

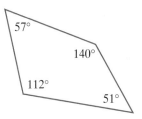

. . . or by dividing the quadrilateral into two triangles . . .

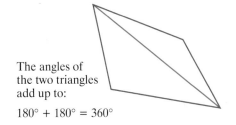

The angles of the two triangles add up to:

$180° + 180° = 360°$

. . . or by tearing off the four corners:

Put the angles together. They make a full turn of 360°.

Example 8

(a) Work out the missing angle in this triangle.

(b) Find the missing angle of this quadrilateral.

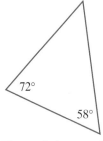

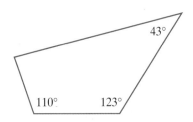

(a) Two of the angles add up to 130°. The third angle must be

$180° - 130° = 50°$.

(b) The 3 angles marked add up to 276°. So the missing angle must be

$360° - 276° = 84°$.

Exercise 3H

Work out the missing angles in these triangles and quadrilaterals.

1

2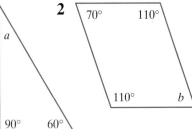

3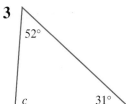

4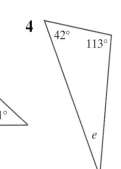

5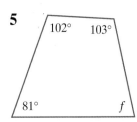

6

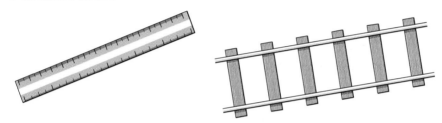

6 g $52°$ $134°$ $108°$

7 h $100°$ h

8 i i $95°$ $95°$

9 j j j

3.9 Alternate and corresponding angles

Parallel lines

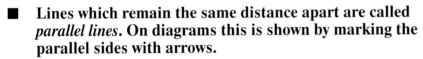

The distance between the two edges of a ruler is the same all the way along it. Similarly the distance between the two rails of a train track is the same wherever it is measured.

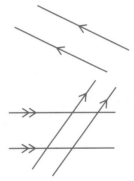

■ **Lines which remain the same distance apart are called** *parallel lines*. **On diagrams this is shown by marking the parallel sides with arrows.**

If there is a second pair of parallel lines in one diagram these are marked with double arrows.

When a straight line crosses a pair of parallel lines it makes angles which are the same size.

Alternate angles

■ **The shaded angles are equal.** **They are called alternate angles.**

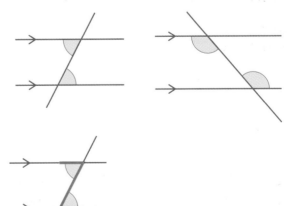

Alternate angles are sometimes called 'Z' angles.

Corresponding angles

■   **The shaded angles are equal.**
     They are called corresponding angles.

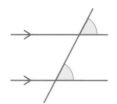
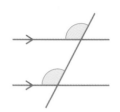

Corresponding angles are sometimes
called 'F' angles.

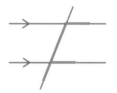

Exercise 3I

1   Find and name as many pairs of parallel lines as you can in
        this diagram.

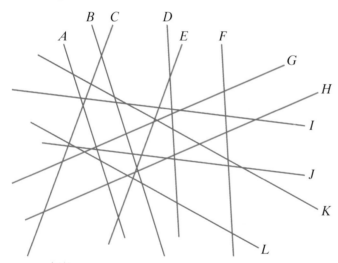

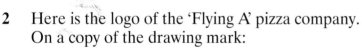

2   Here is the logo of the 'Flying A' pizza company.
        On a copy of the drawing mark:
        (a) a right angle with an R
        (b) two parallel lines each with a P
        (c) an obtuse angle with an O.   [E]

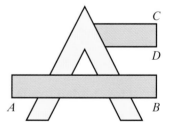

3   In the diagram, which pair of angles
        are alternate angles?

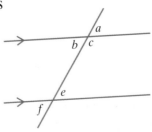

4 In the diagram, which pair of angles are corresponding angles?

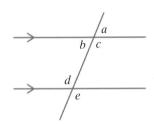

5 Find the size of the angle marked with a letter. Give a reason for your answer.

(a)

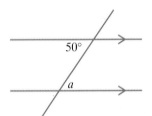

(b)

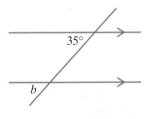

(c)

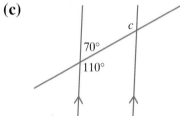

(d)

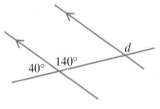

(e)

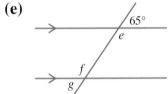

(f)

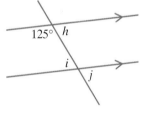

6

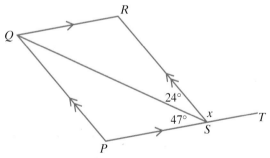

PQRS is a parallelogram:
angle *QSP* = 47°
angle *QSR* = 24°
PST is a straight line.
  (a) **(i)** Find the size of the angle marked *x*.
      (ii) Give a reason for your answer
  (b) **(i)** Work out the size of angle *PQS*.
      (ii) Give a reason for your answer.   [E]

3.10 Proof in geometry

You need to be able to prove that the exterior angle is the sum
of the two interior and opposite angles:

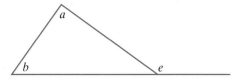

$$e = a + b$$

Through the point of angle e draw a line parallel to the
opposite side of the triangle.
Angle e is now divided into two angles, c and d.

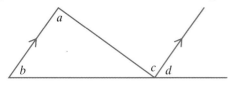

$$a = c \qquad \text{alternate angles}$$
$$b = d \qquad \text{corresponding angles}$$

So $a + b = c + d = e$

Proving the angles of a triangle add up to 180°

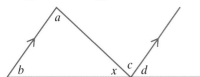

The angles of the triangle are a, b and x.

$$c + d + x = 180° \qquad \text{angles on a straight line}$$

But $\qquad\qquad c + d = a + b \qquad$ exterior angle = sum of interior and
opposite angles.

Therefore $\quad a + b + x = 180°$

3.11 Bearings

Bearings are used to describe directions with angles.

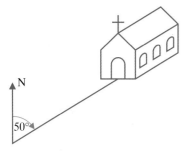

If you begin facing North then turn
clockwise until you face the church
you have passed through 50°.

The **angle** you turn is called the
bearing.

It is always written as a
three-figure number.

When there are less than
three digits in the angle
you need to add zeros to
make a three-figure
number. For example
$$9° = 009°$$

You write the bearing of the church as 050°.

In this diagram the bearing of Birmingham from London is
315°. The bearing of London from Birmingham is 135°.

■ **A bearing is the angle measured from facing North and
turning clockwise.
It is always a three-figure number.**

Exercise 3J

In questions **1–6**, write down the bearing of *B* from *A*.

1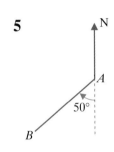

2

3

4

5

6

7 Write down the bearing of
  (a) *A* from *B*
  (b) *C* from *B*
  (c) *B* from *C*.

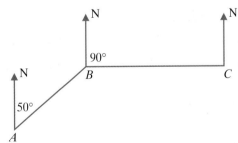

8 Use a protractor to find
  the bearings of:
  (a) *Q* from *P*
  (b) *P* from *R*
  (c) *R* from *Q*.

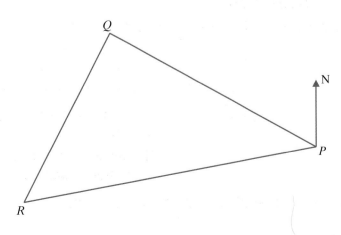

Summary of key points

1 An angle is a measure of turn. It is a change of direction. There is no change of position.

2 An angle which is a $\frac{1}{4}$ turn is called a right angle.

3 An angle which is less than a $\frac{1}{4}$ turn is called an acute angle.

4 An angle which is more than a $\frac{1}{4}$ turn is called an obtuse angle.

5 There are 360 degrees in a full turn. The sign for a degree is °.

6 The angles on a straight line add up to 180°.

7 The angles at a point add up to 360°.

8 When two straight lines cross, the opposite angles are equal. They are called vertically opposite angles.

9 The interior angles of a triangle always add up to 180°.

10 The interior angles of a quadrilateral always add up to 360°.

11 Lines which remain the same distance apart are called parallel lines. On diagrams this is shown by marking the parallel sides with arrows.

12 The shaded angles are equal.
They are called alternate angles.

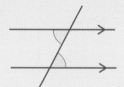

13 The shaded angles are equal.
They are called corresponding angles.

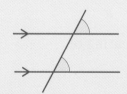

14 A bearing is the angle measured from facing North and turning clockwise. It is always a three-figure number.

4 Fractions

4.1 Fractions from pictures

All these things can be divided into parts called **fractions**:

This football pitch has two halves.

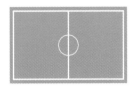

This computer disc has eight sectors.

This chessboard has 64 small squares.

One part is **one half** or $\frac{1}{2}$ of the pitch.

One part is **one eighth** or $\frac{1}{8}$ of the disc.

One part is **one sixty fourth** or $\frac{1}{64}$ of the board.

Using numbers to represent fractions

I am going to eat three quarters or $\frac{3}{4}$ of this cake.

■ The top number shows how many parts I will eat.

The bottom number shows how many parts the cake is divided into.

$$\frac{3}{4}$$

The top number is called the **numerator**.

The bottom number is called the **denominator**.

Two thirds or $\frac{2}{3}$ of these parking spaces are occupied.

Two spaces have cars in them.

The car park is divided into three spaces.

$$\frac{2}{3}$$

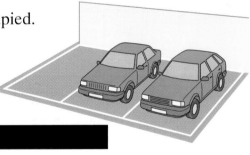

Exercise 4A

1 Copy these shapes into a table like this. The first one is done for you.

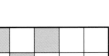 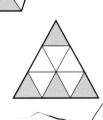

Shape	Fraction shaded	Fraction not shaded
	$\frac{1}{2}$	$\frac{1}{2}$

2   Make four copies of this shape. Shade them to show these fractions:

(a) $\frac{1}{16}$ (b) $\frac{3}{16}$ (c) $\frac{8}{16}$ (d) $\frac{16}{16}$

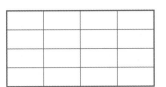

3   Make three copies of this shape. Shade them to show these fractions:

(a) $\frac{1}{6}$ (b) $\frac{3}{6}$ (c) $\frac{1}{2}$

4.2 Fractions from words

Sometimes you will need to use fractions to solve problems given in words.

Example 1

John has a collection of 1000 stamps. 113 are British stamps. The others are foreign.

What fraction of John's collection is:

(a) British   (b) foreign?

(a) 113 of the 1000 stamps are British.

   The fraction is: $\frac{113}{1000}$

(b) 1000 − 113 = 887 stamps are foreign.

   The fraction is: $\frac{887}{1000}$

Exercise 4B

1   There are 28 people on a bus. 13 are female and 15 are male. What fraction of the people are:

(a) male   (b) female?

2   Flora has 28 heavy metal CDs and 47 rock and roll CDs.

(a) How many CDs does she have altogether?

(b) What fraction of her collection is heavy metal?

(c) What fraction of her collection is rock and roll?

3   A transport company has a fleet of 51 vehicles. 37 are lorries, 10 are vans and 4 are cars.
What fraction of the fleet is:

(a) lorries   (b) vans   (c) cars?

4 A plumber's weekly earnings are £217 per week. £198 is basic pay and £19 is bonus pay.
What fraction of weekly earnings is:
(a) basic pay   (b) bonus pay?

5 At a dog show the total entry was 7 Alsatians, 3 Corgis, 5 Spaniels and 1 Labrador.
(a) How many dogs were entered?
What fraction of the entries were:
(b) Alsatians   (c) Corgis
(d) Spaniels   (e) Labradors?

6 An electrical superstore sold 5 CD players, 6 DVD players, 4 refrigerators and 2 deep freezers.
(a) How many items were sold altogether?
What fraction of the items sold were:
(b) CD players   (c) DVD players
(d) refrigerators   (e) deep freezers?

7 A taxi driver works 11 hours a day. She spends 6 hours driving, 4 hours waiting for passengers and 1 hour on paperwork.
What fraction of her day is spent:
(a) driving   (b) waiting   (c) on paperwork?

4.3 Improper fractions and mixed numbers

These fractions are **top heavy**:

$\frac{11}{9}$ is top heavy because 11 is greater than 9

$\frac{18}{5}$ is top heavy because 18 is greater than 5

- **Top heavy fractions are also called improper fractions.**
- **An improper fraction can also be written as a mixed number – a mixture of a whole number and a fraction.**

improper fraction $\frac{11}{9} = \frac{9}{9} + \frac{2}{9} = 1\frac{2}{9}$ mixed number

You say this: "9 goes into 11 one remainder 2 over 9"

- **A mixed number can also be written as an improper fraction:**

mixed number $1\frac{3}{11} = \frac{14}{11}$ improper fraction

Here is how to do this for $1\frac{3}{11}$. Change the mixed number to elevenths.

1 can be written $\frac{11}{11}$ So: $1\frac{3}{11} = \frac{11}{11} + \frac{3}{11} = \frac{14}{11}$

Here is how to write $5\frac{3}{7}$ as an improper fraction. Change the mixed number to sevenths.

5 can be written $\frac{5 \times 7}{7}$ $= \frac{35}{7}$ So: $5\frac{3}{7} = \frac{35}{7} + \frac{3}{7} = \frac{38}{7}$

Exercise 4C

1   Change these improper fractions to mixed numbers.

(a) $\frac{5}{2}$ (b) $\frac{7}{4}$ (c) $\frac{9}{7}$ (d) $\frac{11}{8}$

(e) $\frac{9}{8}$ (f) $\frac{16}{5}$ (g) $\frac{23}{10}$ (h) $\frac{24}{5}$

(i) $\frac{16}{7}$ (j) $\frac{12}{5}$ (k) $\frac{20}{3}$ (l) $\frac{16}{9}$

(m) $\frac{39}{4}$ (n) $\frac{27}{5}$ (o) $\frac{26}{9}$ (p) $\frac{17}{10}$

2   Change these mixed numbers to improper fractions.

(a) $1\frac{1}{2}$ (b) $5\frac{1}{2}$ (c) $2\frac{3}{4}$ (d) $1\frac{2}{3}$

(e) $3\frac{1}{4}$ (f) $4\frac{2}{5}$ (g) $3\frac{7}{10}$ (h) $5\frac{1}{5}$

(i) $7\frac{3}{4}$ (j) $2\frac{1}{4}$ (k) $1\frac{9}{10}$ (l) $9\frac{1}{3}$

(m) $2\frac{5}{6}$ (n) $5\frac{3}{8}$ (o) $3\frac{5}{8}$ (p) $1\frac{9}{100}$

4.4 Simplifying fractions

■ **Fractions can be simplified if the numerator (top) and denominator (bottom) have a common factor.** For example:

$\dfrac{4}{8}$ —————— The numerator will divide by 4
—————— The denominator will divide by 4

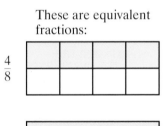

These are equivalent fractions:

$\dfrac{4}{8}$ | $\dfrac{4}{8}$ $\boxed{\text{4 divided by 4 is 1}}$ $\boxed{\text{8 divided by 4 is 2}}$ $\dfrac{1}{2}$ so $\frac{4}{8}$ can be simplified to $\frac{1}{2}$

$\frac{4}{8}$ and $\frac{1}{2}$ are **equivalent fractions**. They represent the same value. To simplify a fraction you find an equivalent fraction that has smaller numbers on the top and bottom.

Example 2

Simplify $\frac{9}{15}$ by finding a common factor.

3 is a common factor of 9 and 15.

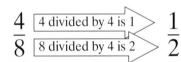

$\dfrac{9}{15}$ $\boxed{\text{9 divided by 3 is 3}}$ $\boxed{\text{15 divided by 3 is 5}}$ $\dfrac{3}{5}$

Factors of 9 are: 1, 3, 9
Factors of 15 are: 1, 3, 5, 15

3 is a common factor

$\frac{9}{15}$ is the same as $\frac{3}{5}$

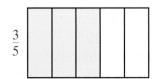
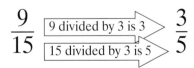

Example 3

Simplify $\frac{24}{30}$ by finding a common factor.

3 is a common factor of 24 and 30.

$$\frac{24}{30} \quad \boxed{24 \text{ divided by 3 is } 8} \quad \boxed{30 \text{ divided by 3 is } 10} \quad \Longrightarrow \quad \frac{8}{10}$$

$\frac{8}{10}$ can also be simplified. 2 is a common factor of 8 and 10.

$$\frac{8}{10} \quad \boxed{8 \text{ divided by 2 is } 4} \quad \boxed{10 \text{ divided by 2 is } 5} \quad \Longrightarrow \quad \frac{4}{5}$$

$\frac{4}{5}$ cannot be simplified any more. This fraction is in its **simplest form**.

Exercise 4D

Simplify these fractions by finding common factors:

1 **(a)** $\frac{4}{6}$ **(b)** $\frac{3}{6}$ **(c)** $\frac{2}{4}$ **(d)** $\frac{3}{9}$

    (e) $\frac{4}{10}$ **(f)** $\frac{8}{12}$ **(g)** $\frac{14}{21}$ **(h)** $\frac{15}{20}$

    (i) $\frac{14}{22}$ **(j)** $\frac{24}{28}$ **(k)** $\frac{27}{36}$ **(l)** $\frac{25}{30}$

2 Copy these shapes into a table like this.
For each shape use two **equivalent fractions** to describe how much is shaded.
The first one is done for you.

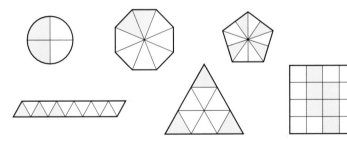

Shape	Fraction	Fraction
	$\frac{2}{4}$	$\frac{1}{2}$

In questions **3**, **4** and **5** write your answers as fractions in their simplest form.

3 A car salesman sold 6 new cars and 15 secondhand cars.
What fraction of the cars were:

    (a) new **(b)** secondhand?

4 Sîan has collected autographs from:
16 pop stars 28 footballers 12 athletes
What fraction of her autographs are:
    (a) pop stars' **(b)** footballers' **(c)** athletes'?

5 Here are Form 11B's traffic survey results:
What fraction of the vehicles were:
  (a) private cars   (b) vans
  (c) lorries   (d) buses?

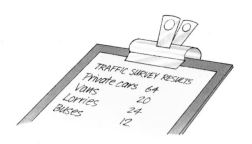

TRAFFIC SURVEY RESULTS
Private cars 64
Vans 20
Lorries 24
Buses 12

4.5 Finding a fraction of a quantity

You need to be able to do calculations like these:

$\frac{3}{4}$ of £60 $\frac{5}{8}$ of 160

Example 4

Find $\frac{3}{4}$ of £60.

Multiply the numerator 3 by the quantity 60: $\frac{3 \times 60}{4} = \frac{180}{4}$

Divide the result 180 by the denominator 4: $\frac{180}{4} = 45$

So $\frac{3}{4}$ of £60 is £45.

Another method:
one quarter is

$\frac{60}{4} = 15$

three quarters is
$3 \times 15 = 45$

Example 5

Find $\frac{5}{8}$ of 160.

$\frac{5 \times 160}{8} = \frac{800}{8}$

$\frac{800}{8} = 100$

So $\frac{5}{8}$ of 160 is 100.

Exercise 4E

1 Find:
  (a) $\frac{1}{2}$ of 70   (b) $\frac{2}{5}$ of £65   (c) $\frac{3}{10}$ of 80 kg
  (d) $\frac{2}{3}$ of 96   (e) $\frac{3}{8}$ of £56   (f) $\frac{3}{4}$ of 60p
  (g) $\frac{1}{4}$ of £6.80   (h) $\frac{7}{10}$ of 90p   (i) $\frac{7}{8}$ of £3.20

Multiplying by $\frac{1}{2}$ is the same as dividing by 2.

2 A superstore employs 85 people. $\frac{2}{5}$ are men.
  (a) How many men does the store employ?
  (b) How many women does the store employ?

3 A mail order company prints 200 000 catalogues. $\frac{13}{20}$ are sent to customers. $\frac{3}{20}$ are sent to agents. $\frac{1}{5}$ are sent to shops.
How many catalogues are sent to:
  (a) customers   (b) agents   (c) shops?

4 A chain store closed $\frac{2}{15}$ of its 345 shops.
How many shops were closed?

5 A factory worker is paid £186 per week. $\frac{2}{5}$ is deducted for tax and national insurance.
How much is deducted?

6 The metal parts of a car weigh 1250 kg. $\frac{3}{10}$ of the metal is recycled.
How much does the recycled metal weigh?

7 Stan sold 560 sandwiches today. $\frac{1}{4}$ were ham sandwiches, $\frac{2}{5}$ were salad, $\frac{1}{8}$ were tuna and the rest were cheese.
How many of each type did Stan sell?

8 A building company built 3050 houses. $\frac{1}{10}$ had four bedrooms, $\frac{7}{50}$ had two bedrooms. The rest had three bedrooms.
How many houses had:
(a) four bedrooms   (b) two bedrooms
(c) three bedrooms?

9 A department store had 480 customers last Saturday. $\frac{2}{3}$ paid by credit card, $\frac{1}{4}$ paid by cheque and the others paid by cash.
How many customers paid by:
(a) credit card   (b) cheque   (c) cash?

4.6 Equivalent fractions

Here is a block of fudge divided in three different ways:

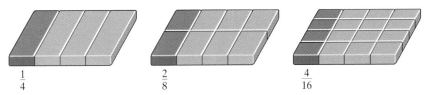

$\frac{1}{4}$ $\frac{2}{8}$ $\frac{4}{16}$

The fractions $\frac{1}{4}$, $\frac{2}{8}$ and $\frac{4}{16}$ all represent the same amount of the fudge.

$\frac{1}{4}$, $\frac{2}{8}$ and $\frac{4}{16}$ are **equivalent fractions**.

■ **Equivalent fractions are fractions that have the same value.**

Simplifying to find equivalent fractions

You can see this by simplifying $\frac{4}{16}$.

Divide the numerator and the denominator by the common factor 2:

Then do this again:

$$\frac{4}{16} \quad \boxed{\begin{array}{l}\text{4 divided by 2 is 2}\\ \text{16 divided by 2 is 8}\end{array}} \quad \frac{2}{8} \qquad\qquad \frac{2}{8} \quad \boxed{\begin{array}{l}\text{2 divided by 2 is 1}\\ \text{8 divided by 2 is 4}\end{array}} \quad \frac{1}{4}$$

So $\frac{4}{16} = \frac{2}{8} = \frac{1}{4}$ These are equivalent fractions.

Multiplying to find equivalent fractions

Another way of finding equivalent fractions is to *multiply* the
numerator and denominator by the same (common) number:

For example:

$$\frac{3}{4} \quad \boxed{\begin{array}{l} \text{3 multiplied by 6 is 18} \\ \text{4 multiplied by 6 is 24} \end{array}} \!\!\!> \quad \frac{18}{24}$$

You can keep doing this:

$$\frac{18}{24} \quad \boxed{\begin{array}{l} \text{18 multiplied by 2 is 36} \\ \text{24 multiplied by 2 is 48} \end{array}} \!\!\!> \quad \frac{36}{48}$$

So $\frac{3}{4} = \frac{18}{24} = \frac{36}{48}$ These are equivalent fractions.

Example 6

Change $\frac{3}{7}$ to any equivalent fraction.

Multiply the numerator and denominator by, say, 5.

$$\frac{3}{7} = \frac{3 \times 5}{7 \times 5} = \frac{15}{35}$$

So $\frac{3}{7}$ and $\frac{15}{35}$ are equivalent fractions.

Remember:

$$\frac{3}{7} \begin{array}{l} \text{—— numerator} \\ \text{—— denominator} \end{array}$$

Example 7

Change $\frac{5}{8}$ to an equivalent fraction with the denominator 48.

To change the denominator from 8 to 48, multiply 8 by 6.

To get an equivalent fraction multiply the numerator 5 by 6 too.

$$\frac{5}{8} = \frac{5 \times 6}{8 \times 6} = \frac{30}{48}$$

So $\frac{5}{8}$ and $\frac{30}{48}$ are equivalent fractions.

Example 8

Complete this set of equivalent fractions: $\frac{2}{5} = \frac{}{10} = \frac{}{15}$

To get from $\frac{2}{5}$ to $\frac{}{10}$ multiply top and bottom by 2:

$$\frac{2}{5} \quad \boxed{\begin{array}{l} \text{2 multiplied by 2 is 4} \\ \text{5 multiplied by 2 is 10} \end{array}} \!\!\!> \quad \frac{4}{10}$$

To get from $\frac{2}{5}$ to $\frac{}{15}$ multiply top and bottom by 3:

$$\frac{2}{5} \quad \boxed{\begin{array}{l} \text{2 multiplied by 3 is 6} \\ \text{5 multiplied by 3 is 15} \end{array}} \!\!\!> \quad \frac{6}{15}$$

So the set of equivalent fractions is $\frac{2}{5} = \frac{4}{10} = \frac{6}{15}$

Exercise 4F

1 Copy these sets of fractions. Fill in the missing numbers to make the fractions equivalent:

(a) $\frac{1}{2} = \frac{}{10} = \frac{}{100} = \frac{}{8} = \frac{}{12} = \frac{3}{}$

(b) $\frac{2}{3} = \frac{}{6} = \frac{}{9} = \frac{}{60} = \frac{}{90} = \frac{20}{}$

(c) $\frac{7}{10} = \frac{}{20} = \frac{}{50} = \frac{}{100} = \frac{}{2000} = \frac{28}{}$

(d) $\frac{3}{4} = \frac{}{16} = \frac{}{20} = \frac{}{28} = \frac{}{400} = \frac{90}{}$

(e) $\frac{1}{8} = \frac{}{24} = \frac{}{32} = \frac{}{64} = \frac{}{480} = \frac{7}{}$

2 Write down five other fractions equivalent to:

(a) $\frac{1}{5}$ (b) $\frac{2}{5}$ (c) $\frac{5}{6}$ (d) $\frac{7}{8}$ (e) $\frac{3}{20}$

3 (a) Give an equivalent fraction to $\frac{1}{2}$ and an equivalent fraction to $\frac{1}{3}$ so that the denominators of the two new fractions are equal.

(b) Repeat part (a) for:

(i) $\frac{2}{5}$ and $\frac{3}{6}$ (ii) $\frac{1}{10}$ and $\frac{1}{7}$ (iii) $\frac{1}{4}$ and $\frac{5}{6}$

(iv) $\frac{1}{2}$ and $\frac{3}{5}$ (v) $\frac{2}{3}$ and $\frac{1}{8}$ (vi) $\frac{3}{4}$ and $\frac{3}{5}$

4.7 Putting fractions in order of size

Which is larger: $\frac{3}{4}$ or $\frac{3}{5}$? Equivalent fractions can help you decide.

First make lists of equivalent fractions for $\frac{3}{4}$ and $\frac{3}{5}$:

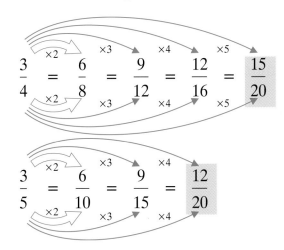

Compare the lists. Find fractions with the same denominator (bottom).

$\frac{15}{20}$ is larger than $\frac{12}{20}$ (Imagine a cake divided into 20 parts and getting either 12 or 15 slices.)

As $\frac{15}{20}$ is equivalent to $\frac{3}{4}$ and $\frac{12}{20}$ is equivalent to $\frac{3}{5}$

$\frac{3}{4}$ must be *larger* than $\frac{3}{5}$

Exercise 4G

1 Which is larger:
   (a) $\frac{2}{5}$ or $\frac{3}{6}$ (b) $\frac{1}{10}$ or $\frac{1}{7}$ (c) $\frac{1}{4}$ or $\frac{5}{6}$
   (d) $\frac{1}{2}$ or $\frac{3}{5}$ (e) $\frac{2}{3}$ or $\frac{1}{8}$ (f) $\frac{3}{4}$ or $\frac{3}{5}$

2 Put these fractions in order of size starting with the smallest:
   $\frac{2}{5}$ $\frac{1}{2}$ $\frac{7}{8}$ $\frac{3}{4}$ $\frac{2}{10}$

Using fractions in your examination

In your examination fractions will usually appear in the context of a number problem or in questions on probability, areas or volumes. You need to be able to add, subtract, multiply and divide fractions in such problems. The rest of this unit shows you how to do this.

4.8 Adding fractions

It is easy to add fractions when the denominator (bottom) is the same:

Easy to add:

$$\frac{1}{4} + \frac{2}{4} = \frac{3}{4}$$

denominators the same

Harder to add:

$$\frac{1}{2} + \frac{2}{3} = ?$$

denominators different

Adding fractions with the same denominator

Add the numerators (top).

$$\frac{7}{10} + \frac{2}{10} = \frac{9}{10}$$

Write them over the **same** denominator (bottom).

Adding fractions with different denominators

$$\frac{1}{2} + \frac{2}{3} = ?$$

Find equivalent fractions to these that have the same denominator (bottom).

Fractions equivalent to $\frac{1}{2}$

Fractions equivalent to $\frac{2}{3}$

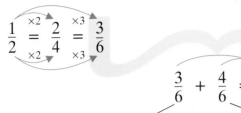

$$\frac{1}{2} = \frac{2}{4} = \frac{3}{6}$$

$$\frac{2}{3} = \frac{4}{6}$$

$$\frac{3}{6} + \frac{4}{6} = \frac{7}{6}$$

These fractions have the same denominators (bottom). Now they are easy to add.

equivalent to $\frac{1}{2}$ equivalent to $\frac{2}{3}$

So $\frac{1}{2} + \frac{2}{3} = \frac{7}{6}$

$\frac{7}{6}$ is top heavy. It is usually written as a mixed number: $1\frac{1}{6}$

■ **To add fractions find equivalent fractions that have the same denominator (bottom).**

Example 9

Work out $\frac{5}{8} + \frac{3}{7}$

Find equivalent fractions to these that have the same denominator (bottom).

An easy way to do this is:

$$\frac{5}{8} \xrightarrow{\times 7} \frac{35}{56} \qquad \frac{3}{7} \xrightarrow{\times 8} \frac{24}{56}$$

same denominators now you can add the numerators

$$\frac{35}{56} + \frac{24}{56} = \frac{59}{56}$$

So $\frac{5}{8} + \frac{3}{7} = \frac{59}{56}$

$\frac{59}{56}$ is top heavy. It is usually written as a mixed number: $1\frac{3}{56}$

Example 10

Work out $\frac{2}{3} + \frac{7}{12}$

Find equivalent fractions to these that have the same denominator (bottom).

$$\frac{2}{3} + \frac{7}{12} = ?$$

Notice that $3 \times 4 = 12$

So you only need to change one of the fractions:

$$\frac{2}{3} \xrightarrow{\times 4} \frac{8}{12}$$

Now both have the same denominator:

$$\frac{8}{12} + \frac{7}{12} = \frac{15}{12}$$

equivalent to $\frac{2}{3}$

So $\frac{8}{12} + \frac{7}{12} = \frac{15}{12}$

$\frac{15}{12}$ is top heavy. It is usually written as a mixed number: $1\frac{3}{12}$
This simplifies to $1\frac{1}{4}$.

Example 11

Work out $2\frac{1}{4} + 3\frac{1}{5}$

These are **mixed numbers**. First add the whole numbers: $2 + 3 = 5$

Then add the fractions:

$$\frac{1}{4} + \frac{1}{5} = ?$$

Change both denominators (bottom) to 20 because $4 \times 5 = 20$:

$$\frac{1}{4} \overset{\times 5}{\underset{\times 5}{=}} \frac{5}{20} \qquad\qquad \frac{1}{5} \overset{\times 4}{\underset{\times 4}{=}} \frac{4}{20}$$

$$\frac{5}{20} + \frac{4}{20} = \frac{9}{20} \qquad \text{same denominator now add the numerators}$$

Now put the whole numbers and the fractions back together:

$5 + \frac{9}{20}$ is $5\frac{9}{20}$

So $2\frac{1}{4} + 3\frac{1}{5} = 5\frac{9}{20}$

Exercise 4H

1   Work out:
   (a) $\frac{3}{8} + \frac{4}{8}$ (b) $\frac{2}{9} + \frac{5}{9}$ (c) $\frac{5}{12} + \frac{1}{12}$ (d) $\frac{5}{18} + \frac{11}{18}$
   (e) $\frac{1}{2} + \frac{1}{4}$ (f) $\frac{1}{4} + \frac{3}{8}$ (g) $\frac{1}{2} + \frac{7}{8}$ (h) $\frac{2}{3} + \frac{1}{6}$
   (i) $\frac{5}{6} + \frac{1}{3}$ (j) $\frac{2}{5} + \frac{3}{10}$ (k) $\frac{7}{12} + \frac{3}{4}$ (l) $\frac{3}{4} + \frac{7}{20}$

2   Work out:
   (a) $\frac{1}{6} + \frac{5}{8}$ (b) $\frac{3}{4} + \frac{1}{10}$ (c) $\frac{4}{9} + \frac{5}{12}$ (d) $\frac{7}{8} + \frac{9}{10}$
   (e) $\frac{3}{10} + \frac{4}{15}$ (f) $\frac{5}{6} + \frac{1}{4}$ (g) $\frac{3}{8} + \frac{7}{12}$ (h) $\frac{1}{6} + \frac{8}{9}$

3 Work out:

(a) $\frac{1}{2} + \frac{1}{3}$ (b) $\frac{2}{5} + \frac{1}{6}$ (c) $\frac{5}{8} + \frac{1}{5}$ (d) $\frac{3}{4} + \frac{1}{9}$

(e) $\frac{5}{6} + \frac{3}{7}$ (f) $\frac{9}{10} + \frac{2}{7}$ (g) $\frac{2}{3} + \frac{7}{10}$ (h) $\frac{3}{5} + \frac{3}{4}$

4 Work out:

(a) $2\frac{1}{2} + \frac{2}{3}$ (b) $3\frac{1}{4} + 2\frac{1}{2}$ (c) $1\frac{1}{4} + 2\frac{7}{8}$ (d) $3\frac{1}{3} + 5\frac{3}{4}$

(e) $3\frac{5}{16} + 1\frac{7}{8}$ (f) $2\frac{11}{12} + \frac{3}{4}$ (g) $\frac{5}{6} + 6\frac{1}{3}$ (h) $2\frac{2}{3} + 4\frac{3}{5}$

5 In a market garden $\frac{1}{4}$ of the garden is used for potatoes, $\frac{3}{20}$ is used for beans and $\frac{1}{10}$ is used for cabbages. What fraction of the garden is used to grow these vegetables altogether?

6 John gave away his old record collection to his brother and two sisters. The elder sister received $\frac{3}{10}$ of them and the younger sister $\frac{5}{16}$ of them. The brother received the rest. What fraction of the collection did the sisters receive altogether?

I said you could have $\frac{5}{16}$ of the records not $\frac{5}{16}$ of each one!

4.9 Subtracting fractions

It is easy to subtract fractions when the denominator (bottom) is the same:

Easy to subtract:

$$\frac{7}{12} - \frac{2}{12} = \frac{5}{12}$$

denominators the same

Harder to subtract:

$$\frac{5}{9} - \frac{1}{4} = \ ?$$

denominators different

Example 12

Work out $\frac{5}{9} - \frac{1}{4}$

Find equivalent fractions to these that have the same denominator (bottom):

An easy way is: change both denominators to 36 because $9 \times 4 = 36$

$$\frac{5}{9} \xrightarrow[\times 4]{\times 4} \frac{20}{36} \qquad \frac{1}{4} \xrightarrow[\times 9]{\times 9} \frac{9}{36}$$

$$\frac{20}{36} - \frac{9}{36} = \frac{11}{36} \qquad \text{same denominators easy to subtract}$$

equivalent to $\frac{5}{9}$ equivalent to $\frac{1}{4}$

So $\frac{5}{9} - \frac{1}{4} = \frac{11}{36}$

Example 13

Work out $4\frac{1}{2} - 1\frac{5}{11}$

These are **mixed numbers**.

First subtract the whole numbers: $4 - 1 = 3$

$4\frac{1}{2} - 1\frac{5}{11}$ is the same as: $3\frac{1}{2} - \frac{5}{11}$

Then find equivalent fractions to $\frac{1}{2}$ and $\frac{5}{11}$ that have the same denominator (bottom).

Change both denominators to 22 because $2 \times 11 = 22$

$$\frac{1}{2} \begin{array}{c} \boxed{\times 11} \\ = \\ \boxed{\times 11} \end{array} \frac{11}{22} \qquad\qquad \frac{5}{11} \begin{array}{c} \boxed{\times 2} \\ = \\ \boxed{\times 2} \end{array} \frac{10}{22}$$

$$\frac{11}{22} - \frac{10}{22} = \frac{1}{22} \qquad \begin{array}{l} \text{same denominators} \\ \text{easy to subtract} \end{array}$$

Now put the whole numbers and fractions back together:
3 and $\frac{1}{22}$ is $3\frac{1}{22}$

So $4\frac{1}{2} - 1\frac{5}{11} = 3\frac{1}{22}$

Example 14

Work out $2\frac{3}{12} - \frac{8}{12}$

$\frac{3}{12}$ is smaller than $\frac{8}{12}$ so you can't just subtract the fractions on their own.

Change the mixed number $2\frac{3}{12}$ into a top heavy (or improper) fraction.

$$2\frac{3}{12} \quad = \quad \frac{24}{12} + \frac{3}{12} = \quad \frac{27}{12}$$

So: $2\frac{3}{12} - \frac{8}{12} = \frac{27}{12} - \frac{8}{12}$

$$= \frac{19}{12} \text{ usually written as a mixed number: } 1\frac{7}{12}$$

■   **To subtract fractions find equivalent fractions that have the same denominator (bottom).**

Exercise 4I

1   Work out:
      (a) $\frac{5}{11} - \frac{3}{11}$ **(b)** $\frac{7}{9} - \frac{5}{9}$ **(c)** $\frac{7}{8} - \frac{1}{8}$ **(d)** $\frac{7}{12} - \frac{5}{12}$

2   Work out:
      (a) $\frac{1}{2} - \frac{1}{4}$ **(b)** $\frac{7}{8} - \frac{3}{4}$ **(c)** $\frac{5}{8} - \frac{1}{2}$ **(d)** $\frac{3}{4} - \frac{1}{8}$

      (e) $\frac{5}{6} - \frac{1}{3}$ **(f)** $\frac{7}{12} - \frac{1}{3}$ **(g)** $\frac{9}{10} - \frac{2}{5}$ **(h)** $\frac{1}{4} - \frac{1}{20}$

3 Work out:

(a) $\frac{5}{8} - \frac{1}{6}$ (b) $\frac{7}{10} - \frac{1}{4}$ (c) $\frac{7}{9} - \frac{5}{12}$ (d) $\frac{5}{6} - \frac{4}{9}$

(e) $\frac{7}{10} - \frac{4}{15}$ (f) $\frac{3}{4} - \frac{7}{20}$ (g) $\frac{3}{4} - \frac{3}{10}$ (h) $\frac{7}{8} - \frac{7}{10}$

4 Work out:

(a) $\frac{2}{3} - \frac{1}{2}$ (b) $\frac{5}{8} - \frac{1}{3}$ (c) $\frac{1}{5} - \frac{1}{6}$ (d) $\frac{3}{5} - \frac{1}{6}$

(e) $\frac{4}{5} - \frac{2}{3}$ (f) $\frac{3}{4} - \frac{3}{5}$ (g) $\frac{7}{10} - \frac{1}{3}$ (h) $\frac{9}{10} - \frac{3}{4}$

5 Work out:

(a) $5\frac{1}{4} - \frac{1}{10}$ (b) $7\frac{1}{2} - \frac{1}{3}$ (c) $6\frac{1}{2} - 5\frac{1}{4}$ (d) $9\frac{1}{2} - 7\frac{3}{10}$

(e) $4 - 1\frac{3}{10}$ (f) $4\frac{4}{5} - 3\frac{9}{10}$ (g) $1\frac{2}{3} - \frac{11}{12}$ (h) $5\frac{3}{4} - 2\frac{19}{20}$

6 In a school, $\frac{7}{16}$ of the students are girls.
What fraction of the students are boys?

7 $\frac{2}{5}$ of the students at Hay College wear contact lenses.
What fraction of the students do not wear them?

8 The garden of Granny Smith's house measures $1\frac{1}{3}$ acres.
Sharky Estates buy $1\frac{1}{4}$ acres of the garden to build new
homes.
How much garden does Granny Smith have left?

4.10 Multiplying fractions

How to multiply two fractions:

■

$$\frac{5}{8} \times \frac{7}{10} = \frac{35}{80}$$

Multiply the numerators (top)

Multiply the denominators (bottom)

So $\frac{5}{8} \times \frac{7}{10} = \frac{35}{80}$ This simplifies to $\frac{7}{16}$ (by dividing top and
bottom of $\frac{35}{80}$ by 5).

Another way of doing this is to simplify the fractions *before* you
multiply them:

$$\frac{5}{8} \times \frac{7}{10} = \frac{5 \times 7}{8 \times 10}$$

You can simplify here by
dividing the top and
bottom by 5.

$$\frac{(5 \times 7) \div 5}{(8 \times 10) \div 5} = \frac{7}{8 \times 2} = \frac{7}{16}$$

This gives the same answer $\frac{7}{16}$ as the first method.

This method is less obvious than the first one. Ask your
teacher if you need help to understand it.

How to multiply a fraction by a whole number:

$$\frac{7}{10} \times 4 = ?$$

You can write 4 as the **top heavy** (or improper) fraction $\frac{4}{1}$

$$\frac{7}{10} \times \frac{4}{1} = \frac{28}{10}$$

Multiply the numerators (top)

Multiply the denominators (bottom)

So $\frac{7}{10} \times 4 = \frac{28}{10}$ This is usually written as a mixed number: $2\frac{8}{10}$

The fraction part of $2\frac{8}{10}$ simplifies (by dividing top and bottom by 2) so $2\frac{8}{10} = 2\frac{4}{5}$

How to multiply two mixed numbers:

$$3\frac{1}{4} \times 2\frac{4}{5} = ?$$

Change both mixed numbers to top heavy (or improper) fractions:

$$3\frac{1}{4} = \frac{12}{4} + \frac{1}{4} = \frac{13}{4} \qquad\qquad 2\frac{4}{5} = \frac{10}{5} + \frac{4}{5} = \frac{14}{5}$$

$$\frac{13}{4} \times \frac{14}{5} = \frac{182}{20}$$

Multiply

Multiply

So $3\frac{1}{4} \times 2\frac{4}{5} = \frac{182}{20}$ This is usually written as a mixed number: $9\frac{2}{20}$

The fraction part of $9\frac{2}{20}$ simplifies (by dividing top and bottom by 2) so $9\frac{2}{20} = 9\frac{1}{10}$

Exercise 4J

1 Work out:

   (a) $\frac{1}{2} \times \frac{3}{4}$ **(b)** $\frac{3}{8} \times \frac{1}{4}$ **(c)** $\frac{2}{5} \times \frac{4}{5}$ **(d)** $\frac{3}{8} \times \frac{3}{4}$

   (e) $\frac{5}{12} \times \frac{1}{3}$ **(f)** $\frac{7}{10} \times \frac{3}{4}$ **(g)** $\frac{3}{10} \times \frac{3}{5}$ **(h)** $\frac{2}{3} \times \frac{2}{3}$

2 Work out:
(a) $\frac{1}{2} \times \frac{4}{5}$ (b) $\frac{3}{4} \times \frac{4}{5}$ (c) $\frac{5}{6} \times \frac{3}{5}$ (d) $\frac{4}{5} \times \frac{3}{10}$
(e) $\frac{5}{6} \times \frac{3}{4}$ (f) $\frac{7}{12} \times \frac{3}{14}$ (g) $\frac{8}{9} \times \frac{3}{10}$ (h) $\frac{3}{4} \times \frac{16}{21}$

3 Work out:
(a) $\frac{1}{2} \times 7$ (b) $\frac{2}{3} \times 5$ (c) $6 \times \frac{4}{5}$ (d) $8 \times \frac{3}{4}$
(e) $\frac{7}{10} \times 20$ (f) $9 \times \frac{2}{3}$ (g) $10 \times \frac{2}{5}$ (h) $\frac{5}{6} \times 12$

4 Work out:
(a) $3\frac{1}{4} \times \frac{1}{2}$ (b) $\frac{2}{3} \times 4\frac{1}{2}$ (c) $\frac{5}{6} \times 1\frac{1}{3}$ (d) $2\frac{1}{2} \times \frac{7}{10}$
(e) $3\frac{1}{2} \times 1\frac{1}{2}$ (f) $2\frac{1}{3} \times 2\frac{3}{8}$ (g) $1\frac{4}{5} \times 2\frac{1}{3}$ (h) $3\frac{3}{4} \times 1\frac{2}{5}$

5 On Monday to Friday inclusive Jamie spends $2\frac{1}{4}$ hours on his homework but his sister Claire spends only $1\frac{3}{4}$ hours each day on hers.
How long in a week does each one spend on homework?

6 A machine takes $5\frac{1}{2}$ minutes to produce a special type of container.
How long would the machine take to produce 15 containers?

Area of rectangle is
length × width
Volume of box is
length × width × depth

7 Calculate the area of a rectangle of length $3\frac{1}{4}$ cm and width $2\frac{1}{4}$ cm.

8 A hand-made chocolate box is 6 inches long by $3\frac{1}{2}$ inches wide by $2\frac{1}{4}$ inches deep.
Calculate the volume of the box.

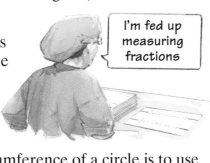

I'm fed up measuring fractions

Roll on metrication

9 One way to calculate the circumference of a circle is to use the formula: Circumference = diameter $\times 3\frac{1}{7}$
Calculate the circumference of a circle with diameter:
(a) 4 cm   (b) $6\frac{1}{2}$ cm   (c) $\frac{3}{4}$ cm

4.11 Dividing fractions

$$\frac{1}{4} \div \frac{3}{5} = ?$$

Turn the ÷ sign into a × sign.

$$\frac{1}{4} \times \frac{5}{3} = \frac{5}{12}$$

So $\frac{1}{4} \div \frac{3}{5} = \frac{5}{12}$

Turn the fraction you are dividing by upside down.

This is called **inverting** the fraction.

Why inverting works
A fraction like $\frac{3}{4}$ is the same as:
$$3 \div 4$$
or $3 \times \frac{1}{4}$
But $\frac{1}{4}$ is $\frac{4}{1}$ inverted.

So dividing is the same as multiplying by the inverted number.
$$12 \div 3 = 4$$
$$12 \times \frac{1}{3} = 4 \quad \text{too}$$

How to divide mixed numbers:

$$3\tfrac{1}{2} \div 4\tfrac{3}{4} = ?$$

Change mixed numbers to top heavy (or improper) fractions.

$$\frac{7}{2} \div \frac{19}{4}$$

Turn ÷ into ×

Invert the dividing fraction.

$$\frac{7}{2} \times \frac{4}{19} = \frac{28}{38}$$

So $3\tfrac{1}{2} \div 4\tfrac{3}{4} = \frac{28}{38}$ This simplifies to $\frac{14}{19}$ (by dividing top and bottom by 2).

How to divide a fraction by a whole number

$$\frac{15}{16} \div 5 = ?$$

Turn ÷ into ×

Invert the dividing number

$$\frac{15}{16} \times \frac{1}{5} = \frac{15}{80}$$

Remember:

$5 = \frac{5}{1}$

so inverting gives

$\frac{1}{5}$

So $\frac{15}{16} \div 5 = \frac{15}{80}$ This simplifies to $\frac{3}{16}$ (by dividing top and bottom by 5).

■ **To divide fractions, invert the dividing fraction (turn it upside down) and multiply.**

Exercise 4K

1 Work out:
   (a) $\frac{1}{3} \div \frac{1}{4}$ **(b)** $\frac{1}{4} \div \frac{1}{3}$ **(c)** $\frac{3}{4} \div \frac{1}{2}$ **(d)** $\frac{1}{2} \div \frac{7}{10}$
   (e) $\frac{2}{3} \div \frac{1}{5}$ **(f)** $\frac{5}{8} \div \frac{1}{3}$ **(g)** $\frac{5}{6} \div \frac{3}{4}$ **(h)** $\frac{7}{10} \div \frac{4}{5}$

2 Work out:
   (a) $2\tfrac{1}{2} \div \tfrac{1}{2}$ **(b)** $3\tfrac{1}{4} \div 2\tfrac{1}{2}$ **(c)** $3\tfrac{3}{4} \div 2\tfrac{1}{4}$ **(d)** $1\tfrac{5}{8} \div 3\tfrac{1}{6}$
   (e) $3\tfrac{2}{3} \div 7\tfrac{1}{3}$ **(f)** $5\tfrac{1}{2} \div 2\tfrac{3}{4}$ **(g)** $1\tfrac{7}{10} \div 2\tfrac{7}{10}$ **(h)** $\tfrac{7}{8} \div 1\tfrac{2}{3}$

3 Work out:
   (a) $\frac{3}{4} \div 8$ **(b)** $\frac{5}{6} \div 2$ **(c)** $\frac{3}{5} \div 6$ **(d)** $\frac{4}{5} \div 5$
   (e) $1\tfrac{1}{3} \div 4$ **(f)** $3\tfrac{1}{4} \div 6$ **(g)** $2\tfrac{5}{6} \div 10$ **(h)** $2\tfrac{1}{2} \div 15$

4 Work out:
   (a) $8 \div \frac{1}{2}$ **(b)** $12 \div \frac{3}{4}$ **(c)** $6 \div \frac{3}{5}$ **(d)** $8 \div \frac{7}{8}$
   (e) $4 \div \frac{4}{5}$ **(f)** $1 \div \frac{7}{12}$ **(g)** $5 \div \frac{1}{3}$ **(h)** $6 \div \frac{1}{4}$

Summary of key points

1 Fraction words you need to know:

$\dfrac{3}{4}$ — numerator
— denominator

2 Top heavy fractions like $\frac{11}{9}$ are also called improper fractions.

3 An improper fraction like $\frac{11}{9}$ can be written as a mixed number: $1\frac{2}{9}$

4 Fractions can be simplified if the numerator (top) and denominator (bottom) have a common factor:

$$\underset{\div 4}{\overset{\div 4}{\dfrac{8}{12}}} \quad \text{simplifies to} \quad \dfrac{2}{3} \qquad \text{the common factor is 4}$$

5 Equivalent fractions are fractions that have the same value:

$$\dfrac{8}{12} = \dfrac{4}{6} = \dfrac{2}{3}$$

6 To add fractions find equivalent fractions that have the same denominator (bottom):

$$\times 3 \left(\dfrac{1}{2} + \dfrac{4}{6} \right) \times 3$$
$$\dfrac{3}{6} + \dfrac{4}{6} = \dfrac{7}{6}$$

7 To subtract fractions find equivalent fractions that have the same denominator (bottom):

$$\dfrac{4}{6} \underset{\times 3}{-} \left(\dfrac{1}{2} \right) \times 3$$
$$\dfrac{4}{6} - \dfrac{3}{6} = \dfrac{1}{6}$$

8 To multiply two fractions:

Multiply numerators (top)

$$\dfrac{3}{4} \times \dfrac{4}{7} = \dfrac{12}{28}$$

Multiply denominators (bottom)

9 To divide fractions, invert the dividing fraction (turn it upside down) and multiply:

$$\dfrac{1}{4} \div \dfrac{2}{5}$$

Turn ÷ into ×

Invert (turn upside down)

$$\dfrac{1}{4} \times \dfrac{5}{2}$$

5 Two-dimensional shapes

This design for a bathroom tile is made from three different types of shape: triangles, squares and rectangles.

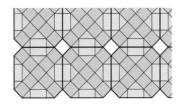

On paper the design is flat and has no thickness. It is a two-dimensional (or 2-D) shape.

■ **2-D shapes are flat; they have no thickness.**

This unit shows you how to recognise and draw some common 2-D shapes.

5.1 Recognising triangles and quadrilaterals

■ **A triangle is a three-sided shape.**

You need to be able to recognise these special types of triangles:

Shape					
Name	**Isosceles triangle**	**Equilateral triangle**	**Right-angled triangle**	**Obtuse-angled triangle**	**Scalene triangle**
Properties	2 equal sides 2 equal angles	3 equal sides all angles equal 60°	one angle 90°	one angle greater than 90°	no special features

Dashes show equal sides

These show equal angles

This is a right angle (90°)

■ **A quadrilateral is a four-sided shape.**

You need to be able to recognise these special types of quadrilateral.

Name	Shape	Properties
Trapezium		1 pair of parallel sides.
Parallelogram		2 pairs of parallel sides, opposite sides equal, opposite angles equal.
Rhombus		2 pairs of parallel sides, all sides equal, diagonals cross at right angles.
Rectangle		All angles are 90°, opposite sides equal and parallel, diagonals equal.
Square		All angles 90°, all sides equal, opposite sides parallel, diagonals equal and cross at right angles.
Kite		Two pairs of adjacent sides (next to each other) equal, 1 pair opposite angles equal, diagonals cross at 90°.

Sides marked with the same number of arrows are parallel.

Sides marked with the same number of dashes are equal.

Exercise 5A

1   Copy and complete these sentences. Use numbers and/or
    the correct word chosen from:

    equal   opposite   parallel   sides

    The first two have been done for you.

    (a) A **triangle** has __3__ sides and __3__ angles.

    (b) An **equilateral triangle** has <u>equal</u> sides. Each angle
    measures <u>60</u>°.

    (c) An **isosceles triangle** has _____ equal sides and _____
    angles which are equal.

    (d) The largest angle in a **right-angled triangle** always
    measures _____ .

    (e) **Quadrilaterals** have _____ sides and angles.

    (f) The angles in a **rectangle** are all _____ . Opposite sides
    are _____ .

    (g) All the sides of a **square** are _____ and the angles are
    all _____ .

    (h) A **kite** has 2 pairs of equal _____ and one pair of _____
    angles.

    (i) _____ sides and angles of a **parallelogram** are _____ .

    _____ sides are parallel.

    (j) A **rhombus** has 4 equal _____ and opposite angles are

    _____ .

    _____ sides are parallel.

    (k) A **trapezium** has 1 pair of _____ sides.

2   Name these special quadrilaterals:

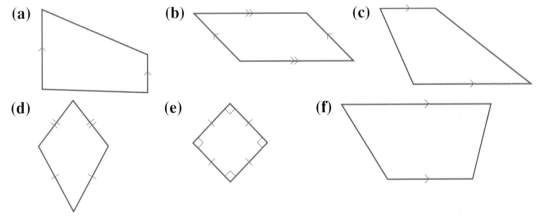

3 You will need a ruler and a protractor.

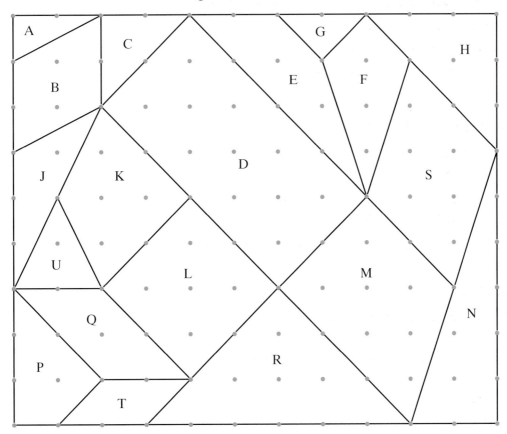

Copy and complete these statements about the 2-D shapes in the picture.
Use a ruler and protractor to help you.
The first one has been done for you.

(a) Shapes <u>A, C, G, H, N and R</u> are all right-angled triangles.

(b) Shape ____ is a square.

(c) Shapes ____ are all parallelograms.

(d) Shape D is a ____ .

(e) Shapes ____ and ____ are trapeziums.

(f) The kites are shapes ____ and ____ .

(g) Shape ____ is an ordinary quadrilateral.

(h) Shape U is an ____ triangle.

(i) The only square in the picture is shape ____ .

(j) Shapes ____ are isosceles triangles.

5.2 Drawing 2-D shapes on squared paper

Squared paper makes it easier to draw some 2-D shapes. Parallel lines and lines at right angles are easily drawn, and lengths are easily measured.

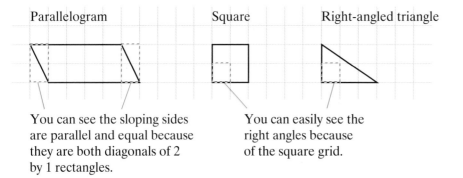

Parallelogram   Square   Right-angled triangle

You can see the sloping sides are parallel and equal because they are both diagonals of 2 by 1 rectangles.

You can easily see the right angles because of the square grid.

Exercise 5B

1   On squared paper draw:
   (a) a rectangle with sides 2 cm and 3 cm
   (b) a right-angled triangle with base 3 cm and height 2 cm
   (c) a square with sides 3 cm long.

2   On squared paper draw:
   (a) an isosceles triangle with a base of 5 cm and a height of 3 cm
   (b) a parallelogram with the longest sides 5 cm
   (c) a trapezium with parallel sides that add up to 9 cm and with height 3 cm.

3   Design some square tiles using quadrilaterals and triangles. Do not use too many different shapes in any one design. Here are two designs from the floor of a Roman Villa:

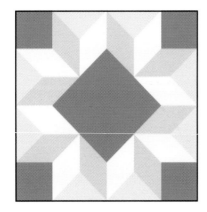
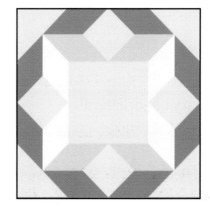

5.3 Drawing more complicated shapes

Sometimes you will need to draw more complicated shapes accurately using a ruler, protractor and compasses.

Example 1

Draw a triangle PQR with sides $PQ = 7\,cm$, $PR = 5\,cm$ and $QR = 9\,cm$.

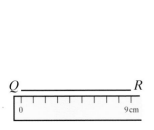 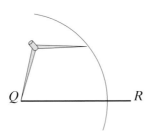 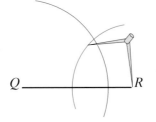

Point P is 7 cm from Q and 5 cm from R.

Draw the line $QR = 9\,cm$. Use compasses to draw the points 7 cm from Q. Use compasses to draw points 5 cm from R. Join P to Q and P to R to make a triangle PQR.

Exercise 5C

Use a protractor and compasses to help you do these questions.

1 Draw:

   (a) an equilateral triangle with sides 5 cm long

   (b) a triangle with sides 6 cm, 5 cm and 7 cm long

   (c) an isosceles triangle with sides 6 cm, 5 cm and 5 cm.

2 Draw these shapes accurately:

(a) **(b)** 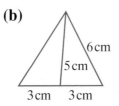 **(c)**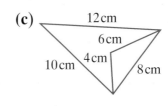

Make accurate drawings of:

3 **(a)** 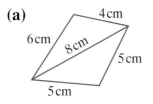 **(b)** 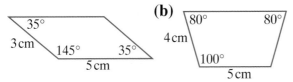 **(c)**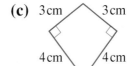

4 A parallelogram $PQRS$ where $PQ = RS = 8\,cm$, $PS = QR = 5\,cm$, angle $QPS = 60°$ and angle $PQR = 120°$.

5   A rhombus *DEFG* with sides of 6 cm and angle
    DEF = 35°.

6   A right-angled triangle *EFG* with the right angle at *F* and
    with *EF* = 4 cm and *FG* = 3 cm. Measure *EG*.

7   A triangle *XYZ* where *XY* = 6 cm, angle *XYZ* = 65° and
    angle *YXZ* = 45°. Measure *XZ* and *YZ* as accurately as
    you can.

5.4 Congruent shapes

These three shapes are facing different ways but their sides are
the same length and their angles are the same size.

■   **Shapes which are exactly the same size and shape are
    congruent.**

Reflected shapes are still
the same size and shape.
These shapes are
congruent.

Example 2

Which of these shapes are congruent?

A B C D

Shapes **A** and **C** are **congruent**. They have the same length
sides and the same size angles.

Exercise 5D

Write down the letters of the shapes which are congruent.

1     2

   A   B   C   D   A   B   C   D

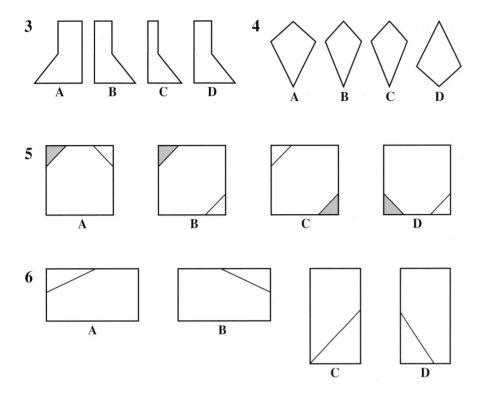

3 A B C D   **4** A B C D

5 A B C D

6 A B C D

5.5 Congruent triangles

You may be asked to identify congruent triangles from the information you are given about their sides and angles.

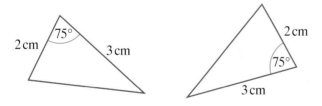

> You can measure the missing sides and angles to check these triangles are congruent.

These two triangles are congruent.

This is because they have two sides the same *and* the angle between the two sides is the same. This is known as side, angle, side or **SAS**.

Other ways of identifying congruent triangles are:

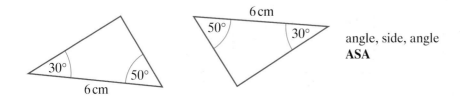

angle, side, angle
ASA

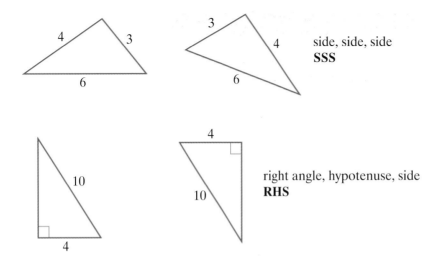

side, side, side
SSS

right angle, hypotenuse, side
RHS

■ **For triangles to be congruent they must demonstrate SAS, ASA, SSS or RHS.**

Example 3

From the information given say which pairs of triangles are congruent and give a reason for your answer.

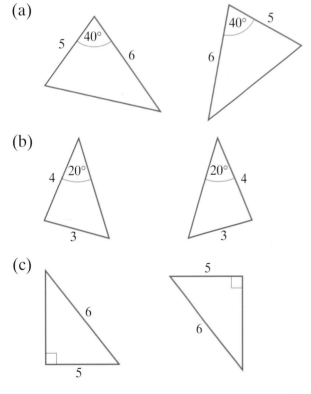

(a)

(b)

(c)

(a) Congruent (SAS)
(b) Not necessarily congruent
(c) Congruent (RHS)

Exercise 5E

1   For each pair of triangles say **(i)** if they are congruent
    (ii) give a reason if they are congruent.

(a)
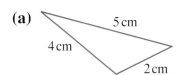

(b)
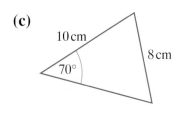
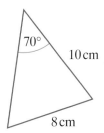

(c)
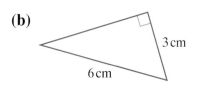
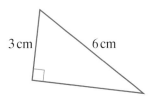

(d)
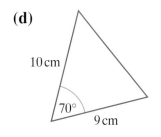
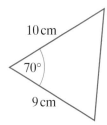

(e)
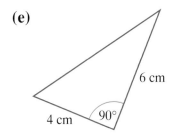
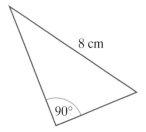

(f)
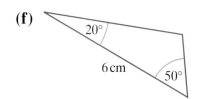
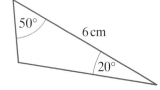

Write down the letters of the congruent triangles in
questions **2–5**, and give a reason for your answer.

2 **A** 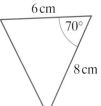 **B** 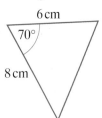 **C**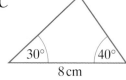

3 **A** 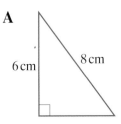 **B** 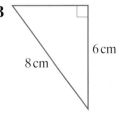 **C**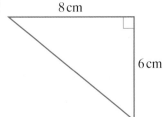

4 **A** 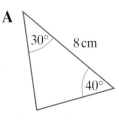 **B** 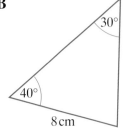 **C**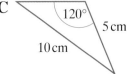

5 **A** 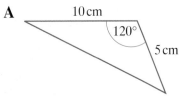 **B** 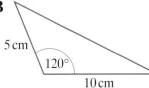 **C**

6 Draw two triangles to show why the triangles in these pairs
might not be congruent.

(a)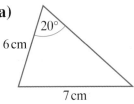

(b)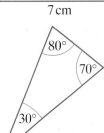

5.6 Polygons

All these shapes are polygons:

Not all sides of a polygon need to be the same.

Polygons can have sides which point inwards.

Polygons are closed shapes – there are no gaps in the perimeter.

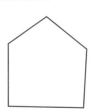

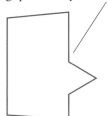

- ■ **Polygon is a 2-D shape with straight sides.**

You need to know the names of some special polygons:

- A polygon with 5 sides is a **pentagon**. (penta = 5; a modern pentathlon has 5 events)
- A polygon with 6 sides is a **hexagon**. (hex = 6)
- A polygon with 8 sides is an **octagon**. (octo = 8; an octopus has 8 tentacles)
- A polygon with 10 sides is a **decagon**. (deca = 10; a decathlon has 10 events)

Regular polygons

- ■ **A polygon is a *regular polygon* if its sides are all the same length and its angles are all the same size.**

- ■ **Regular polygons can be constructed by equal division of a circle.**

Equilateral triangles and squares are regular polygons. Here are some others you will use:

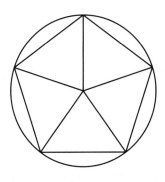

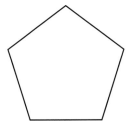
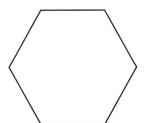
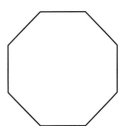

Interior and exterior angles of polygons

When you extend a side of a polygon you get two angles. One inside the polygon and one outside.

Learn these two rules:

- ■ **For a polygon: interior angle + exterior angle = 180°.**

- ■ **The exterior angles of a polygon always add up to 360°.**

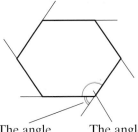

The angle outside is called the **exterior angle**.

The angle inside is called the **interior angle**.

Example 4

Find the sum of:
  (a) interior angles of a pentagon
  (b) the exterior angles of a pentagon.

(a) The inside of the pentagon can be divided into three triangles. The angles of a triangle always add up to 180°, so the angles of a pentagon add up to 3 × 180° = 540°.

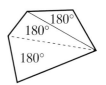

(b) The exterior angles are marked on this pentagon. Start at one corner and work clockwise ↻ around the pentagon.

At each corner you turn clockwise.

By the time you have gone past every corner you will have turned one full turn, so the sum of the exterior angles is 360°.

Exercise 5F

1 Draw the different types of polygons listed in this table. For each one extend the sides to show the exterior angles. Then:

  • Measure the interior and exterior angles and record them in your table.
  • Complete the table to show the sum of the interior angles and the sum of the exterior angles.
  The first one is done for you.

Type of polygon	Interior angles	Sum	Exterior angles	Sum
Triangle	43° + 122° + 15° =	180°	137° + 58° + 165° =	360°
Equilateral triangle				
Isosceles triangle				
Parallelogram				
Pentagon				

Example 5

Work out the exterior angle of a regular hexagon and use it to find the interior angle.

The hexagon has 6 sides.
The exterior angles add up to 360°
So each exterior angle is 360° ÷ 6 = 60°

Each interior angle is 180° − 60° = 120°

There are six equal exterior angles of 60°

Each interior angle is 180° − 60° = 120°

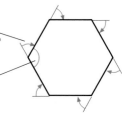

Exercise 5G

1 Work out:
   (a) the exterior angle of a regular octagon
   (b) the interior angle of a regular pentagon
   (c) the interior angle and exterior angle of a regular decagon.

2 A regular polygon has an exterior angle of 24°. Work out how many sides it has.

3 A regular polygon has an interior angle of 172°. Find the exterior angle and work out how many sides it has.

4 Draw a regular hexagon with sides 6 cm long.
   ● Join all the vertices (corners).
   ● Identify as many of the special types of quadrilaterals from page 85 as you can find in your hexagon.

5 Repeat question **4** for a regular pentagon.
   ● Identify different congruent shapes and say how many of each of them there are.
   ● Present your answers in a table.
   ● Find and mark this shape in your pentagon: ▲
     It is called an arrowhead.

5.7 Tessellations

■ **A pattern of shapes which fit together without leaving any gaps or overlapping is called a tessellation.**

You can tessellate some shapes to cover a flat surface. It is easier to show this if you work on dotted grid paper.

Example 6

Tessellate this trapezium on a grid.

There are three possible ways of doing this. They are:

  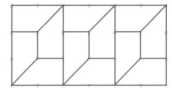

Exercise 5H

You will need dotted grid paper.

1 Show how each of these shapes tessellates.

(a)   (b)   (c)  (d)

(e)   (f) 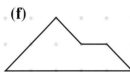 (g)  (h)

2 Make some interesting patterns using combinations of different shapes that tessellate together.

3 Invent some shapes with curved sides that tessellate.

Summary of key points

1 2-D shapes are flat. They have no thickness.

2 A triangle is a three-sided shape.

3 A quadrilateral is a four-sided shape.

4 Shapes which are exactly the same size and shape are congruent.

5 For triangles to be congruent they must demonstrate SAS, ASA, SSS or RHS.

6 Polygon is the general name for a 2-D shape with straight sides.

7 A polygon is a regular polygon if its sides are all the same length and its angles are all the same size.

8 For a polygon: interior angle + exterior angle = 180°.

9 The exterior angles of a polygon always add up to 360°.

10 A pattern of shapes which fit together without leaving gaps or overlapping is called a tessellation.

6 Decimals

6.1 Understanding place value

Some things in life have whole number values.

There are a *whole* number of people and animals

Other items do not have whole number values.
The weight of a packet of sugar is 2.2 lbs.
The height of a person may be 1.76 metres.

2.2 is read as two point two.
1.76 is read as one point seven six.

These are *decimal* numbers. Decimals are used when you need to record a value more accurately than a whole number.

- **In a decimal number the decimal point separates the whole number from the part that is smaller than 1.**

Example 1

A Formula One Grand Prix driver has his lap time recorded as 53.398 seconds.

You can understand better what 53.398 seconds really means by drawing a decimal place value diagram.

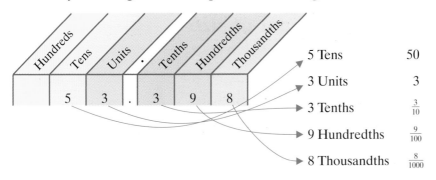

Read the whole number and then read the digits in order.
Fifty three point three nine eight

Hundreds	Tens	Units	.	Tenths	Hundredths	Thousandths		
	5	3	.	3	9	8		

5 Tens   50

3 Units   3

3 Tenths $\frac{3}{10}$

9 Hundredths $\frac{9}{100}$

8 Thousandths $\frac{8}{1000}$

Example 2

A woman 400 m hurdler's time is 56.08 seconds. Draw up a place value table.

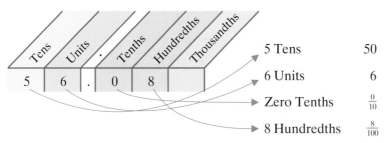

					5 Tens	50
					6 Units	6
					Zero Tenths	$\frac{0}{10}$
					8 Hundredths	$\frac{8}{100}$

Even though there are zero *tenths* the 0 has to be recorded to keep the 8 in its correct place value position.

Example 3

Write down the face and place value of the digit underlined in each number

(a) 3<u>2</u>.8 (b) 0.38<u>5</u> (c) 10.<u>0</u>3
(d) 4.2<u>9</u>0

(a) 2 units (b) 5 thousandths (c) 0 tenths
(d) 9 hundredths

Exercise 6A

1   Draw a place value diagram like the one in Examples 1 and 2 and write in these numbers.
   (**a**) 41.6 (**b**) 4.16 (**c**) 34.6 (**d**) 1.463
   (**e**) 0.643 (**f**) 1.005 (**g**) 5.01 (**h**) 0.086

2   What is the place value of the digit underlined in each number?
   (**a**) 2<u>5</u>.4 (**b**) 2.<u>5</u>4 (**c**) 25.<u>4</u>6 (**d**) 3.5<u>4</u>6
   (**e**) <u>1</u>8.07 (**f**) 9.66<u>9</u> (**g**) 216.0<u>3</u>1 (**h**) 2.135<u>7</u>
   (**i**) 9.1<u>0</u>2 (**j**) 3.<u>3</u>36 (**k**) 2.59<u>1</u> (**l**) 0.0<u>2</u>7

6.2 Writing decimal numbers in order of size

To arrange decimal numbers in order of size you need to have a good understanding of place value.

■   **You can sort decimal numbers in order of size by first comparing the whole numbers, then the digits in the tenths place, then the digits in the hundredths place, and so on.**

Example 4

Write these decimal numbers in order of size, starting with the largest: 3.069, 5.2, 3.4, 3.08, 7.0

Step 1	**Step 2**	**Step 3**
Whole numbers:	Tenths place:	Hundreths place:
7 is bigger than **5**	**4** is bigger than **0**	**8** is bigger than **6**
5 is bigger than **3**		
7.0	7.0	7.0
5.2	5.2	5.2
3.069	3.**4**	3.**4**
3.**4**		
3.**08**	3.**069**	3.**08**
	3.**08**	3.**069**

So the order is 7.0, 5.2, 3.4, 3.08, 3.069.

Exercise 6B

1 Rearrange these decimal numbers in order of size starting with the largest:

   (a) 0.62, 0.71, 0.68, 0.76, 0.9

   (b) 3.4, 3.12, 3.75, 2.13, 2.09

   (c) 5.2, 3.6, 5.04, 5.16, 3.47

   (d) 0.42, 0.065, 0.407, 0.3, 0.09

   (e) 3.0, 6.52, 6.08, 3.58, 3.7

   (f) 0.06, 0.13, 0.009, 0.105, 0.024

   (g) 0.08, 0.8, 0.05, 0.2, 0.525

   (h) 2.09, 1.08, 2.2, 1.3, 1.16

2 Put these decimal numbers in order of size, smallest first:

   (a) 4.85, 5.9, 5.16, 4.09, 5.23

   (b) 0.34, 0.09, 0.37, 0.021, 0.4

   (c) 5, 7.23, 5.01, 7.07, 5.009

   (d) 1.001, 0.23, 1.08, 1.14, 0.07

3 The table gives the price of a pack of Sudso soap powder in different shops:

Shop	Stall	Corner	Market	Main	Store	Super
Price	£1.29	£1.18	£1.09	£1.31	£1.20	£1.13

Write the list of prices in order starting with the lowest price.

4 The table gives the height in metres of six girls:

Rachel	Ira	Sheila	Naomi	Latif	Jean
1.56	1.74	1.78	1.65	1.87	1.7

Write the list of names in descending order of height, starting with the tallest.

5 The fastest lap times (in minutes) of six drivers was:

Ascarino	53.072	Bertolini	53.207
Rascini	52.037	Alloway	57.320
Silverman	53.027	Killim	53.702

Write down the drivers' times in order, fastest first.

6 A new cereal gives the weights per 100 g of vitamins and minerals:

Fibre	1.5 g	Iron	0.014 g
Vitamin B6	0.002 g	Thiamin B1	0.0014 g
Riboflavin B2	0.0015 g	Sodium	0.02 g

Write down the weights in order starting with the lowest.

6.3 Rounding decimal numbers

As with ordinary numbers, it is sometimes helpful to round a decimal number and give the result correct to the nearest whole number or correct to so many decimal places (d.p.).

Example 5

Round £5.11 to the nearest pound.

■ **To round to the nearest whole number, look at the figure in the tenths column or first decimal place. If it is 5 or more round the whole number up. If it is less than 5 do not change the whole number.**

In this example the first decimal place is **less** than 5 so you do not change the whole number

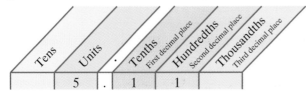

£5.11 to the nearest pound is £5

Example 6

Round 7.815 to the nearest whole number.

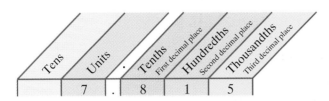

7 is in the units column
8 is in the first decimal place
1 is in the second decimal place
5 is in the third decimal place

8 in the first decimal place is more than 5 so you round up.

7.815 is rounded to 8.

Example 7

Terry spent £47.52 on a new video game. Round the cost to the nearest pound.

Because the figure in the first decimal place is 5 (or more), the whole number is rounded to £48.

■ **To round a decimal correct to one place of decimals (1 d.p.) you look at the second place of decimals. If it is 5 or more you round up, if it is less you leave this and any remaining numbers out.**

To correct to *two* places of decimals you look at the third place.
To correct to three places of decimals you look at the fourth place and so on …

Example 8

Round the following numbers correct to *one* place of decimals.

(a) 25.27   (b) 25.72

(c) 25.55   (d) 25.528

(a) The second decimal place is 7 which is 5 or more so round the 2 up to 3. The answer is 25.3.

(b) The second decimal place is 2 which is less than 5 so leave this number out. The answer is 25.7.

(c) The second decimal place is 5 which is 5 or more so round the 5 up to 6. The answer is 25.6.

(d) The second decimal place is 2 which is less than 5 so leave this and any other numbers out. The answer is 25.5.

Exercise 6C

1   Round to the nearest whole number:
   (a) 7.8   **(b)** 13.29   **(c)** 14.361   **(d)** 5.802
   (e) 10.59   **(f)** 19.62   **(g)** 0.771   **(h)** 20.499
   (i) 0.89   **(j)** 100.09   **(k)** 19.55   **(l)** 1.99

2   Round these numbers to one place of decimals:
   (a) 3.6061   **(b)** 5.3391   **(c)** 0.0901   **(d)** 9.347
   (e) 10.6515   **(f)** 7.989   **(g)** 2.0616   **(h)** 0.4999
   (i) 2.45   **(j)** 125.67   **(k)** 0.05   **(l)** 9.890

3   Round:
   (a) 13.6 mm to the nearest mm
   (b) 80.09 m to the nearest m
   (c) 0.907 kg to the nearest kg
   (d) £204.49 to the nearest £
   (e) 3.601 lb to the nearest lb
   (f) 0.299 tonne to the nearest tonne
   (g) 10.5001 g to the nearest g
   (h) 8.066 mins to the nearest min

6.4 Adding and subtracting with decimals

Example 9

Two children weigh 24.5 kg and 35.75 kg. What is their combined weight?

Combined weight is 24.5 kg + 35.75 kg

Keep numbers in their columns like in a place value diagram.

Put the decimal points under each other.

24.5

35.75

Decimal point in the answer will be in line.

Then add:
$$
\begin{array}{r}
24.5 \\
+\ 35.75 \\
\hline
60.25
\end{array}
$$

Example 10

Add 13.6 and 125.403

$$
\begin{array}{r}
13.6 \\
+\ 125.403 \\
\hline
139.003
\end{array}
$$

Exercise 6D

Work these out **without a calculator**, showing all your working.

1	1.5 + 4.6	**2**	3 + 0.25
3	26.7 + 42.2	**4**	125.7 + 0.32
5	0.1 + 0.9	**6**	16.1 + 2.625
7	9.9 + 9.9	**8**	10 + 1.001
9	156.3 + 1.84	**10**	0.005 + 1.909
11	117 + 1.17	**12**	4.56 + 0.751

Exercise 6E

Work these out **without a calculator**, showing all your working.

1	6.3 + 17.2 + 8.47	**2**	13.08 + 9.3 + 6.33
3	0.612 + 3.81 + 14.7	**4**	8.6 + 3.66 + 6.066
5	7 + 3.842 + 0.222	**6**	23.43 + 5.36 + 2.216
7	3.07 + 12 + 0.0276	**8**	5.02 + 31.5 + 142.065

Example 11

Bill earns £124.65 per week but loses £33.40 in tax and national insurance. What does he take home?

£124.65 − £33.40

$$
\begin{array}{r}
124.65 \\
-\ 33.40 \\
\hline
\ .\ \\
\end{array}
$$

Remember to put the decimal points under each other.

$$
\begin{array}{r}
1^{1}24.65 \\
-\ 33.40 \\
\hline
91.25
\end{array}
$$

Bill takes home £91.25

Exercise 6F

Work these out **without a calculator**, showing all your working.

1	19.9 − 13.7	**2**	5.84 − 1.7	**3**	23.5 − 9.4
4	100.7 − 3.4	**5**	0.59 − 0.48	**6**	1 − 0.65
7	16.9 − 10.71	**8**	21.64 − 10.5	**9**	2.5 − 1.6
10	5.84 − 1.77	**11**	23.5 − 9.47	**12**	14 − 0.75
13	6.125 − 4.9	**14**	14.01 − 2.361	**15**	3.29 − 1.036
16	204.06 − 35.48				

■ **When working out a decimal addition or subtraction sum always put the decimal points under each other.**

6.5 Multiplying decimals

Example 12

Find the cost of 5 books at £4.64 each.

$$
\begin{array}{r}
464 \\
\times\ \ 5 \\
\hline
{}_{3\ 2} \\
\hline
2320
\end{array}
$$

Multiply the numbers together ignoring the decimals.

$$5 \quad \times \quad 4.64$$

0d.p.　+　　2d.p.　= 2d.p.

Count the total number of decimal places (d.p.) in the numbers you are multiplying.

2320 must have 2d.p.

Answer is £23.20

The answer must have the same number of decimal places.

Example 13

0.52 × 0.4

$$
\begin{array}{r}
52 \\
\times\ \ 4 \\
\hline
208
\end{array}
$$

$$0.52 \quad \times \quad 0.4$$

2d.p.　+　1d.p.　= 3d.p.

The answer must have 3d.p. so is 0.208

■ **When multiplying decimals the answer must have the same number of decimal places as the total number of decimal places in the numbers being multiplied.**

Exercise 6G

Work these out **without a calculator**, showing all your working.

1 Find the cost of:
   (a) 6 books at £2.25 each
   (b) 4 tins of biscuits at £1.37 each
   (c) 8 ice creams at £0.65 each
   (d) 1.5 kilos of pears at £0.80 per kilo.

2 **(a)** 7.6×4 **(b)** 0.76×4
   (c) 0.76×0.4 **(d)** 2.25×5
   (e) 2.25×0.5 **(f)** 0.225×0.5
   (g) 22.5×0.05 **(h)** 2.25×0.005
   (i) 0.225×0.005

3 **(a)** 24.6×7 **(b)** 3.15×0.03
   (c) 0.12×0.12 **(d)** 0.2×0.2
   (e) 1.5×0.6 **(f)** 0.03×0.04

4 **(a)** 6.42×10 **(b)** 64.2×10
   (c) 0.642×10 **(d)** 56.23×10
   (e) 5.623×10 **(f)** 0.05623×10

   Look carefully at your answers to question **4**. What do you notice?

5 **(a)** 0.045×100 **(b)** 0.45×100
   (c) 4.5×100 **(d)** 0.0203×100
   (e) 0.203×100 **(f)** 2.03×100

   What do you notice about your answers to question **5**?

6.6 Division with decimals

Example 14

Five friends win £216.35 in a charity lottery. They share the money equally. How much do they each get?

$216.35 \div 5$

Put the decimal points in line.

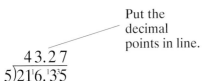

Because 5 is a whole number divide straight away.

Exercise 6H

Work out **without a calculator**, showing all your working.

1. **(a)** $64.48 \div 4$ **(b)** $3.165 \div 5$ **(c)** $133.56 \div 9$
    (d) $205.326 \div 6$ **(e)** $35.189 \div 7$ **(f)** $0.0368 \div 8$

2. **(a)** $34.5 \div 10$ **(b)** $3.45 \div 10$ **(c)** $0.345 \div 10$
    (d) $78 \div 10$ **(e)** $7.8 \div 10$ **(f)** $0.78 \div 10$
    (g) $65 \div 10$ **(h)** $65 \div 100$ **(i)** $65 \div 1000$

    Write down anything you notice about your answers to question **2**.

3. How many 3 litre jugs would be needed to hold 43.5 litres of lemonade?

4. Seven people share £107.80 equally. How much will each receive?

Example 15

1.2 metres of fabric costs £1.56. What is the cost per metre?

£1.56 ÷ 1.2 — not a whole number

> If the number you are dividing by is **not** a whole number change it to a whole number.

To change 1.2 to a whole number multiply by 10

$1.2 \times 10 = 12$

Do the same to £1.56 $1.56 \times 10 = £15.6$
The sum becomes   £15.6 ÷ 12

$$12\overline{)15.\overset{3}{6}}^{\,1.3}$$ Answer is 1.3 which in money terms is £1.30

Exercise 6I

Work out **without a calculator**, showing all your working.

1. $7.75 \div 2.5$ 2. $7.92 \div 2.2$ 3. $9.86 \div 5.8$

4. $18.9 \div 12.6$ 5. $0.129 \div 0.03$ 6. $0.27 \div 0.1$

7. $6.634 \div 0.62$ 8. $0.2121 \div 0.21$ 9. $3.5 \div 1.4$

10. $353.6 \div 6.8$

6.7 Fractions and decimals

Fractions can be changed into decimals by dividing the
numerator by the denominator.

Example 16

Change $\frac{3}{4}$ into a decimal.

$\frac{3}{4}$ means $3 \div 4$ (numerator $\div$ denominator)

$$4\overline{)3.^30^20} \quad \underset{0.7\,5}{}$$ ———— Put extra '0' here.

Answer is 0.75

To remind yourself about
numerators and
denominators look back
at Unit 4.

Some fractions which you often use are shown in this table
along with their decimal conversion equivalents:

Fraction	Decimal
$\frac{3}{10}$	0.3
$\frac{2}{5}$	0.4
$\frac{1}{4}$	0.25
$\frac{1}{2}$	0.5
$\frac{3}{4}$	0.75
$\frac{1}{8}$	0.125
$\frac{1}{3}$	0.3333 . . .

Decimals can be changed into fractions by using a place value
table.

Example 17

Change 0.763 into a fraction.

The place value table shows 0.763 is 7 tenths, 6 hundredths and
3 thousandths which is the same as 763 thousandths, so you
can write the decimal as a fraction like this:

$$\frac{763}{1000}$$

Units	.	Tenths	Hundredths	Thousandths
0	.	7	6	3

Example 18

Write as a fraction: (a)   0.7   (b)   0.59   (c)   0.071

Answers using the place value table:
(a) 7 tenths $= \frac{7}{10}$
(b) 59 hundredths $= \frac{59}{100}$
(c) 71 thousandths $= \frac{71}{1000}$

■ **Fractions can be changed into decimals by dividing the numerator by the denominator. Decimals can be changed into fractions by using a place value table.**

Exercise 6J

1   Change these fractions to decimals. Show your working.

(a) $\frac{3}{5}$ (b) $\frac{1}{2}$ (c) $\frac{7}{10}$ (d) $\frac{7}{20}$

(e) $\frac{4}{25}$ (f) $\frac{3}{50}$ (g) $\frac{7}{8}$ (h) $\frac{9}{20}$

(i) $\frac{19}{25}$ (j) $\frac{5}{16}$ (k) $\frac{1}{8}$ (l) $\frac{27}{50}$

(m) $\frac{9}{100}$ (n) $\frac{13}{200}$ (o) $\frac{2}{3}$ (p) $\frac{19}{20}$

2   Change these decimals to fractions.

(a) 0.3   (b) 0.37   (c) 0.93   (d) 0.137

(e) 0.293   (f) 0.07   (g) 0.59   (h) 0.003

(i) 0.00003   (j) 0.0013   (k) 0.77   (l) 0.077

(m) 0.39   (n) 0.0041   (o) 0.019   (p) 0.031

Summary of key points

1   In a decimal number the decimal point separates the whole number from the decimal places.

2   You can sort decimal numbers in order of size by first comparing the whole numbers, then the digits in the tenths place, then the digits in the hundredths place, and so on.

3   To round a decimal to the nearest whole number, look at the first decimal place. If it is 5 or more round the whole number up, if it is less than 5 do not change the whole number.

4   To round a decimal correct to one place of decimals look at the second place. If it is 5 or more round up, if it is less leave this and any remaining numbers out.

5   When making a decimal addition or subtraction sum always put the decimal points under each other.

6   When multiplying decimals the answer must have the same number of decimal places as the total number of decimal places in the numbers being multiplied.

7   Fractions can be changed into decimals by dividing the numerator by the denominator. Decimals can be changed into fractions by using a place value table.

7 Measure 1

7.1 Estimating

In real-life people estimate all the time. How long will it take me to walk to the shops? Have I got time for another cup of tea? Is there enough milk in the fridge for the rest of the week?

Here are some measures that you have to estimate in real life:

You need to be able to estimate measurements in metric units and the old style imperial units such as gallons, miles, and pounds.

Some imperial measures
12 inches	= 1 foot
3 feet	= 1 yard
1760 yards	= 1 mile
8 pints	= 1 gallon

7.2 Estimating lengths

■ **Lengths and distances are measured in these metric and imperial units:**
   metric: kilometres (km), metres (m), centimetres (cm), millimetres (mm)
   imperial: miles, yards, feet, inches

Here are some estimates of distances in real life:

   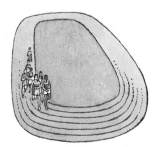

A 30 cm ruler is about 1 foot long.

A door is 2 m high or about $6\frac{1}{2}$ feet.

A long stride is 1 m long or about 3 feet.

$2\frac{1}{2}$ times around the track is 1 km or about $\frac{5}{8}$ mile.

Exercise 7A

Look at this picture, then write down an estimate for each of the following:

1 The height of the:
 (a) man **(b)** girl
 (c) bus **(d)** car

2 The length of the:
 (a) car **(b)** bus
 (c) wall

3 The height of the:
 (a) house **(b)** tree
 (c) wall

4 The width of the:
 (a) drive **(b)** garage
 (c) house

7.3 Estimating capacities

■ **Capacity is a measure of the amount a container can hold.
It is measured in these units:**
metric: litres (*l*), centilitres (*cl*), millilitres (*ml*)
imperial: gallons, pints, fluid ounces

20 fluid ounces = 1 pint
8 pints = 1 gallon

A milk carton holds
1 pint or 570 m*l*.

A petrol can holds
1 gallon or 4.5 litres.

A mug and a can of
cola hold about $\frac{1}{2}$ pint
or about 300 m*l*.

Exercise 7B

Look at this picture, then copy and complete the table with your estimates for each of the following.
Give your answers in metric and imperial units.

	Amount of	Metric	Imperial
1	milk in a full carton		
2	milk in the open carton		
3	cola in a full can		
4	cola in the glass		
5	coffee in the jug		
6	petrol in a full tank		
7	water in a full bucket		
8	water in a full paddling pool		

7.4 Estimating weights (masses)

| 16 ounces = 1 pound |
| 14 pounds = 1 stone |
| 8 stones = 1 hundredweight |
| 20 hundredweights = 1 ton |

■ **Weight is measured in these units:**
  metric: tonnes (t), kilograms (kg), grams (g), milligrams (mg)
  imperial: tons, hundredweight, stones, pounds, ounces

A $\frac{1}{4}$ pound packet of tea weighs about 100 g.

A 1 kg bag of sugar weighs about 2.2 pounds.

A 50 kg bag of cement weighs about 110 pounds.

Exercise 7C

Look at this picture, then copy and complete the table with your estimates.

Give your answer in metric and imperial units.

	Weight of the	Metric	Imperial
1	bag of potatoes		
2	block of butter		
3	bag of apples		
4	packet of coffee		
5	loaf of bread		
6	packet of cereal		
7	bottle of squash		
8	packet of biscuits		
9	packet of crisps		
10	box of soap powder		

Example 1

My car is 20 m long.

This is not a sensible statement because it means the car would be over 60 feet long or about as long as an articulated lorry.

A sensible answer would be about 4 or 5 m.

Exercise 7D

For each of these statements say whether the measurements are sensible or not. If the statement is not sensible then give a reasonable estimate for the measurement.

1 (a) My teacher is 20 m tall.
   (b) My father is 20 cm tall.
   (c) The classroom measures 2 m by 3 m.
   (d) I bought 2 g of potatoes at the supermarket.
   (e) A can of cola holds 3 *l* of liquid.
   (f) A house is 10 m high.

2 (a) The tallest boy in school is 2 m tall.
   (b) John can just lift 50 kg.
   (c) Jane has to walk 1 km to school each day.
   (d) A cup full of tea contains 2 *l* of liquid.
   (e) The river Thames is 20 km long.
   (f) A 50 p piece weighs 0.5 kg.

3 (a) A box of chocolates weighs 500 g.
   (b) A 50 cc moped can travel at 150 miles per hour.
   (c) A pint glass will hold 1 *l* of liquid.
   (d) The capacity of the petrol tank in my car is 5 *l*.

4 (a) My new sports car can travel at 100 miles per hour.
   (b) The Eiffel Tower in Paris is more than 200 m high.
   (c) The capacity of the petrol tank in my car is 50 gallons.
   (d) A packet of tea weighs 50 g.
   (e) A kilo bag of sugar weighs 2 pounds.

7.5 Choosing appropriate units of measure

When you want to measure something you have to choose the most appropriate units to use. For example, to measure how long it takes to run 100 m you use seconds. To measure how long it takes to run a 26 mile marathon you would need to use hours, minutes and seconds.

SHOULD I USE HOURS, MINUTES OR SECONDS TO TIME THE 400 METRE RACE

Exercise 7E

Copy and complete this table with appropriate units for each measurement. Give both metric and imperial units of measurement.

		Metric	Imperial
1	The length of your classroom		
2	The width of this book		
3	The distance of Edinburgh from London		
4	The length of a double decker bus		
5	The weight of a sack of potatoes		
6	The weight of a packet of sweets		
7	The weight of a lorry full of sand		
8	The amount of petrol in a car's petrol tank		
9	The amount of liquid in a full cup of tea		
10	The amount of medicine in a medicine spoon		
11	The amount of water in a raindrop		
12	The amount of water in a reservoir		
13	The time it takes to boil an egg		
14	The time it takes to run 400 metres		
15	The time it takes to walk 20 miles		
16	The time it takes to sail from Southampton to New York		
17	The length of a pencil		
18	The thickness of a page in this book		
19	The weight of 30 of these books		
20	The time it takes to travel from Earth to Mars		

7.6 Measuring time

You need to be able to:
- read the time using digital and analogue clocks
- use 12-hour and 24-hour clock times and convert from one type to the other.

■ **Digital clocks have a number display:**

■ **Analogue clocks have hands:**

Reading the time from an analogue clock

When you *say* the time you use phrases such as 'half past four' and 'ten to five'.

This clock shows the key phrases you need to know:

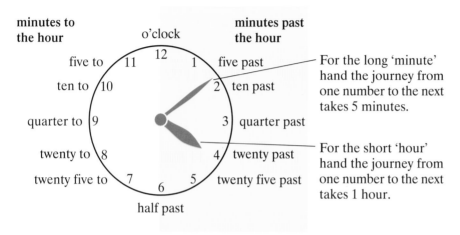

minutes to the hour

o'clock

minutes past the hour

five to 11
ten to 10
quarter to 9
twenty to 8
twenty five to 7

12

five past
ten past
quarter past
twenty past
twenty five past

6
half past

For the long 'minute' hand the journey from one number to the next takes 5 minutes.

For the short 'hour' hand the journey from one number to the next takes 1 hour.

Exercise 7F

1   Write down the times shown by these clocks as you would say them.

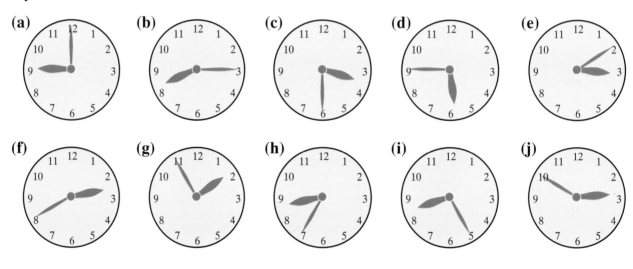

2   Draw six clock faces and mark these times on them:
   (a) seven o'clock
   (b) ten past eight
   (c) quarter to five
   (d) twenty to three
   (e) quarter past nine
   (f) twenty five past four

Reading the time from a digital clock

Digital clocks make it easy to say the time:

The colon : keeps
the hours and minutes apart.

9:20   means 20 minutes past 9.
You can also say: nine twenty.

9:40   means 40 minutes past 9, or twenty minutes to 10.
You can also say: nine forty.

Exercise 7G

1 Write these times as you would *say* them:

    (a) 8:30   **(b)** 10:10   **(c)** 11:05

    (d) 4:45   **(e)** 3:55

2 Draw five digital watches and show the following times on
them:

    (a) quarter past 9   **(b)** half past 3   **(c)** twenty to 5

    (d) quarter to 7   **(e)** five o'clock

You need to be able to tell the difference between times such
as 2 o'clock in the morning and 2 o'clock in the afternoon.
Here are two ways of doing this.

12-hour clock times use **am** or **pm** to show whether a time is before or after midday:

		ante meridiem before midday am		midday or noon today			3 pm	post meridiem after midday pm		12 midnight today
12 midnight yesterday	3 am									
12	1 2 3 4 5 6 7 8 9 10 11 12 1 2 3 4 5 6 7 8 9 10 11 12									

24-hour clock times number the hours from **1** to **24**:

3 o'clock in
the morning

3 o'clock in
the afternoon

00 | 00 1:00 2:00 3:00 4:00 5:00 6:00 7:00 8:00 9:00 10:00 11:00 12:00 13:00 14:00 15:00 16:00 17:00 18:00 19:00 20:00 21:00 22:00 23:00 24 | 00

        Morning   *Afternoon*

The time shown by these clocks is
    9:25 am or 09:25

The time shown by these clocks is
    1:35 pm or 13:35

Changing 12-hour clock times to 24-hour clock times

Up to 12 noon the times are the same:

12-hour → 24-hour

9:35 am → 09:35

A digital clock
shows a zero here.

After 12 noon add 12 to the hour number:

12-hour → 24-hour

1:45 pm → 13:45

+ 12

Exercise 7H

1 Change these times from 12-hour clock times (am and pm)
to 24-hour clock times.
(a) 10:00 am   (b) 10:00 pm   (c) 9:30 am
(d) 9:30 pm   (e) 8:20 pm   (f) 8:20 am
(g) 7 am   (h) 8 pm   (i) 3:30 pm
(j) 4:40 am   (k) 1:08 am   (l) 1:08 pm
(m) 5:50 pm   (n) 5:50 am   (o) 11 pm
(p) 8 am
(q) Quarter past 8 in the morning
(r) Quarter to 9 in the evening
(s) Five to three in the afternoon
(t) Twenty to seven in the morning

2 Change these times from 24-hour clock times to 12-hour
clock times (am or pm).
(a) 08:00   (b) 09:20   (c) 21:30   (d) 13:10
(e) 12:10   (f) 00:20   (g) 01:40   (h) 08:00
(i) 15:45   (j) 18:00   (k) 16:30   (l) 21:10
(m) 23:55   (n) 14:02   (o) 06:25   (p) 00:00
(q) 24:00   (r) 12:00   (s) 10:55   (t) 20:55

7.7 Reading scales

You need to be able to read these types of scales:

... a ruler to measure
lengths.

... weighing scales

... a measuring jug to measure
capacity

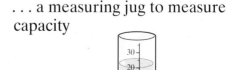

This line is 4 cm long.   This scale shows 30 g.   This jug contains 20 ml of liquid.

Exercise 7I

1 Write down the readings on these scales.

(a)

(b)

(c)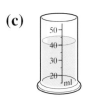

(d)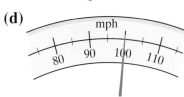

(e)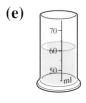

(f)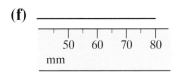

2 Draw diagrams to show these readings on a scale.

    (a) 5 cm   **(b)** 20 m*l*   **(c)** 50 g

    (d) 3 cm   **(e)** 25 m*l*   **(f)** 250 g

Using the marks on a scale

Here is how to read measurements that are in between the numbers marked on a scale:

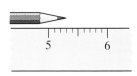

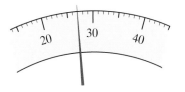

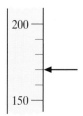

This pencil ends between the 5 and the 6.

There are 10 spaces between the 5 and the 6 so each mark shows $\frac{1}{10}$ or 0.1

As the pencil ends on the third mark it must be 0.3 or $\frac{3}{10}$ more than 5.

The pencil is 5.3 units long.

This pointer is between 20 and 30.

There are 10 spaces between 20 and 30 so each mark shows 1 unit.

As the pointer is on the seventh mark it must be 7 more than 20.

The reading is 27 units.

This reading is between 150 and 200.

There are 5 spaces marked between 150 and 200 so each mark shows 10 units.

The reading is 170 units.

Exercise 7J

Write down the reading on these scales.

1

2

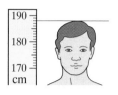

3

4

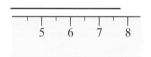

5

6

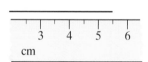

Estimating from a scale

Some scales have no helpful marks on them so you have to estimate a reading.

The only helpful mark on this scale is halfway between 4 and 5 at 4.5.

As the pointer is a little more than halfway between 4.5 and 5 a good estimated reading is 4.8 units.

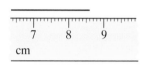

Exercise 7K

Write down the measurements on these scales.

1

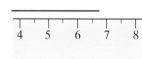

2

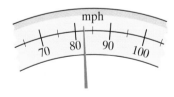

3

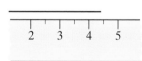

4

5

6

Post Office scales

Post Office scales weigh letters and parcels and also show you how much it costs to post them.

This scale shows that for a letter weighing up to about 60 g the cost of postage is 19p second class and 27p first class.

Exercise 7L

1 Write down the cost of posting these letters:

(a) 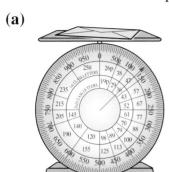   **(b)** 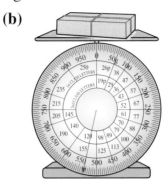   **(c)**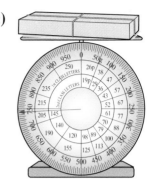

2 Draw a scale showing letters with the following weights.
   (a) 30 g   **(b)** 50 g   **(c)** 100 g   **(d)** 200 g

Measuring lines accurately

When you measure the length of a line remember to start measuring from the 0 on the scale you are using, *not the end* of the ruler.

Start measuring from the zero mark.

This line measures 3.7 cm.

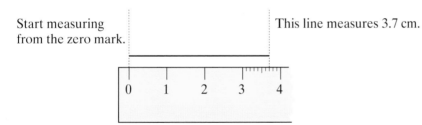

Exercise 7M

1 Measure and write down the lengths of these lines in centimetres:

   (a) _____

   (b) _____

   (c) _____

   (d) _____

   (f) \

   (g) _____

   (e)

   (h) _____

2 Draw and label lines with lengths:
  (a) 4 cm   **(b)** 6 cm   **(c)** 2.5 cm   **(d)** 5.7 cm
  (e) 4.8 cm   **(f)** 3.2 cm   **(g)** 8.3 cm   **(h)** 10.2 cm
  (i) 4.6 cm   **(j)** 3.9 cm   **(k)** 6.4 cm   **(l)** 7.2 cm

3 Draw and label lines with lengths:
  (a) 20 mm   **(b)** 35 mm   **(c)** 55 mm   **(d)** 100 mm
  (e) 74 mm   **(f)** 8 mm   **(g)** 18 mm   **(h)** 68 mm

Exercise 7N

1 **(a)** What metric unit of length would you use to measure the length of a large coach?
  (b) Using the unit you gave in part **(a)** estimate the length of a large coach.

[E]

2 The world record time for running 100 metres is about 10 seconds.
  (a) Estimate the fastest time a sixteen-year-old boy could run 100 metres.
  (b) How long would it take the Prime Minister to *walk* 400 metres? Give your answer in minutes.

[E]

3 The scale diagram shows a man and a dinosaur called a velociraptor.

The man is 6 ft tall.

Estimate the height of the velociraptor:
  (a) in feet
  (b) in metres.

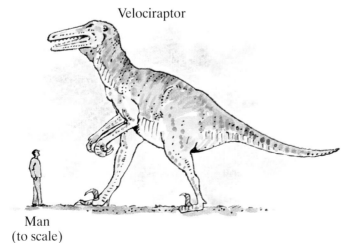

Velociraptor

Man
(to scale)

[E]

Summary of key points

1   Lengths and distances are measured in these metric and imperial units:

   metric: kilometres (km), metres (m), centimetres (cm), millimetres (mm)

   imperial: miles, yards, feet, inches

2   Capacity is a measure of the amount a container can hold. It is measured in these units:

   metric: litres (*l*), centilitres (c*l*), millilitres (m*l*)

   imperial: gallons, pints, fluid ounces

3   Weight is measured in these units:

   metric: tonnes (t), kilograms (kg), grams (g), milligrams (mg)

   imperial: tons, hundredweight, stones, pounds, ounces

4   Digital clocks have a number display:

5   Analogue clocks have hands:

8 Collecting and recording data

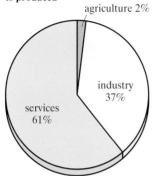

UK – how the country's wealth is produced

agriculture 2%

industry 37%

services 61%

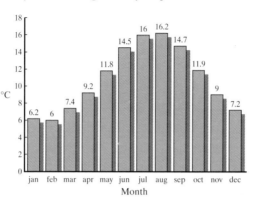

Plymouth – average monthly temperature

°C

6.2, 6, 7.4, 9.2, 11.8, 14.5, 16, 16.2, 14.7, 11.9, 9, 7.2

jan feb mar apr may jun jul aug sep oct nov dec

Month

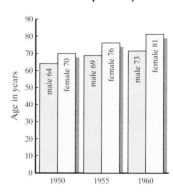

France – life expenctancy

Age in years

male 64, female 70, male 69, female 76, male 73, female 81

1950 1955 1960

You often see charts and tables like this in the press and on TV. They give you information or **data** as it is usually called. Sometimes you can use this data to make forecasts and plan for the future.

For example, the number of babies born this year helps us plan the number of school places needed in five years' time.

Statistics is the branch of mathematics concerned with:

- **collecting** and **recording** data
- **sorting** and **tabulating** that data
- presenting data visually in **charts** and **diagrams**
- making **calculations**
- **interpreting** results.

This unit shows you how to start collecting and recording data.

8.1 Ways of collecting data

You can collect data:

- using questionnaires
- by making observations and recording results
- by carrying out experiments
- from databases or records.

You must be careful how you collect data. If you want to find out what people think of school dinners, don't just ask those going up for 'seconds'. They are the keen ones, so you might be led to the wrong conclusions!

8.2 Designing questions to collect data

When you are writing questions for a questionnaire:

■ **Be clear what you want to find out, and what data you need.**

■ **Ask short, simple questions.**

Here are some good examples:

Are you:

Male ☐   This has a clear choice
Female ☐   of **two** answers.

What age are you: Which of these do you like:

under 21 ☐   Cinema ☐
21–40 ☐   Theatre ☐
41–60 ☐   Opera ☐
over 60 ☐   Ballet ☐

These both offer **four** choices.

■ **Avoid questions which are too vague, too personal, or which may influence the answer.**

Do you go swimming:

sometimes ☐   occasionally ☐   often ☐

Sometimes, occasionally and often may mean different things to different people.

Have you ever stolen anything from a shop:

Yes ☐   No ☐

Even a hardened criminal is unlikely to answer this question honestly!

Do you agree that sport is good for you:

Yes ☐   No ☐

This question suggests the right answer is Yes. It is **biased**.

Test your questionnaire on a few people first to see if it works or needs to be improved. This is called a **pilot survey**.

Exercise 8A

1 Look at these pairs of questions. Choose whether Question X or Question Y is best to do the job in the *To find out* column.

	Question X	Question Y	To find out:
(a)	You like vegetables don't you?	Do you like vegetables?	if people like vegetables
(b)	How old are you?	Are you under 21, 21 to 40, 40 to 60, over 60?	the age group a person is in
(c)	Do you usually watch BBC or ITV?	Do you watch BBC?	the most popular channel
(d)	Do you use brand K?	Which soap powder do you normally use?	the most popular soap powder
(e)	Do you usually walk or ride to school?	Did you walk or ride to school this morning?	how pupils get to school
(f)	Did you enjoy your stay?	Was everything all right?	if the hotel was satisfactory
(g)	Did you see the film Henry the Fifth?	Do you like Shakespeare?	if they saw the film
(h)	Do you often have an egg for breakfast?	Do you usually have a cooked breakfast?	what sort of breakfast pupils have

Example 1

Decide if the following question is suitable for use in a questionnaire. If not, give a reason and rewrite the question to improve it.

> How much pocket money do you get:
>
> a little ☐ some ☐ a lot ☐

A little to some people may be *a lot* to others. It would be better to be more precise. Also the word *some* means different amounts to different people. A better question would be:

> How much pocket money do you get:
>
> ≥ 0 and < £2 ☐ ≥ £2 and < £4 ☐ ≥ £4 and < £6 ☐
>
> ≥ £6 and < £8 ☐ ≥ £8 ☐

Remember:
< less than
≥ greater than or equal to

Exercise 8B

1   Here are some questions that are not suitable for use in a questionnaire. For each one say why and write a more suitable question.

   (a) You don't like rice pudding, do you?
   Yes ☐   No ☐

   (b) How well did you do in your test?
   Badly ☐   Well ☐   OK ☐

   (c) How many hours television do you watch?
   1 ☐   2 ☐   3 ☐
   4 ☐   5 ☐

   (d) What was the weather like?
   Lousy ☐   Quite good ☐
   OK ☐

   (e) What type of job do you have?
   Clerical ☐   Manual ☐
   Other ☐

   (f) Have you ever cheated in an examination?
   Yes ☐   No ☐

   (g) You will vote for Williams in the election won't you?
   Yes ☐   No ☐

2   Draw up a questionnaire to find out what kind of holiday people had last year. (Hint: where, when, age, length, accommodation.)

8.3 Collecting data by sampling

If you carry out a survey in a mixed school, but only question the first five students on each form register, you could end up asking all boys or all girls.

Ideally you should ask everyone in the school, but this is usually not practical. Instead you ask a limited number of students – a **sample**.

Make sure each student in the school has an equal chance of being picked to be part of the sample. You might question six pupils from each class, drawing their names out of a hat. This is called a **random sample**. Then questioning your sample should give a similar result to questioning the whole school.

If you do not choose a random sample your answers may be **biased**. To find out which sports a typical teenager watches you need a random sample of teenagers. Choosing a sample from teenagers at a football match will **bias** your answers – there will be more football lovers than in a random sample.

For a fair survey you may need to ask people of different ages, sexes, jobs, nationalities, and so on.

A **random sample** helps this market researcher find out what a typical person thinks without asking them all.

■   **When you carry out a survey select a random sample to avoid bias.**

Exercise 8C

Which of these methods of collecting information could be biased? Give your reasons.

Data needed:	Who to ask:
1 How people get to work	(a) Every fifth person near a bus stop (b) A group of people arriving together (c) A group of people during a tea break
2 If people like gardening	(a) People in a garden centre (b) Every tenth person passing in a street (c) People queuing for a train
3 If people think Superstore is a good shop	(a) People in a public library (b) People going into the shop (c) People coming out of the shop

4 Nadia wants to find out which computer games people play.
   She has designed a questionnaire with only boxes to tick.

> **QUESTIONNAIRE: COMPUTER GAMES**
>
> 1 Do you have a computer game at home? YES/NO
>
> 2 Tick the computer you have games for:
>
>   ☐   ☐   ☐
>   ZEGA   GAMESTATION   PC
>
> 3 Tick the type of computer game you like the best:
>
>   ☐   ☐   ☐   ☐   ☐
>   Arcade   Beat-em-up   Fantasy Factual Educational
>
> 4 How many computer games have you bought in the last 12 months?
>
>   ☐   ☐   ☐   ☐
>   None   1   2   3

(a) How could you improve part 4 of the questionnaire?
(b) Nadia also wants to find out how much time people spend playing computer games each week.
     Design a question that she could use. Include tick boxes for a response.   [E]

5 Write six questions you might include in a survey to decide on school uniform.

6 (a) Draw up a short questionnaire to find out how students in school spend their leisure time.
  (b) Describe briefly how you would collect your data.

7 Laurie is designing a survey to find out about people who use a superstore near his home. One of the things he wants to know is how far people have travelled to get to the superstore.

Decide which of these questions is best to ask. Give two reasons for your decision.

(a) How far have you travelled to get here today?

(b) Where do you live?

(c) Do you live far from here?

(d) Please show me on this map where you have travelled from.

8.4 Collecting data by observation

You could do a traffic survey by counting vehicles and recording what type they are as they pass you.

You would have to decide where, how and for how long to carry out your survey.

For example, if you did your survey during rush hour the results would be different from a survey on a Sunday.

Here is a **data capture sheet** from a traffic survey:

Traffic survey by H.Long on 5/6/96 at Main Street Ash 9.00–9.30 am.

Type of vehicle	Tally	How many
Bus	‖‖‖ ‖‖‖ ‖	12
Car	‖‖‖ ‖‖‖ ‖‖‖ ‖‖‖ ‖	22
Lorry	‖‖‖ ‖	7
Van	‖‖‖ ‖‖‖	10
M/cycle	‖‖‖ ‖‖‖ ‖‖‖ ‖	16

You record 5 in a tally chart like this:

‖‖‖

Exercise 8D

1 Prepare data capture sheets for surveys to find out two of the following by observation:

(a) the colours of cars, clothes or shoes

(b) the types and numbers of birds in a park

(c) people using a pedestrian crossing

(d) the age groups and sexes of people entering a library or supermarket.

8.5 Collecting data by experiment

■ **When you do an experiment you can use a data capture sheet to record your results.**

Example 2

A game starts when you throw a six on a dice.

Greta throws a dice sixty times to see if it is a fair dice. Here are her results in a data capture sheet:

Score	Tally	How many
1	‖‖‖ ‖‖	9
2	‖‖‖ ‖‖‖ ‖	11
3	‖‖‖ ‖‖‖ ‖‖	12
4	‖‖‖ ‖‖	9
5	‖‖‖ ‖‖‖	10
6	‖‖‖ ‖‖	9

The dice seems to be fair as the results for each score are about the same.

Exercise 8E

Carry out one or more of the following experiments

1   Find out how accurate people are at estimating.
    Ask people to estimate:
    (a) the length in centimetres of a piece of wood
    (b) the number of sweets in a jar
    (c) the weight in grams of a piece of metal.

2   Find out whether a typical science textbook has longer words than a typical English textbook.
    ● Choose two passages of about 50 words from each book.
    ● For each passage make a data collection table like this and complete it:

Number of letters in a word	Tally	How many
1		
2		
3		
4		
5		
6		
7		
8 or more		

8.6 Obtaining data from a database

■ **A database is an organised collection of information. It can be stored on paper or in a computer.**

Here is a table of data from a database of exam results:

Name	Maths		English		Science	History	French
	paper 1	paper 2	paper 1	paper 2			
Peter	34	41	53	46	57	29	43
Ahmed	73	85	60	58	63	52	46
Lucy	49	60	74	83	48	60	52
Chandra	55	38	46	40	62	45	36
Cita	61	65	50	44	71	58	52
Glenys	17	24	38	55	30	47	19

A computer database allows you to obtain information quickly and in a variety of forms, for example:

- in alphabetical order
- in numerical order
- girls' results only
- pupils scoring over 60 in one subject only . . .

Exercise 8F

1 Here is some information from a table in a gardening book.

Plant	Use	Colours	Flowering period	Where to plant
Alyssum	Edging	White, pink or purple	June–Sept	Full sun
Begonia	Small bedding	Pink, red or white	June–Sept	Any position
Convolvulus	Medium bedding	Red	July–Sept	Full sun or partial shade
Dahlia	Large bedding	Pink, red or white	July–Sept	Full sun or partial shade

(a) Which plant *must* be planted in full sun?
(b) Write down the colour of a Convolvulus.
(c) Write down the name of a pink bedding plant. [E]

2 This database contains details of some second-hand cars.

Make	Model	Colour	Insurance group	Number of doors	Year	Price (£)
Ford	Sierra 2.0	Red	10	5	1990	3200
Ford	Escort 1.3	Blue	4	3	1992	4050
Honda	Civic 1.4	Blue	9	3	1991	5000
Renault	Clio 1.4	Silver	5	5	1992	4750
Peugeot	205 1.4GR	Red	7	5	1990	3150
Volvo	740 GL	Silver	12	4	1991	5950
Rover	1.6 LX	Green	9	4	1990	2500
Peugeot	205 1.4 XT	Silver	8	3	1992	3850
Vauxhall	Cavalier 1.6	Black	7	4	1989	3000
VW	Polo 1.3	White	6	3	1990	2300

Use the database to answer the following questions.

(a) Which is the oldest car?

(b) Which is the most expensive car?

(c) Which car has the highest insurance group?

(d) How many cars have 5 doors?

(e) Karen has £3000. Which cars could she buy?

(f) Gareth bought a silver 4-door car. What year was it?

(g) Which is the cheapest 5-door red car?

(h) How many cars are less than £3500?

(i) List the models of cars in order of price, most expensive first.

(j) Write a question of your own using at least two pieces of data.

3 Draw up a database to provide information about pupils in your class which might be useful for a new teacher.

4 This computer database has been asked to print out details of crops which are harvested from June to October.

Crop	Form of tree	Space in metres between plants	Time in years before fully grown	Harvest time
Apple	Bush	3	3–4	Aug–Oct
Apple	Cordon	1	2–5	Aug–Oct
Apple	Standard	6	5–8	Aug–Oct
Blackberry	Fan or rod	3	2–3	Aug–Sep
Cherry	Fan	4.5	4–5	Jun–Sep
Cherry	Standard	7.5	5–7	Jun–Sep
Gooseberry	Bush	1.8	3–4	Jun–Jul
Gooseberry	Cordon	0.75	2–3	Jun–Jul
Pear	Cordon	1	3–4	Aug–Oct
Pear	Espalier	4.5	4–5	Aug–Oct
Pear	Dwarf	1	4–5	Aug–Oct
Plum	Bush	5	3–5	Aug–Oct
Plum	Fan	5	3–6	Aug–Oct

(a) Which crop has a space of 7.5 metres between plants?

(b) Which fruits can be harvested in July?

(c) Which two Cordon trees should be planted 1 metre apart?

(d) Which Fan type tree takes the longest before it is fully grown?

(e) Name two crops which grow on Standard trees.

(f) Name the crop and form of tree which is planted with a space of 4.5 metres and can be harvested in October. [E]

8.7 Secondary data

■ **Data you collect is called primary data. Data that may have been collected by other people is called secondary data.**

The National Census carried out every 10 years provides information about people in the UK. The Central Statistics Office publishes monthly and annual figures on a wide range of subjects.

Example 3

Below is an extract from a table in the 1991 Census which shows the percentage of different groups of residents in some boroughs in London:

	Bexley	Brent	Bromley	Camden	Croydon	Newham	Greenwich	Total London
Retired	17.4	14.3	19.6	17.3	16.1	14.2	17.6	16.8
Birth rate	13.8	17.0	12.8	13.5	15.1	20.2	15.9	15.4
Unemployed	38.1	41.6	35.5	40.6	40.2	43.6	44.6	40.8
2-car families	26.1	16.0	28.9	9.4	24.6	8.8	14.8	18.2
No car families	26.7	43.4	25.6	55.8	30.5	53.5	43.6	40.7

(a) Which borough has the highest percentage (i) birth rate (ii) unemployment?
(b) What percentage of families have 2 cars in (i) Bromley (ii) Brent?
(c) Give a reason why Camden has a large percentage of families with no car.
(d) Which boroughs have a higher percentage of families with 2 cars than the total London figure?

(a) (i) Newham   (ii) Greenwich
(b) (i) 28.9   (ii) 16.0
(c) High unemployment
(d) Bexley, Bromley and Croydon

Exercise 8G

1   The table shows the percentage unemployment figures for August:

Year	1987	1988	1989	1990	1991	1992	1993
Male	7.9	5.2	4.0	2.5	4.8	7.1	9.2
Female	5.8	3.9	2.6	2.0	2.2	3.2	4.4

(a) What was the female unemployment rate in 1992?
(b) Which year had the (i) highest (ii) lowest percentage total unemployment?
(c) Between which two years did male unempolyment rise the most?

2 This table shows money spent in £ per person in different boroughs of London.

	Camden	Barnet	Haringay	Islington	Lambeth	Redbridge	Richmond	Southwark
Average weekly rent	53	51	56	50	41	61	48	44
Management	18	13	14	16	16	16	12	23
Repairs	9	11	10	13	14	20	12	14
Bad debts	4	0.04	3	0.77	0.48	0.23	0.14	1.76
Rent rebates	32	31	41	31	24	43	27	26

(a) Which borough has (i) the highest (ii) the lowest average weekly rent?

(b) Which borough has a bad debts figure of 0.48?

(c) Which borough has (i) the highest (ii) the lowest repairs figure?

(d) What is the rent rebate per person in Richmond?

3 The annual traffic figures for peak hour travel (8–9 am and 5–6 pm) are shown in this table.

Location/Year	1987	1988	1989	1990	1991	1992	1993	1994	1995	
Burnt Ash Lane am	1426	1478	1501	1476	1503	1017	1416	1539	1806	
Burnt Ash Lane pm	1436	1417	1382	1416	1387	1261	1430	1448	1530	
Location/Year	1982	1983	1984	1985	1986	1987	1988	1989	1990	1991
Mottingham Lane am	616	845	842	1079	935	838	553	401	622	485
Mottingham Lane pm	661	710	679	840	810	783	297	284	250	183

(a) In 1995, was Burnt Ash Lane busier in the morning or the evening?

(b) Which year did Burnt Ash Lane have the least traffic in the afternoon?

(c) In which year do you think Mottingham Lane introduced speed humps? Give your reason.

4 This table shows the amount of money spent per person in the London boroughs.

	Bromley	Ealing	Greenwich	Hackney	Havering	Kingston	Lambeth	Lewisham
Education	200	487	508	556	450	341	445	480
Social services	104	133	200	278	200	138	327	200
Libraries	19	15	15	30	16	14	19	17
Highways	40	37	33	38	24	37	17	37
Public transport	19	17	17	17	0	19	16	16
Environmental health	7	14	12	20	7	11	43	12
Planning	1	8	14	20	6	10	15	15
Refuse & cleaning	15	23	26	24	15	15	16	26
Sport & recreation	27	24	33	47	18	9	25	29
Housing	10	13	31	54	24	6	78	14

(a) Which borough spends the most on (i) education (ii) highways (iii) housing?

(b) Which borough spends least on (i) libraries (ii) sport & recreation (iii) social services?

(c) How much more per person did Lambeth spend than Ealing on (i) social services (ii) housing?

Summary of key points

1 When you are writing questions for a questionnaire:
- be clear what you want to find out and what data you need
- ask short, simple questions
- avoid questions which are too vague, too personal, or which may influence the answer.

2 When you carry out a survey select a random sample to avoid bias. This is a sample in which everyone has an equal chance of being chosen.

3 When you do an experiment you can use a data capture sheet to record your results.

4 A database is an organized collection of information. It can be stored on paper or in a computer.

5 Data you collect is called primary data. Data that may have been collected by other people is called secondary data.

9 Algebra 2

9.1 Coordinates in the first quadrant

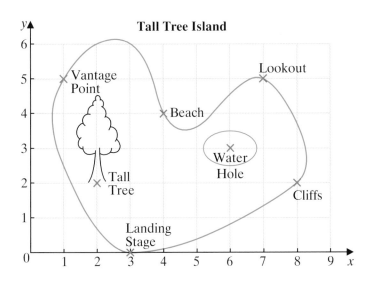

You can describe the position of a place on a grid by using two numbers.

The number of units across is put first and the number of units up is put second.

On this map 6 across, 3 up gives the position of the Water Hole. You can write this as the point (6,3). The numbers 6 and 3 are the **coordinates**.

Exercise 9A

1   Write down the names of the places on the map with these coordinates:

   (a) (8,2)   **(b)** (7,5)   **(c)** (1,5)
   (d) (3,0)   **(e)** (2,2)   **(f)** (4,4)

2   Write down the coordinates of these places on the map of Tall Tree Island:

   (a) Tall Tree   **(b)** Landing Stage   **(c)** Cliffs
   (d) Vantage Point **(e)** Lookout   **(f)** Beach

Here is a straight line *AB*.
The straight line is called a line segment *AB*.

A has coordinates (1, 1)
B has coordinates (3, 6)

Halfway along line segment *AB* is *M*.
M is called the **mid-point of the line segment *AB***.

You can find the coordinates of the mid point:

1 add the *x*-coordinates and divide by 2:

$$\frac{1 + 3}{2} = \frac{4}{2} = 2$$

2 add the *y*-coordinates and divide by 2:

$$\frac{1 + 6}{2} = \frac{7}{2} = 3\tfrac{1}{2}$$

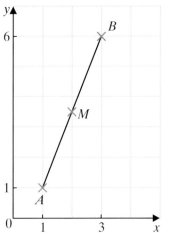

So the coordinates of the mid-point is $(2, 3\tfrac{1}{2})$.

■ **The mid-point of the line segment $A(x_1, y_1)$ and $B(x_2, y_2)$ is**
$$\left(\frac{x_1 + x_2}{2}, \frac{y_1 + y_2}{2}\right).$$

Example 1

Work out the coordinates of the mid-point of the line segment
AB where *A* is (2, 3) and *B* is (7, 11).

Mid-point *x*-coordinate is $\dfrac{2 + 7}{2} = \dfrac{9}{2} = 4\tfrac{1}{2}$

Mid-point *y*-coordinate is $\dfrac{3 + 11}{2} = \dfrac{14}{2} = 7$

So mid point is $(4\tfrac{1}{2}, 7)$.

Exercise 9B

1 On squared paper draw a coordinate grid and number it
from 0 to 12 across the page and 0 to 6 up the page. Join
the points in order given:

   (2,3) (3,1) (10,1) (11,3) (7,3) (9,5) (6,5) (4,3) (2,3)

2 Write down the
coordinates of all
the points
marked that
make up the
shape in the
diagram.

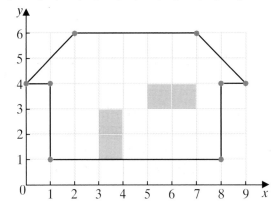

3 Write down the coordinates of all the corner points that make up this ship.

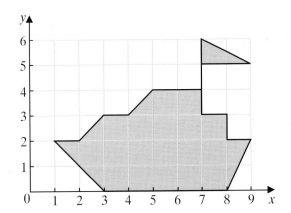

4 Draw a coordinate grid from 0 to 10 in both directions. On your grid plot the points and draw the following shapes by joining the points in order.

(a) (1,1) (4,1) (4,4) (1,4) (1,1)
(b) (6,1) (9,1) (9,6) (6,6) (6,1)
(c) (0,6) (5,6) (0,10) (0,6)
(d) (5,7) (5,10) (8,10) (5,7)

5 Draw a coordinate grid from 0 to 10 in both directions. On your grid draw a shape of your own and label the coordinates of each point.

6 Invent an island of your own and mark positions on it using a coordinate grid. List all the places with their coordinates.

7 Work out the coordinates of the mid-point of the line segments:

(a) *AB*
(b) *CD*
(c) *EF*
(d) *GH*

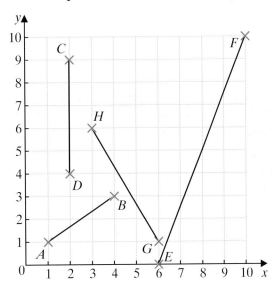

8 Work out the coordinates of the mid-point of the line segments:

(a) $A(0, 0)$ $B(4, 5)$ (b) $C(3, 1)$ $D(4, 9)$
(c) $E(5, 7)$ $F(1, 6)$ (d) $G(1, 0)$ $H(5, 8)$
(e) $I(9, 10)$ $J(3, 7)$

9.2 Linear graphs

You can use graphs to show relationships. For example, if you buy several packets of crisps the price you pay will be related to the number of packets you buy.

Example 2

Stan sells packets of crisps at 15p each. He wants a quick way of remembering how much different amounts of packets cost.

He makes a table:

Number of packets	1	2	3	4	5	6
Cost in pence	15p	30p	45p	60p	75p	90p

Notice that the cost goes up by 15p for each extra packet.

A pattern like the one in the table is known as a **linear relationship**. Linear relationships can be spotted very easily by drawing a graph.

Here is a graph of Stan's table.

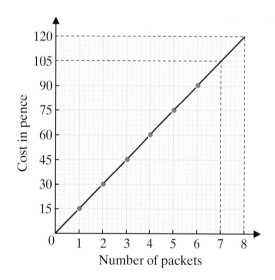

The points plotted make a straight line. This will be true for any linear relationship.

■   **A graph representing a linear relationship will always have a straight line.**

If the line of Stan's graph is made longer the cost of 7 and 8 packets can be read off. Check the dotted lines to see that the cost of 7 and 8 packets is 105p and 120p.

Example 3

Sharon charges £1 for the use of her taxi and 50p per mile after that.

You can use the information to draw up a table and then plot the points on a graph.

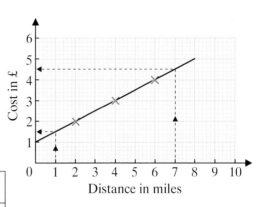

The table gives the charges for some journeys up to 6 miles long.

Distance in miles	0	2	4	6
Cost in £	1	2	3	4

Notice that the graph gives a straight line.

Because the relationship is linear you can use the graph to find the cost of a journey more than 6 miles, for example 7 miles, and for journeys between the values that you have worked out.

The cost of a 7 mile journey is £4.50

The cost of a 1 mile journey is £1.50

Exercise 9C

1 The table shows the cost of potatoes per kg.
   (a) Draw a graph for this table.

Weight in kg	1	2	3	4	5
Cost in pence	30	60	90	120	150

   (b) Work out how much 2.5 kg of potatoes would cost.
   (c) Extend the graph to work out the cost of 6 kg of potatoes.

2 The table shows the cost of ice lollies.
   (a) Draw a graph for the cost of ice lollies from the table.

Number of ice lollies	1	2	3	4	5
Cost in pence	25	50	75	100	125

   (b) Extend the graph and then use it work out the cost of
       (i) 8 ice lollies   (ii) 6 ice lollies

3   The table shows the number of litres of petrol left in a
car's petrol tank on a journey.

Number of hours travelled	1	2	3	4	5	6	7	8
Number of litres left	55	50	45	40	35	30	25	20

(a) Draw a graph from the information given in the table.

(b) How many litres were in the tank at the start of the
journey (after 0 hours)?

(c) How many litres were in the tank after $5\frac{1}{2}$ hours?

4   A car uses 2 litres of petrol for every 5 km it travels.

(a) Copy and complete the table showing how much
petrol the car uses.

Distance travelled in km	0	5	10	15	20	25
Petrol used in litres	0	2	4			

(b) Draw a graph from the information given in your
table.

(c) Work out how much petrol was used in travelling 4 km.

(d) Work out how many kilometres were travelled after
15 litres of petrol were used.

5   The water in a reservoir is 144 m deep. During a dry
period the water level falls by 4 m each week.

Copy and complete this table showing the expected depth
of water in the reservoir.

Weeks	0	1	2	3	4	5	6	7	8
Expected depth of water in m	144	140							

(a) Draw a graph from the information given in your
table.

(b) How deep would you expect the reservoir to be after
10 weeks?

If the water level reaches 96 m the water company will
divert water from another reservoir.

(c) After how long will the water company divert water?

9.3 Conversion graphs

Another use of linear graphs (straight line graphs) is for converting one measurement into another measurement. These graphs are usually called **conversion graphs**.

Here is an example of a conversion graph that relates temperatures in degrees Fahrenheit (°F) to temperatures in degrees Celsius (°C).

To draw a conversion graph you need to know two pairs of values which link the two measurements you want to convert between.

In this case use the fact that

$$0°C = 32°F \text{ (the freezing point of water)}$$
$$\text{and } 100°C = 212°F \text{ (the boiling point of water)}$$

To make a conversion graph you draw axes from 0 to 100 and label them °C and from 0 to 212 and label them °F. The two points are plotted and joined with a straight line.

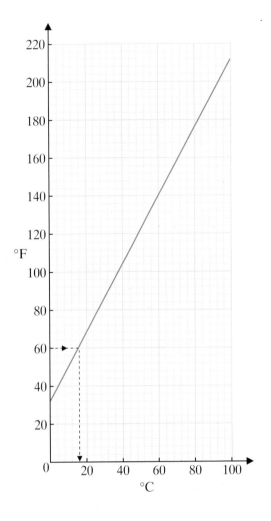

Example 4

Change 60° Fahrenheit to °Celsius.

- Draw a horizontal line from 60 °F across to the straight line on the graph.
- From the line draw a vertical line down to the °C axis.
- Read off the scale.

The answer is 16 °C.

■ **A conversion graph is used to convert one measurement into another measurement.**

Sometimes you may be given a graph where the line is not straight.

Example 5

The distance travelled by a stone when it is dropped from a cliff is shown on this graph.

(a) What distance did the stone drop in 2 seconds?

(b) How long did the stone take to drop 32 metres?

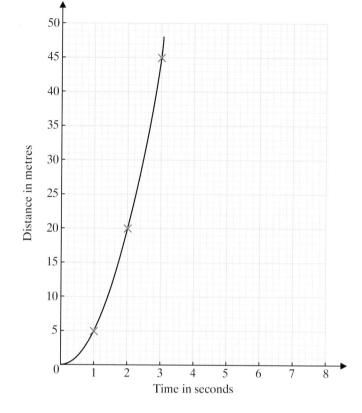

(a) Draw a line up from 2 seconds to meet the line. Then draw it across to the distance axis. It cuts at 20 metres. The stone drops 20 metres in 2 seconds.

(b) Draw a line across from 32 metres to the line. Then draw it down to the time axis. It cuts it at about 2.6 seconds. The stone drops 32 metres in 2.6 seconds.

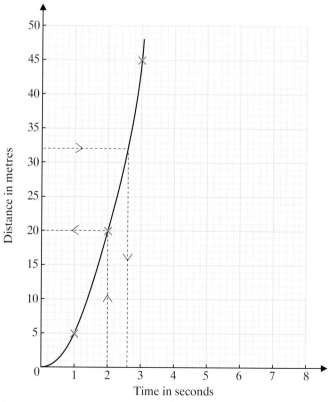

Exercise 9D

1   Copy the table and use the temperature conversion graph
    to complete it.

°C	5	20		28			35	80		40
°F			80		50	100			200	

2   **(a)** Draw a conversion graph from pounds to kilograms.
        Use the fact that 0 pounds is 0 kilograms and
        50 kilograms is 110 pounds. On your graph draw scales
        for kilograms and pounds using a scale of
        1 cm = 10 pounds and 1 cm = 10 kg.
        Plot the points (0,0) and (50,110) and join them with
        a straight line.
    (b) Copy and complete this table using your conversion
        chart to help you.

Kilograms	0			45	30	15			35	50
Pounds	0	10	20				50	14		110

3   Copy this table and then use the information in the table
    to draw a conversion graph from inches into centimetres.
    Use your graph to help you fill in the missing values.

Inches	0	1	2				9	8		12
Centimetres	0			10	15	20			25	30

4   Copy this table then use the information in the table to
    draw a conversion graph from miles into kilometres. Use
    your graph to help you fill in the missing values.

Miles	0	5		40		30			24	50
Kilometres	0		16		36		72	20		80

5   Copy this table then use the information in the table to
    draw a conversion graph from acres into hectares. Use
    your graph to help you fill in the missing values.

Hectares	0			12	15	17			3	20
Acres	0	20	30				24	45		50

6   Use the graph in Example 5 to find:

  (a) the distance travelled by the stone in:
  - **(i)** 1.5 seconds
  - **(ii)** 3 seconds

  (b) the time taken for the stone to drop:
  - **(i)** 40 metres
  - **(ii)** 25 metres.

7   The speed of a ball when it is dropped is given by the following table of values.

Distance in metres	0	5	10	15	20	25
Speed in metres per second	0	10	14	17	20	22

  (a) Draw a graph using the information given in the table. You should join up your points with a **smooth** curve.

  (b) Use the graph to work out the speed when the distance dropped is 12 metres.

  (c) Use the graph to work out the distance dropped when the speed is 18 metres per second.

9.4 Distance-time graphs

Distance-time graphs give information about journeys.
We always use the horizontal scale for time and the vertical scale for distance.

Example 6

This is a graph showing a car journey.

On the graph from O to A the car travels 10 km in 10 minutes. From A to B the car travels 20 km in 10 minutes. From B to C the car does not go anywhere for 5 minutes. The journey home takes 15 minutes for the 30 km.

Suppose that the speed of the car needed to be worked out for each part of the journey.

speed = distance travelled ÷ time taken

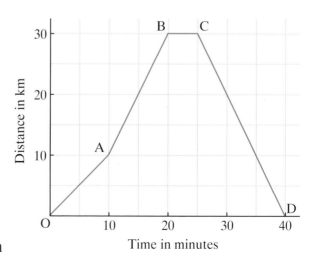

Between O and A

speed = 10 km ÷ 10 minutes
      = 1 km per minute

There are 60 minutes in 1 hour, so speed = 60 km per hour.

From the graph check that the speed:

from A to B = 120 km per hour
from B to C = 0 km per hour
from C to D = 120 km per hour

■ **Distance-time graphs are used to relate the distance travelled to the time taken, and to calculate speeds.**

Exercise 9E

1   Jane walks to the shops, does some shopping then walks home again.

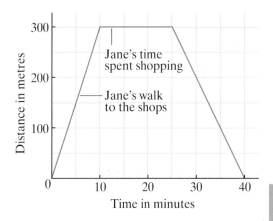

   (a) How many minutes did it take for Jane to walk to the shops?

   (b) How far away were the shops?

   (c) How many minutes did Jane spend shopping?

   (d) How many minutes did it take for Jane to walk home?

   (e) Work out the speed at which Jane walks to the shops. First give your answer in metres per minute, then change it to km per hour.

   (f) Work out the speed at which Jane walks back from the shop. First give your answer in metres per minute, then change it to km per hour.

2   Here is a graph of David's car journey to see his aunt.

   (a) Write a story of the journey explaining what happened during each part of it.

   (b) Work out David's speed during each part of the journey. Give your answer in km per hour.

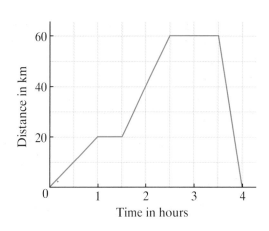

3 This graph shows Wayne's and Tracey's journeys. Wayne sets off from London at 08:00 and travels to a town 150 km away to meet his girlfriend Tracey. He stops for a rest on the way. Once he gets to Tracey's he turns around and drives straight home because he discovers that Tracey set off for London some time ago to see Wayne.

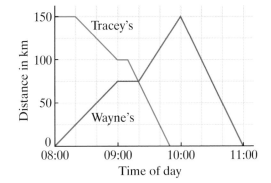

   (a) Describe Wayne's journey in detail explaining after what distance he stopped on the way and for how long.

   (b) Describe Tracey's journey in detail explaining after what distance she stopped on the way and for how long.

   (c) At what time did Tracey and Wayne pass each other and what distance were they from London when it happened?

4 Karen skis down a mountain. This graph shows her journey.

   (a) From the graph write down the height Karen was at after:

      (i) 1 minute

     (ii) 2 minutes 30 seconds

    (iii) 4 minutes 15 seconds.

   (b) Use the graph to write down the time Karen was at the following heights:

      (i) 900 m

     (ii) 750 m

    (iii) 625 m.

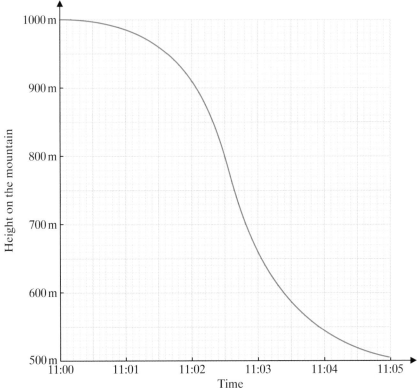

5 Imran has a bath. The graph shows the height of the bath water.

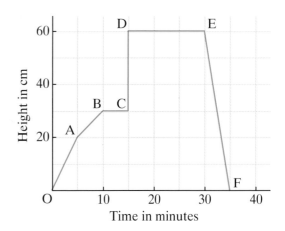

He starts at O by turning the hot and cold water taps on.

Between O and point A on the graph the height of water goes up to 20 cm in 5 minutes.

Explain what happens between letters AB, BC, CD DE and EF on the graph and how long each part of the process takes.

6 Gerald walked to the telephone box near his house to make a phone call. It took him 4 minutes to walk to the phone box that was 400 m away. Gerald talked on the phone for 2 minutes and he walked home in 3 minutes. Use graph paper to draw a distance-time graph for this journey.

7 Kirsti took up a trip in a hot-air balloon. The balloon rose 400 metres in the air in one hour and stayed at this height for two-and-a-half hours. The balloon then came back to earth in half-an-hour. Use graph paper to draw a distance-time graph for this balloon flight.

8 Annabel travels to school. She walks the 250 metres to the bus stop in 4 minutes, waits at the bus stop for 5 minutes then travels the remaining 1000 metres by bus. She arrives at the bus stop outside the school 15 minutes after she set off from home.

(a) Draw a distance-time graph of the journey.

(b) Work out the speed of the bus, first in m per min, then in km per hour.

9 Kam went shopping by car. She drove the 10 miles to the shops in 30 minutes. She stayed at the shops for 30 minutes and then started to drive home. The car then broke down after 5 minutes when she had travelled 4 miles from the shops. It took 10 minutes to repair the car and another 5 minutes to get home. Draw a distance-time graph for Kam's journey.

9.5 Graphs of simple algebraic equations

Section 9.1 shows how to plot points on a grid with positive coordinates only. You also need to be able to plot coordinates that include negative numbers:

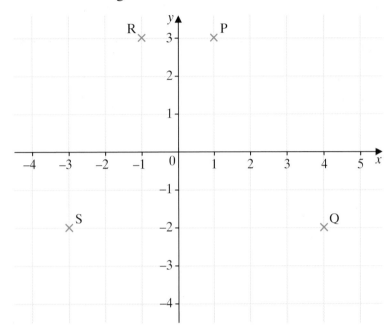

Point P is (1,3).

Point R is at −1 along the x-axis and 3 up the y-axis. R is (−1, 3).

Point Q is (4,−2) and point S is (−3,−2).

Exercise 9F

1   Write down the coordinates of all the points A to L marked on the coordinate grid.

2   Draw a coordinate grid with the horizontal axis (the x-axis) marked from −4 to +4 and the vertical axis (the y-axis) marked from −10 to +10.

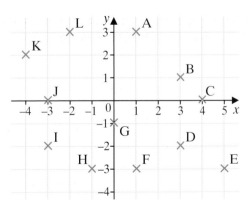

Plot the following points and join them in the order of the following list:

(−1,6) (−2,6) (−4,5) (−4,6) (−2,7) (0,8) (1,7)
(1,−2) (2,−6) (1,−8) (0,−9) (−1,−9) (−2,−8)
(−3,−6) (−2,−5) (−1,−5) (0,−6) (−2,−6) (−1,−8)
(0,−8) (1,−6) (−1,−2) (−3,0) (−3,2) (−1,6).

Using algebra to describe lines

This graph shows a linear (straight line) relationship between the numbers on the horizontal and vertical axes.
You can use algebra to describe the relationship and the line.

Remember: The horizontal axis is called the **x-axis** and the vertical axis is called the **y-axis**.

In a pair of numbers the x-coordinate is given first; the y-coordinate is given second:

$$(x, y)$$

For example in the point $(3, 2)$ the x-coordinate is 3 and the y-coordinate is 2.

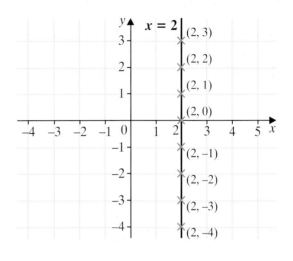

Look for a number pattern connecting the coordinates to give the line a name. On this grid all the x-coordinates are 2 so the line is called $x = 2$

The **equation** of the line is $x = 2$

In this graph all the y-coordinates are –2 so the line is called $y = -2$

The **equation** of the line is $y = -2$

■ **The equation of a line is a way of using algebra to show a relationship between x- and y-coordinates.**

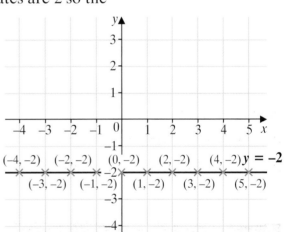

Exercise 9G

1 Write down the equations of the lines marked **(a)** to **(d)** in this diagram.

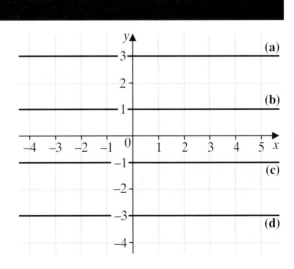

2 Write down the equations of the lines labelled **(a)** to **(d)** in this diagram.

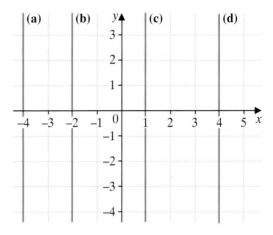

3 Draw a coordinate grid with x- and y-axes labelled from –5 to 5. On the grid draw and label the graphs of:
    (a) $x = 4$ **(b)** $x = -2$ **(c)** $x = -4$ **(d)** $x = 1$

4 Draw a coordinate grid with axes labelled from –5 to 5. On the grid draw and label the graphs of:
    (a) $y = 4$ **(b)** $y = -2$ **(c)** $y = -4$ **(d)** $y = 1$

5 Draw a coordinate grid with axes labelled from –5 to 5. On the grid draw and label the graphs of:
    (a) $y = 3$ **(b)** $x = -1$
    (c) Write down the coordinates of point where the two lines cross.

Example 7

Draw the graph of $y = x - 1$

Step 1: Choose some values for x, for example let $x = -2, 0$ and $+2$

Step 2: Put these values of x into the equation $y = x - 1$

if $x = -2$: $y = -2 - 1$ = –3
if $x = 0$: $y = 0 - 1$ = –1
if $x = 2$: $y = 2 - 1$ = 1

Step 3: These are three pairs of values which make the equation $y = x - 1$ true.

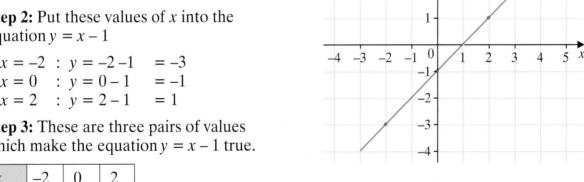

x	–2	0	2
y	–3	–1	1

Step 4: Plot the points $(-2, -3)$, $(0, -1)$ and $(2, 1)$ and join them to give the line.

Step 5: Label the line.

Worked examination question 1

Draw and complete a table of values for the graphs of
$y = 2x - 1$ and $y = -x + 1$.

Draw the graphs on graph paper and write down the
coordinates of the point where they cross.

Step 1
Choose three values for x.
For example

$x = -1, 0, 2$

Step 2
Put these values of x into each equation.

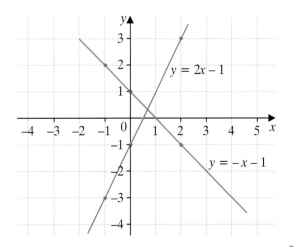

$x = -1$	$y = -2 - 1$	$= -3$
$x = 0$	$y = 0 - 1$	$= -1$
$x = 2$	$y = 4 - 1$	$= 3$

$x = -1$	$y = 1 + 1$	$= 2$
$x = 0$	$y = 0 + 1$	$= 1$
$x = 2$	$y = -2 + 1$	$= -1$

These pairs of values make
$y = 2x - 1$ true

x	−1	0	2
y	−3	−1	3

These pairs of values make
$y = -x + 1$ true

x	−1	0	2
y	2	1	−1

Plot the points $(-1, -3)$, $(0, -1)$ and $(2, 3)$ and join them with a
straight line.

Do the same for $(-1, 2)$, $(0, 1)$ and $(2, -1)$.

The lines cross at about

$$\left(\tfrac{2}{3}, \tfrac{1}{3}\right)$$

Make sure you read the coordinates accurately from your
graph.

Sometimes you can draw a straight line graph without using a table of values. This is called drawing the graph **by inspection**.

Example 8

Draw the graph of $x + y = 4$.

This means that the x-coordinate and the y-coordinate when added make 4.

Therefore the points on the line will be:

$(0, 4)$ $(1, 3)$ $(2, 2)$ $(3, 1)$ $(4, 0)$

as well as:

$(-1, 5)$ $(5, -1)$ $(-2, 6)$ $(6, -2)$

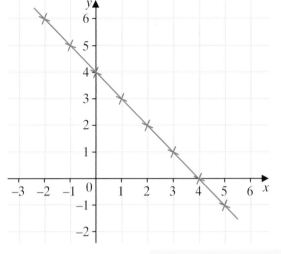

The coordinates of the points on the line $x + y = 4$ always add up to 4.

Exercise 9H

1   On a coordinate grid with the x-axis labelled from –5 to 5 and the y-axis labelled from –5 to 5 draw the following graphs.

  (a) $y = x + 2$ **(b)** $y = x + 4$ **(c)** $y = x + 1$

  (d) $y = x - 2$ **(e)** $y = x - 4$ **(f)** $y = x - 1$

2   On a coordinate grid with the x-axis labelled from –4 to 4 and the y-axis labelled from –10 to 10 draw the following graphs.

  (a) $y = x$ **(b)** $y = 2x$

  (c) $y = 3x$ **(d)** $y = \frac{1}{2}x$

3   On a coordinate grid with the x-axis labelled from –4 to 4 and the y-axis labelled from –6 to 6 draw the following graphs.

  (a) $y - 1 = 2x$

  (b) $y + x = -2$

  Write down the coordinates of the point there they cross.

  Hint: Write these equations in the form $y =$ before drawing your graph.

4   On a coordinate grid with the x-axis labelled from –4 to 4 and the y-axis labelled from –6 to 6 draw the following graphs.

  (a) $y = -2x - 1$

  (b) $y = x + 2$

  Write down the coordinates of the point where they cross.

5 On a coordinate grid with the *x*-axis labelled from –4 to 4 and the *y*-axis labelled from –10 to 10 draw the following graphs.

(a) $y = x + 1$ (b) $y = -x - 2$ (c) $y = 2x - 3$

Write down the coordinates of the points where they cross as accurately as you can.

6 (a) Copy and complete the table of values for the graphs below:

$y = 3x - 2$ and $y = 5 - x$

x	–1	0	1	2
y				

x	–1	0	1	2
y				

(b) Draw the graphs on graph paper and write down the coordinates of the point where they cross.

Using algebra to describe curves

Algebra can also be used to describe curved line graphs.

The following table shows *x*- and *y*-coordinates related by the equation $y = x^2$

Remember:

$x^2 = x$ times x

and

$-2 \times -2 = + 4$

$-3 \times -3 = + 9$ and so on

x	$y = x^2$
–4	16
–3	9
–2	4
–1	1
0	0
1	1
2	4
3	9
4	16

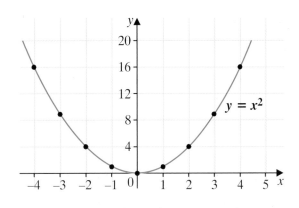

■ **Graphs of equations which have an x^2 term do not have straight lines.**

Exercise 9I

1   Copy and complete this table of values for the graph of
    $y = x^2 - 2$.

x	−4	−3	−2	−1	0	1	2	3	4
x^2	16		4			1			
−2	−2		−2			−2			
y	14		2			−1			

Draw a coordinate grid with values for x from −4 to 4 and
values of y from −2 to 15. Plot the points from the table
and join them with a smooth curve.

2   Copy and complete this table of values for the graph of
    $y = x^2 + 1$.

x	−4	−3	−2	−1	0	1	2	3	4
$y = x^2 + 1$	17				1			10	

Draw a coordinate grid with values for x from −4 to 4 and
values of y from 0 to 17. Plot the points from the table and
join them with a smooth curve.

3   On a coordinate grid with values for x from −3 to 3 and
    values of y from −4 to 16 draw the curve of $y = x^2$ and
    $y = 4x$. Write down the coordinates of *both* points where
    the two lines cross.

4   On a coordinate grid with values of x from −3 to 3 and
    values of y from −4 to 16 draw the curve of $y = x^2 - 2$ and
    $y = 2x + 3$. Write down the coordinates of *both* points
    where the two lines cross.

5   Draw a coordinate grid with values for x from −6 to 6 and
    values of y from −5 to 25. On the grid draw the graphs of:
    (a) $y = 3x - 1$
    (b) $x = 5$
    (c) $y = x^2 - 3$

9.6 1-D, 2-D or 3-D

The number line goes in one direction.

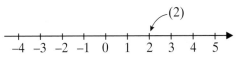

■ **The number line is 1-dimensional or 1-D. You can describe position on the number line using one number or coordinate, for example (2).**

Coordinate grids and flat shapes go in two directions.

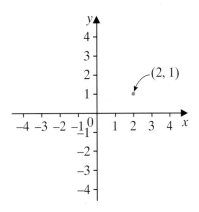

■ **Flat shapes are 2-dimensional or 2-D. You can describe position on a flat shape using two numbers or coordinates, for example (2, 1).**

Solid shapes go in three directions.

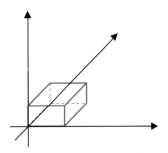

■ **Solid shapes are 3-dimensional or 3-D. You can describe position in a solid shape using three numbers or coordinates. 3-D coordinates look like (1, 2, 3).**

Example 9

Say whether each shape is 2-D or 3-D.

(a)

Cone

(b)

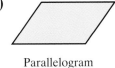

Parallelogram

(c)

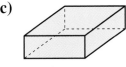

Cuboid

(a) 3-D   **(b)** 2-D   **(c)** 3-D

Exercise 9J

1   Say whether each shape is 1-D, 2-D or 3-D.

(a)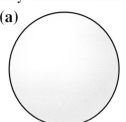

Sphere

(b)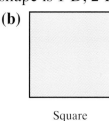

Square

(c)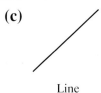

Line

(d)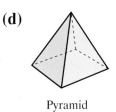

Pyramid

2   Say whether each shape is 1-D, 2-D or 3-D.
   (a) Hexagon   (b) Cylinder
   (c) Circle   (d) Cube

3   Write down all the 3-dimensional coordinates from this list:
   (6)   (3, 1, 1)   (4, 2)   (6, 6, 9)   (3)
   (12, 8)   (4, 1, 16)   (9, 3)   (28, 1, 8)   (126)

Summary of key points

1   A graph representing a linear relationship always has a straight line.

2   A conversion graph is used to convert one measurement into another measurement.

3   Distance-time graphs are used to relate the distance travelled to the time taken, and to calculate speeds.

4   The equation of a line is a way of using algebra to show a relationship between x- and y-coordinates.

5   Graphs of equations which have an x^2 term do not have straight lines.

6   The number line is 1-dimensional or 1-D. You can describe position on the number line using one number or coordinate, for example (2).

7   Flat shapes are 2-dimensional or 2-D. You can describe position on a flat shape using the numbers or coordinates, for example (2, 1).

8   Solid shapes are 3-dimensional or 3-D. You can describe position in a solid shape using three numbers or coordinates. 3-D coordinates look like (1, 2, 3).

10 Sorting and presenting data

10.1 Sorting and presenting data

Here are sixty pupils' marks out of 10 for a spelling test.

3 6 2 5 4 7 7 6 7 9 5 7 8 6 6 5 6 8 6 3 7 8 6 1 8 4 5 5 4 7 6 4 7 3 5
10 3 7 6 8 9 5 8 4 7 6 1 9 8 6 8 2 6 5 4 5 7 6 7 9

To see how well the pupils did sort them like this:

Mark	Tally	Frequency
0		0
1				2
2				2
3						4
4							6
5										9
6													13
7											11
8									8
9						4
10			1

This table, or tally chart, shows the frequency of each mark (how often each mark occurred). It is easier to count tally marks later on if you draw the 5th tally mark through the other four, like this:

|||| |||| |||| |||

is better than

||||||||||||||||||

Another way of showing up any pattern in data is to draw a **bar chart**, sometimes called a bar graph.

This bar chart shows the data from the spelling test.

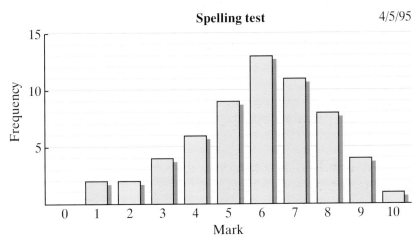

Spelling test   4/5/95

When drawing a bar chart you must make sure that:

● the horizontal and vertical axes are clearly labelled
● the chart is named and dated
● the scale is designed so that the bar chart is a sensible size.

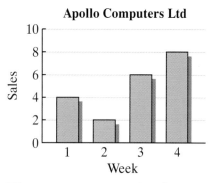

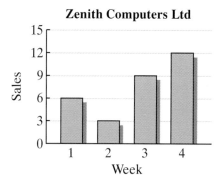

These computer sales figures look the same until you look at the scale on the vertical axes.

For Apollo Computers Ltd the number of computers sold in a month was $4 + 2 + 6 + 8 = 20$

For Zenith Computers Ltd the number sold was $6 + 3 + 9 + 12 = 30$

■ **Tally charts and bar charts are two ways of displaying data that can be counted.**

Exercise 10A

For each of the following sets of data recorded at an island weather station, draw up:
(a) a tally and frequency chart
(b) a bar chart
to make it easier to study the data.

1   Hours of sunshine

5	6	0	1	3	1	4	7	5	6
6	2	4	3	1	0	7	10	9	11
5	4	7	6	9	9	11	12	12	7
9	10	11	10	9	7	8	4	6	5
7	8	10	8	6	3	6	8	3	3
4	1	10	9	11	7	2	6	10	7

2   Force of the wind measured on the Beaufort scale

3	1	4	4	3	5	2	6	4	2
2	2	0	1	4	2	5	3	3	4
4	3	6	7	5	4	0	1	2	3
1	5	4	3	2	2	8	10	9	7
7	8	6	5	4	3	5	2	6	7
5	5	1	2	6	4	3	4	6	5

3 **Maximum temperature, in degrees Fahrenheit**

64	66	66	69	67	70	68	65	67	65
63	64	63	65	64	63	66	60	65	62
69	71	69	70	69	71	71	74	72	69
74	70	70	73	70	73	69	67	68	67
66	67	65	67	66	62	67	63	68	72
69	70	72	72	71	73	70	69	67	68

> Think carefully about your horizontal scale on the bar chart for this data.

10.2 Grouping data

When the range of data is wide it is usual to group it so that the charts are more compact and it is easier to spot patterns.

Here are the marks in an exam:

15	9	34	28	17	44	11	17	21	39
26	38	14	23	47	32	27	19	23	41
17	25	16	30	26	26	42	35	24	19
16	27	33	22	12	19	31	27	36	24
32	36	23	29	37	40	33	16	47	10
34	29	24	18	31	36	26	28	33	22
27	36	34	22	43	18	24	31	26	7

■ **The range is the highest mark − lowest mark.**

Check from the table that the **range** of this data is $47 - 7 = 40$

First group the marks in fives and make a frequency table. Then draw a bar chart to illustrate the data.

Mark	Tally	Frequency
0–4		0
5–9	||	2
10–14	||||	4
15–19	ⅢⅢ ||	12
20–24	ⅢⅢ |	11
25–29	ⅢⅢ ||||	14
30–34	ⅢⅢ ||	12
35–39	Ⅲ |||	8
40–44	Ⅲ	5
45–49	||	2
50–		0

The intervals $0-4$, $5-9$, etc are called **class intervals**.

The **modal class** is 25–29 because this group has the greatest frequency.

■ **The modal class is the group which has the greatest frequency.**

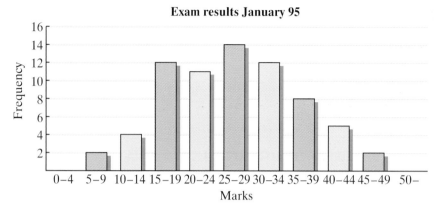

Exam results January 95

Worked exam question 1

Thirty-five people took part in an archery competition. The points they scored are shown below.

18 24 19   3 24 11 25 10 25 14   9 16 26 25
21 27 13   5 26 22 12 27 20   7 28 21 20 23
22 16 12 25   7 25 19

(a) Work out the range of points scored.
   Range is highest − lowest = 28−3. Range is 25.

(b) Complete the frequency table.

Class interval	Tally	Frequency
1–5	||	2
6–10	|| ||	4
11–15	|||||	5
16–20	|||| ||	7
21–25	|||| |||| ||	12
26–30	|||||	5
Total =		35

Exercise 10B

1 In a music contest the marks awarded were:

15 21 13 18 22 17   9 12   7 19
24 16 11   8 14 28 17 15 18   7
 5 17 10 26   7 16 23 14 11 20
12   6 26 16 10 19 13 29 17   8

(a) Using class intervals 0−4, 5−9, etc draw up a frequency table.

(b) Illustrate your answer with a bar chart.

(c) Write down the range of marks.

2   The number of questions answered correctly in a general
    knowledge quiz were:

11	8	23	17	15	31	29	17	9	28
33	26	14	24	35	30	16	8	27	16
35	27	38	22	11	32	25	36	14	20
28	34	21	17	30	6	14	25	28	31
29	17	5	18	22	31	24	35	17	26

(a) Draw up a frequency table, using class intervals 0–4, 5–9,
    and so on.
(b) Illustrate your data with a bar chart.
(c) Give the range of correct answers.
(d) Write down the modal class.

10.3 Comparing data

You can use bar charts to compare different sets of data.

The number of patients who attended morning and evening
surgeries in a doctor's practice one week are shown in the table
and the following bar chart.

	Mon	Tues	Wed	Thur	Fri	Sat	Sun
Morning	145	120	96	116	125	136	27
Evening	81	65	43	55	64	28	–

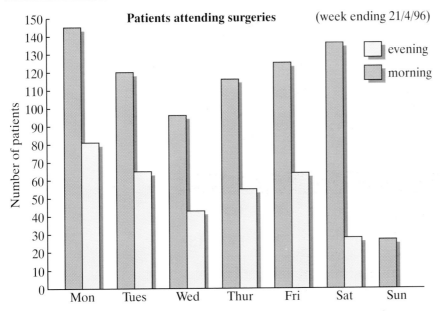

This graph is called a **dual bar chart** because it compares **two**
sets of data. Often it is easier to see a pattern by looking at the
chart rather than the table.

■   **A dual bar chart is used to compare two sets of data.**

Example 1

Use the table and bar chart on page 163 to answer the following questions.

(a) Which day did most patients attend?
(b) Which day did least patients attend?
(c) On which day was there no evening surgery?
(d) On which day did 171 patients visit?
(e) How many more patients attended on Tuesday morning than evening?

(a) Monday (This is clearer from the bar chart.)
(b) Sunday (This is clearer from the bar chart.)
(c) Sunday (Use the table or the bar chart.)
(d) Thursday (Use the table for exact numbers.)
(e) 55 (Use the table.)

Exercise 10C

1 The average daily temperature, in °F, in London and Majorca is recorded in the table.

	Oct	Nov	Dec	Jan	Feb	Mar	Apr
London	58	50	45	42	44	50	56
Majorca	72	65	58	55	58	62	66

(a) Draw a dual bar chart to illustrate the table.

(b) Write down three things you notice from your chart.

2 The cost of holiday accommodation in Majorca per person, in £ sterling, per week is given in the table.

	Oct	Nov	Dec	Jan	Feb	Mar	Apr
Hotel	260	290	270	280	295	315	330
Self catering	190	150	140	110	125	150	180

(a) Draw a dual bar chart to illustrate the table. (Think carefully about the vertical scale you choose.)

(b) Which months have the greatest difference in price between hotel and self catering?

(c) What is the hotel price range?

(d) Write down three things you notice from your chart.

3   The amount of money, in pesetas, spent by Norma and Adrian buying presents was:

	Mon	Tues	Wed	Thur	Fri	Sat
Norma	560	0	940	120	1500	350
Adrian	320	460	150	1260	380	260

(a) Represent this data by drawing a dual bar chart.

(b) Write down three statements about Norma's and Adrian's spending.

4   Record the times you and a friend spent doing homework on each day during one week.

(a) Enter the results on a table.

(b) Draw a dual bar chart of the data.

(c) Comment on the result.

10.4 Pictograms

■ **Pictograms are a quick, visual way of showing information by using symbols to represent amounts.**

For example, a school office secretary recorded the number of Year 10 students who were late to school one week and drew a pictogram like this:

Monday	🧍 🧍 🧍 🧍
Tuesday	🧍 🧍
Wednesday	🧍 🧍 🧍 🧍 🧍 🧍
Thursday	🧍 🧍 🧍
Friday	🧍 🧍 🧍 🧍 🧍

Each 🧍 represents one student.

Rashid went round a factory car park and made a note of the colours of the cars. His findings were:

Black	35	White	20
Red	10	Grey	15
Silver	15	Beige	2
Green	5	Other	14

He decided to use the symbol to represent 5 cars and drew this pictogram.

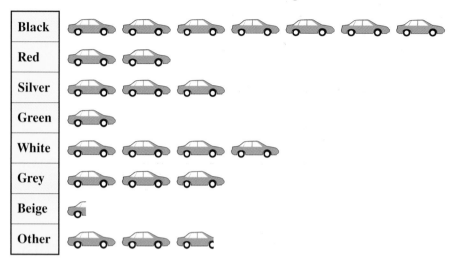

Car colours in car park

Exercise 10D

1 The pictogram shows how the staff at an office travel to work.

Travel

Walk	
Train	
Bus	
Car	
M/cycle	
Bicycle	
Other	

$\dot{\wedge}$ = 10 people

> You can use parts of a symbol to represent smaller numbers. Here each line represents 2 people:
>
> e.g. $\dot{\uparrow}$ = 6 people
>
> $\uparrow$ = 2 people

(a) How many people travel by car?

(b) Which was the most popular way to travel?

(c) How many people travel by bicycle?

2 Draw a pictogram to illustrate each of the tables. First decide what symbol you will use and what it will represent.

(a) This table shows the number of telephone calls to an answering service between midnight and 7:00 am during a certain week.

	Mon	Tues	Wed	Thur	Fri	Sat	Sun
Calls	24	20	32	48	28	32	12

(b) This table shows the number of cars using a car-park in a week.

	Sun	Mon	Tues	Wed	Thur	Fri	Sat
Cars	40	160	120	150	140	180	210

3 The pictograms show how many drinks were sold from two machines.

Drinks machine sales

Dining room

Coffee	🥤🥤
Tea	🥤🥤🥤🥤
Hot chocolate	🥤🥤🥤
Soup	🥤
Hot blackcurrant	🥤🥤🥤

Staff room

Coffee	🥤🥤🥤🥤🥤
Tea	
Hot chocolate	🥤
Soup	🥤
Hot blackcurrant	🥤

Key: 🥤 10 drinks sold
     5 drinks sold

(a) **(i)** Which drink was most popular in the dining room?
  (ii) How many hot blackcurrants were sold in the dining room?
  (iii) How many hot chocolates were sold in the staff room?

(b) The staff room machine sold 45 teas.
  Copy and complete the pictogram.

(c) Work out the total number of drinks sold in the dining room.

(d) 64 people used the machine in the dining room.
  Work out the average number of drinks sold per person.

(e) Comment on the differences in sales from the two machines. [E]

4 **(a)** Draw up a table to show the amount of time, in minutes, you spent each day last week:
  (i) doing homework
  (ii) watching television.

(b) Draw a pictogram to illustrate your result.

10.5 Line graphs

The table shows the temperature, in °Centigrade in Leeds at midday during the first week in May.

May	1	2	3	4	5	6	7
Temperature (°C)	12	16	14	11	12	15	13

We could show this data on a **line graph**. Here are two ways of doing so.

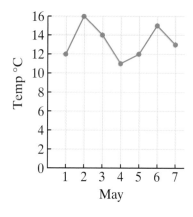

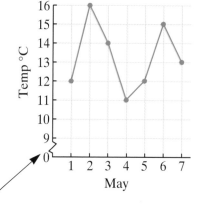

The temperature scale on this graph starts at 0 °C.

The break in the axis shows the scale goes straight from 0 to 9.

This graph gives more space on the *y*-axis to values between 10 and 16 °C. This gives a more representative line graph.

10.6 Discrete and continuous data

It is important, when drawing graphs, to know if the data is discrete or continuous.

Data which can be *counted* is called **discrete data**.
For example, George has three sisters.
               Portsmouth beat Charlton 3–1.
               There are six tennis balls in the box.

Data which is *weighed* or *measured* is called **continuous data**.
For example, George is 1.82 m tall.
               The athlete ran 100 metres in 10.3 seconds.
               The weight of the six tennis balls was 372.5 grams.

■ **Line graphs can be used to show continuous data.**

This table shows the number of passengers that got off a train at each stop on a line in SE London.

London Bridge (B)	Hither Green (G)	Lee (L)	Mottingham (M)
30	19	28	24

If you drew a graph like this it might give the impression that people got off the train between stations!

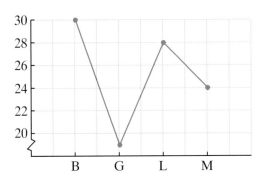

The data is *discrete* so you do not join the points.

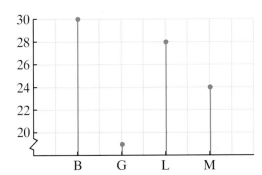

This is called a vertical line graph. This is used to display data that can be counted in a similar way to bar charts.

Exercise 10E

1 A class of 30 pupils listed their favourite hobby. The results are shown in this table.

Sport	Music	Reading	Model making	Computer games
8	5	3	2	12

Illustrate this data by drawing:

(a) a bar chart

(b) a pictogram using suitable symbols.

2   This table shows the temperatures recorded in °F one day
in July.

Washington	Vancouver	Sydney	Luxor	Hong Kong	London
93	64	52	104	88	72

(a) Draw a bar chart for this data.

(b) Draw a pictogram using a suitable symbol.

(c) Which did you find easier to draw? Give your reasons.

(d) Which do you think is more accurate? Say why.

3

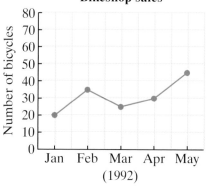

The graphs show the sales of bicycles during January to May.

(a) Look quickly at the graphs. Which shop appears to
sell the most bicycles?

(b) How many bikes did Cycleshop sell in March?

(c) How many bicycles were sold altogether by:
  (i) Cycleshop
  (ii) Bikeshop?

(d) Which shop sold more bicycles?

(e) In which month did they both sell the same number of
bicycles?

4   Which of the following are discrete data and which are
continuous?

(a) The number of pages in a book

(b) The weight of a bag of sweets

(c) The number of sweets in a tube of Tasties

(d) The cost of a tube of Tasties

(e) The time of a telephone call

(f) The height of a flagpole

(g) The number of peas in a pod

(h) The length of the pod

(i) The distance to the moon

(j) The number of these questions you get right

(k) The time it took you to do this exercise

5 Draw line graphs to represent each of the following:

(a) The number of letters delivered to a guest house one week:

Mon	Tues	Wed	Thur	Fri	Sat	Sun
12	25	15	19	23	20	0

(b) The noon temperature, in °F, in Cambridge:

July	1	2	3	4	5	6	7
Temperature	73	69	65	70	75	79	76

Think carefully about your vertical scale.

(c) The longest rivers in Great Britain, in km to the nearest 10 km:

River	Severn	Towy	Shannon	Tay	Thames
Length	365	100	390	190	340

6 Muriel recorded the maximum and minimum temperatures over a period of ten days. Her results are given in this table.

April 1995	4th	5th	6th	7th	8th	9th	10th	11th	12th	13th
Max (°C)	12	9	10	10	13	17	15	19	16	18
Min (°C)	5	2	4	5	6	7	5	8	6	7

(a) On the same graph illustrate these temperatures.

(b) Write a comment about each set of data.

7   The amount of petrol (in litres) in the storage tank at a garage was measured every hour between 7 am and 7 pm on one day. This is the shape of the line graph showing the results.

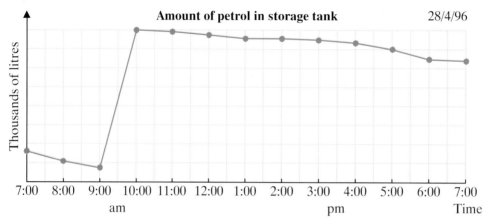

(a) When was the amount of petrol in the tank at its lowest?

(b) What do you think happened between 9 am and 10 am?

(c) What were the sales like between 1 pm and 5 pm?

(d) Give a reason for your answer to part **(c)**.

8   The graphs show the number of petrol sales at two garages during a week.

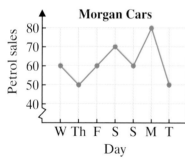 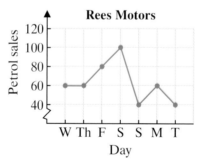

(a) What is different about the scales?

(b) Copy and complete the table below.

Petrol sales	Wed	Thur	Fri	Sat	Sun	Mon	Tue
Morgan Cars							
Rees Motors							

(c) Draw a dual bar chart for this information.

(d) Use your chart to make three comments about the sales.

(e) On which day was the most petrol sold at each garage?

10.7 Time series

A line graph used to illustrate data collected at intervals in time (e.g. weekly, monthly or annually) is called a **time series**. By observing the results over a long period, you can predict what may happen in the future.

■ **A line graph used to illustrate data collected at intervals in time is called a time series.**

Example 2

This graph shows the temperature, in ° Centigrade, at noon during the first ten days of June in Llangrannog.

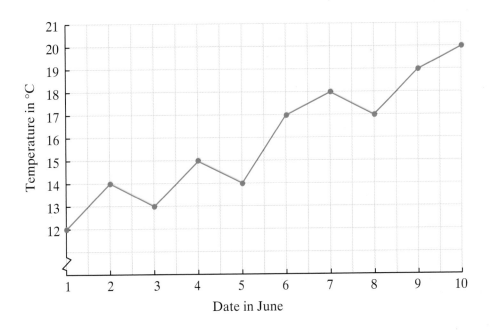

(a) What was the highest temperature recorded?

(b) What do you think the temperature might be in the next few days?

(c) Give a reason why your prediction might be wrong.

(a) 20 °C

(b) 20–21 °C

(c) The weather may change very suddenly.

Exercise 10F

1   This table shows the average temperature, in ° Fahrenheit, in Berne and Lugano in Switzerland.

	Jan	Feb	Mar	Apr	May	Jun	Jul	Aug	Sept	Oct	Nov	Dec
Berne	35	40	48	56	64	70	74	73	66	55	44	36
Lugano	43	48	56	63	70	78	83	82	75	63	52	45

(a) Draw a time series line graph to illustrate the data for each of these cities.

(b) What was the highest temperature and in which month was it recorded?

(c) Where was the lowest temperature recorded?

2   The value, in pence, of shares in two companies on the last day of the month last year is shown in this table.

	Jan	Feb	Mar	Apr	May	Jun	Jul	Aug	Sept
Engico	74	75	74	72	70	68	69	67	66
Mechacorp	35	36	40	41	39	42	41	43	44

(a) Draw time series for the share price of each company on the same graph.

(b) What do you think the value of each share might have been in October? Give your reason.

3   This table shows the annual turnover of Angel Exports Ltd in millions of Euros.

Year	1991	1992	1993	1994	1995	1996	1997	1998	1999
Turnover	12	9	11	12	14	15	13	16	17

(a) Draw a time series graph to represent this data.

(b) Comment on your graph. (Hint: you could say what the highest and lowest values were, whether the graph has gone up or down, or what you think the next value might be.)

4 This table shows the quarterly sales of cars at Autobuy garages.

Year	1998				1999				2000			
Quarter	1	2	3	4	1	2	3	4	1	2	3	4
Sales	90	86	82	77	94	92	88	85	100	95	92	

(a) Draw a time series graph to represent this data.

(b) Comment on your graph.

(c) Make a prediction of sales for the last quarter of 2000. Give your reason.

5 This table shows the annual number of anglers fishing in Wey Pond.

Year	1991	1992	1993	1994	1995	1996	1997	1998	1999
Anglers	279	268	272	240	228	212	209	195	

(a) Draw a time series graph to represent this data.

(b) Comment on your graph.

(c) Make a prediction for the number of anglers in 1999.

10.8 Histograms

You usually draw bar charts or line graphs to represent frequency distributions. These diagrams use the height of bars or lines to represent the frequency. If the data is continuous and is grouped as described in Section 10.2 you can use a **histogram**.

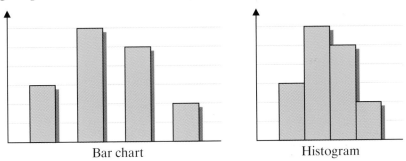

| Bar chart | Histogram |

You will notice that a histogram looks like a bar chart but there are no gaps between the bars in a histogram. Later you will find that when the intervals are not equal the area has to be considered.

■ **Histograms are used to display grouped data that is continuous.**

10.9 Frequency polygons

Another useful way of displaying data is a **frequency polygon** in which the midpoints of the class intervals are joined by straight lines.

The table shows the frequency distribution of the ages of members of a swimming club in 1990 and 1991.

Age	0–9	10–19	20–29	30–39	40–50	over 50
1991	5	15	21	30	19	15
1992	10	24	28	22	10	6

You can draw bar charts and join the midpoints to get frequency polygons.

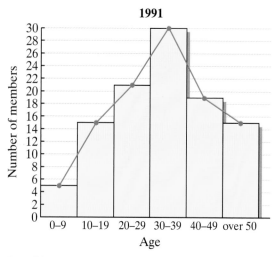

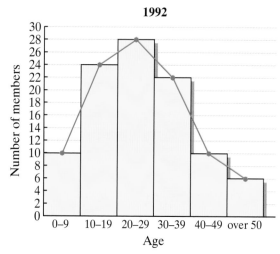

■ **Frequency polygons can show the general pattern of data represented by bar charts or histograms.**

It is often easier to compare data like this by placing one polygon on top of the other:

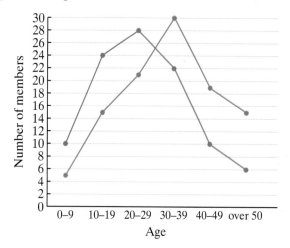

Exercise 10G

1   The number of pupils late for school are given in this table.

	Mon	Tues	Wed	Thur	Fri
Week 1	6	8	14	10	12
Week 2	9	13	8	6	10

(a) Draw bar graphs of this data.

(b) Join the midpoints to make frequency polygons.

(c) Use the frequency polygons to compare the two weeks and write down three observations you have found.

2   The table shows the points scored by 75 pupils in a fitness programme held in September, January and May.

Mark	0–4	5–9	10–14	15–19	20–24
Sept	10	12	22	16	15
Jan	6	15	26	18	10
May	2	8	25	27	13

(a) Draw frequency polygons for each set of results.

(b) Make a tracing of the January and May polygons.

(c) Comment on the changes you find.

Summary of key points

1   A tally chart can be used to display data that can be counted.

2   A bar chart can be used to display data that can be counted.

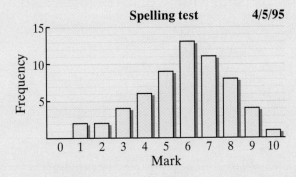

3   The range of a set of data = highest value − lowest value.

4   The modal class is the group with the greatest frequency.

5   A dual bar chart is used to compare two sets of data.

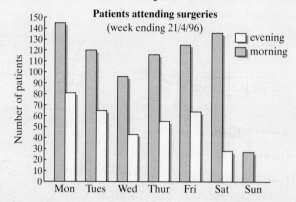

6   Pictograms are a quick, visual way of showing information by using symbols to represent amounts.

7   Line graphs can be used to display continuous data.

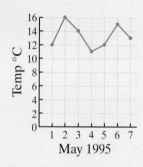 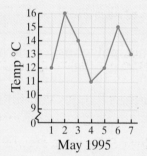

8   A line graph used to illustrate data collected at intervals is called a time series.

9   Histograms are used to display data that is continuous and which is grouped.

10   Frequency polygons can show the general pattern of data represented by bar charts or histograms.

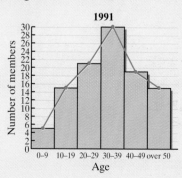 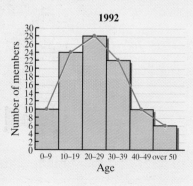

11 Three-dimensional shapes

All the boxes and packets shown on the supermarket shelves
in the picture are 3-dimensional or 3-D. They have height,
width and depth. This unit is all about 3-D shapes.

11.1 Horizontal and vertical surfaces

In the picture at the top of the page the shelves are horizontal
surfaces. The floor and ceiling are also horizontal surfaces.

The walls and the sides of the packets are vertical surfaces.

■ **A flat surface is called a plane.**

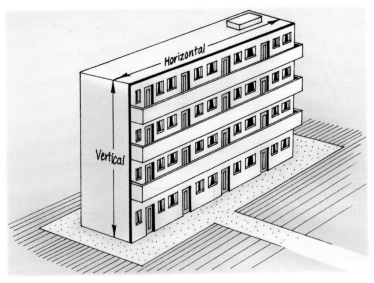

Exercise 11A

Identify and list 5 horizontal and 5 vertical surfaces in your classroom. You can include the floor and the walls as part of your answer.

11.2 Faces, edges and vertices

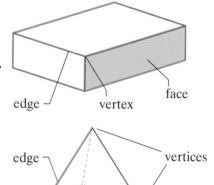

■ **Surfaces of three-dimensional shapes are called faces.**

■ **The shaded surface on the box is called a** *face.*
  The box has 6 faces.

■ **The line where two faces meet is called an** *edge.*
  The box has 12 edges.

■ **The point where three edges meet is called a** *vertex.*
  The box has 8 vertices.

Exercise 11B

Find or make as many different 3-D shapes as you can. Try to get some shapes with 4, 5, 6, 7 and 8 faces. Using straws for the edges is a quick way of getting new and unusual shapes. Count the faces, edges and vertices of your shapes and write them in a copy of this table.

Description of shape	Number of faces	Number of vertices	Number of edges

There is a rule which connects the number of faces and the number of vertices with the number of edges. Use your results to find this rule.

A **sketch** is a drawing which does not pretend to be exact. It should be good enough not to be misleading.

In sketches parallel lines are drawn parallel. Vertical lines always look vertical. Horizontal lines may go in any direction.

Exercise 11C

Draw sketches for your shapes in Exercise 11B.

11.3 Looking at shapes

The chart on this page shows pictures, sketches, names and some of the properties of some 3-D shapes.

Name and properties	Picture	Sketch
Cube 6 square faces		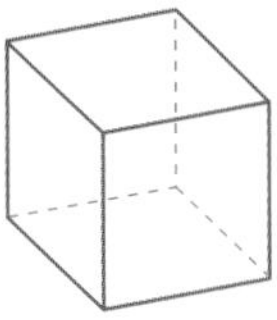
Cuboid 6 rectangular faces	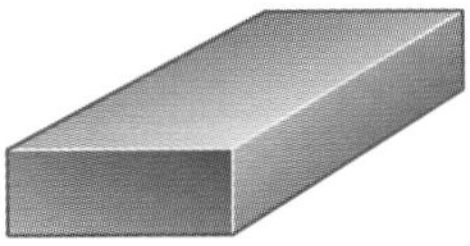	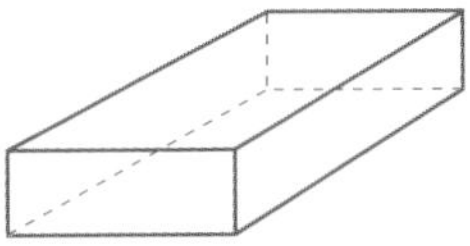
Sphere		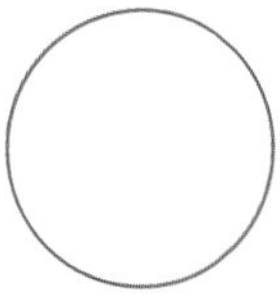
Square-based pyramid Square base, 4 triangular faces	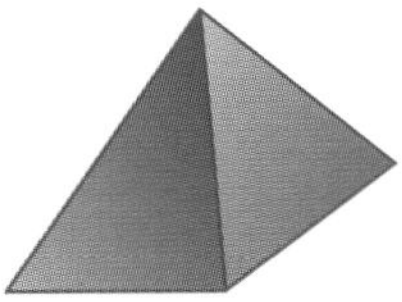	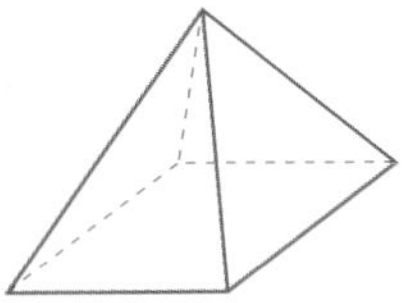
Triangular-based pyramid (Tetrahedron) 4 triangular faces	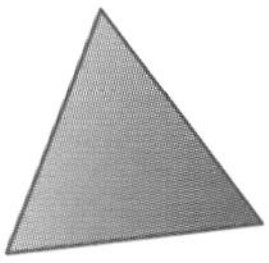	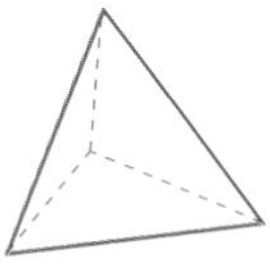
Cone Pyramid with a circular base		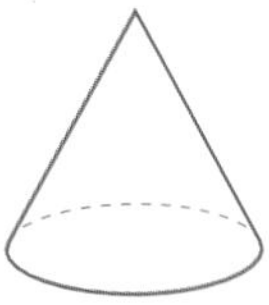
Cylinder 2 circular faces	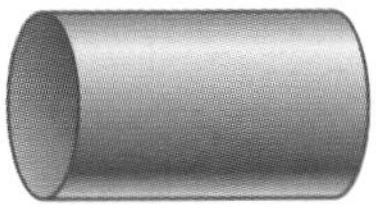	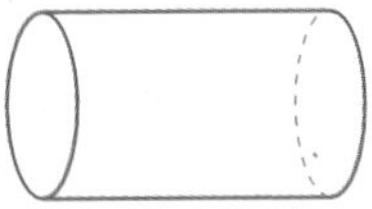

To sketch a cube:

Stage 1
Draw a face
It is a parallelogram

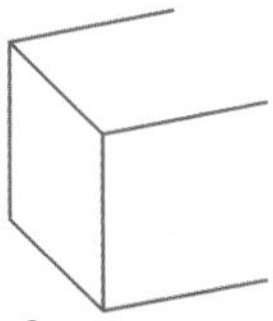

Stage 2
Draw 3 parallel edges

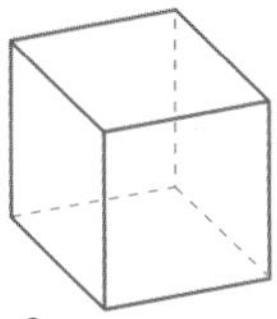

Stage 3
Complete the cube

To sketch a square-based pyramid:

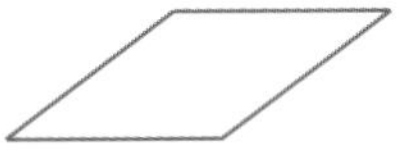

Stage 1
Draw the base

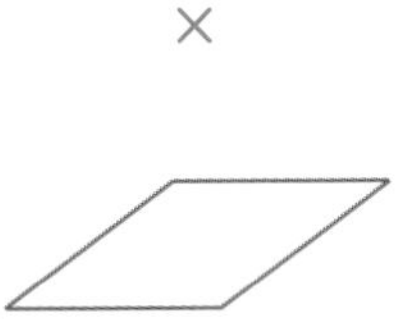

Stage 2
The top is above the middle of the base

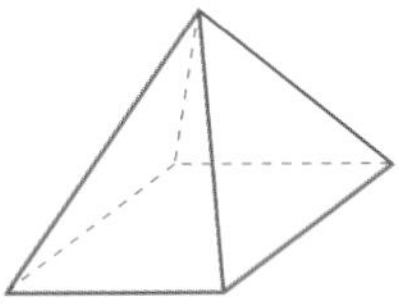

Stage 3
Join the top to the other corners

Exercise 11D

Look at the picture below and make a list of as many 3-D shapes as you can. For example, the football is a sphere.

There are at least 12 3-D shapes.

11.4 Prisms

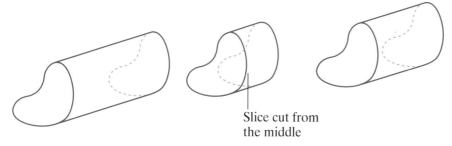

Slice cut from
the middle

The faces at either end of this shape are identical and parallel. Parallel means the faces are the same distance from each other at all points.

If the shape is cut anywhere parallel to these faces then the cut surface will be identical to the end faces. This type of shape is called a **prism**.

■ **A prism is a shape whose cross section is constant wherever it is cut.**

Some other prisms are drawn below. Where the shape of the cross section is a known 2-D shape then it is used to describe the type of prism.

Drawing a prism:
These are difficult to draw. They can be made to look reasonable by drawing the front and back faces and then joining the vertices.

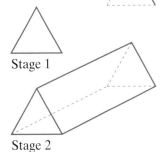
Stage 1

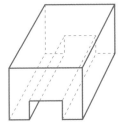
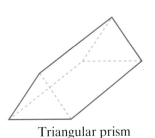
Triangular prism

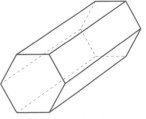

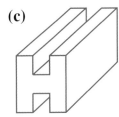
Hexagonal prism

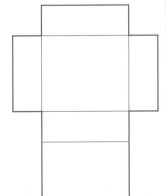
Stage 2

Exercise 11E

1 Which of the following shapes are prisms?

(a)
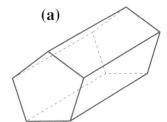

(b)
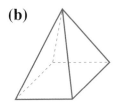

(c)

2 What name is usually given to a circular prism?

3 Write down the name of two other 3-D shapes that are also prisms.

4 Write down 5 things in your classroom which are prisms.

11.5 Nets

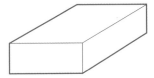

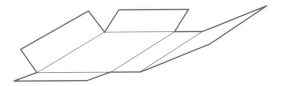

The box (cuboid) in the pictures has been opened out to make a 2-D shape. This 2-D shape is called the **net** of the box.

■ **A net is a 2-D shape that can be made into a 3-D shape.**

Example 1

Draw the net of this triangular prism.

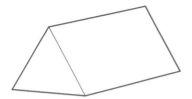

The prism has three rectangular faces measuring 6 cm by 3 cm which are joined along their long sides. These faces can be drawn as:

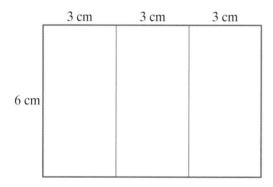

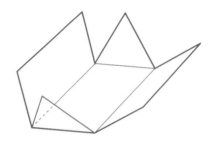

The faces at either end of the prism are equilateral triangles with 3 cm sides. These must join to the short side of one of the rectangles. This makes the complete net.

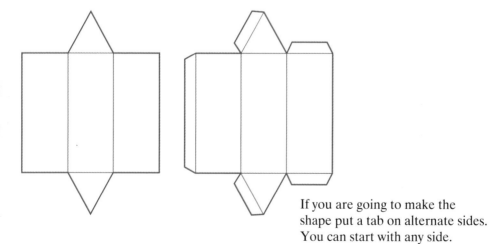

If you are going to make the shape put a tab on alternate sides. You can start with any side.

You need to be able to construct accurate drawings of 2-D shapes if your nets are to fold together to make good solids. There is more about this in Unit 5 on 2-D shapes.

Exercise 11F

You need a ruler, pencil and pair of compasses for some of the questions in this exercise.

1   Sketch the nets of these solids:

(a) 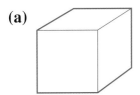   **(b)**   **(c)**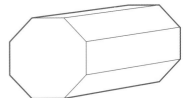

2   These nets will form a 3-D solid. Draw a sketch of each solid.

(a) 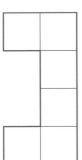   **(b)**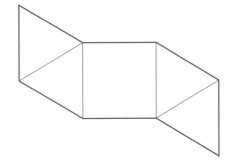

3   Which of the following are nets of a cube?

(a) 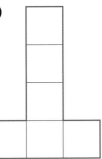   **(b)**   **(c)**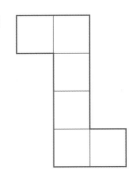

4   Which of the following are nets of a tetrahedron (triangular-based pyramid)?

(a)    **(b)** 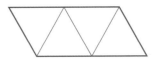   **(c)**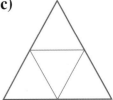

5 Draw accurate nets for these shapes. You could do this on card and make the shapes (remember to add tabs).

(a)

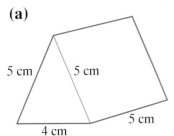

5 cm   5 cm

5 cm

4 cm

(b)

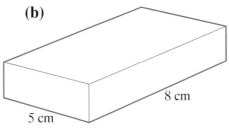

2 cm

8 cm

5 cm

(c)

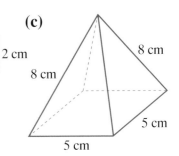

8 cm

8 cm

5 cm

5 cm

6 Draw an accurate net for the following shapes:

(a) a cube with sides of 5 cm

(b) a regular tetrahedron with sides of 4 cm.

Exercise 11G Mixed questions

1 For each shape in this question write down:

  (i) the name   **(ii)** the number of edges

  (iii) the number of faces   **(iv)** the number of vertices

(a)

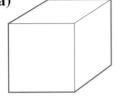

(b)

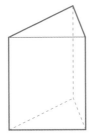

(c)

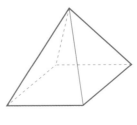

2 The base of each shape in question **1** is horizontal.

  (i) Write down the number of horizontal edges each shape has.

  (ii) Write down the number of vertical edges each shape has.

3 Which of these shapes are prisms?

C

A

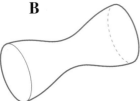

B

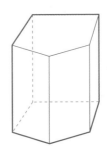

4 Sketch the following solids and their nets:

  (a) cuboid   **(b)** cylinder   **(c)** cone

  (d) tetrahedron   **(e)** hexagonal prism.

5 Sketch the solids that these nets form. Mark the
measurements on your sketches.

(a)

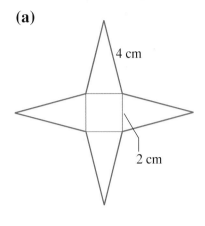

4 cm

2 cm

(b)

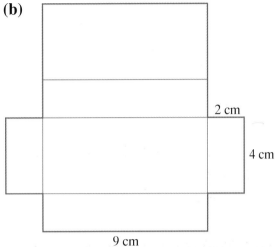

2 cm

4 cm

9 cm

6 Draw accurate nets for these solids.

(a)

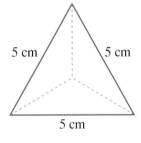

5 cm   5 cm

5 cm

(b)

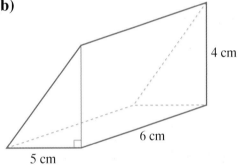

4 cm

6 cm

5 cm

7 Here is a list of the names of some shapes:

Square   Cube   Kite   Rectangle
Pyramid   Cylinder Triangle   Circle

Use the list to help you write down the names of these
shapes.

(a)

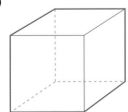

(b)

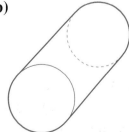

8 The diagram shows part of an accurately drawn net for a cuboid.

One face of the cuboid is missing.

The cuboid is 2 cm by 3 cm by 4 cm.

(a) Write down the size of the missing face.

(b) Copy and complete the net by drawing in the missing face.

(c) Work out the area of the whole net. [E]

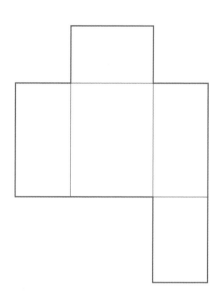

9 Draw an accurate net for the following shapes:

(a)

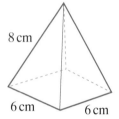

8 cm

6 cm   6 cm

(b) a regular tetrahedron with sides of 5 cm.

11.6 Plan and elevation

Architects and designers often represent three-dimensional objects with two-dimensional drawings.

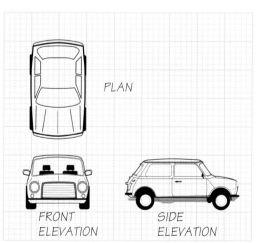

PLAN

FRONT ELEVATION

SIDE ELEVATION

■ **The plan of a solid is the view when seen from above.**

■ **The front elevation is the view when seen from the front.**

■ **The side elevation is the view when seen from the side.**

Example 2

Draw the plan and elevations of this shape.

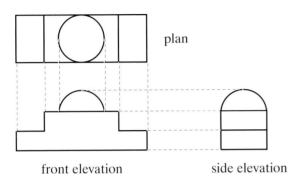

plan

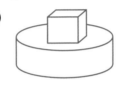

front elevation   side elevation

> You should draw plans and elevations using dotted lines to show how the different drawings match up.

Exercise 11H

1 Sketch the plans and elevations of these shapes.

(a) **(b)** **(c)**

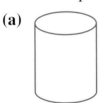  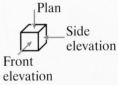

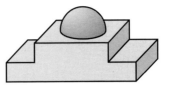

Plan
Side elevation
Front elevation

(d) **(e)** **(f)**

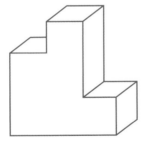 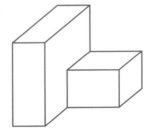

2 Use multilink cubes to construct these solids. Sketch each one.

(a)

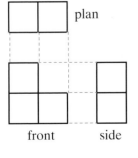

plan

front   side

(b)

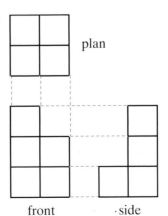

plan

front   ·side

(c)

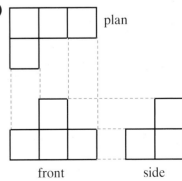

plan

front   side

3 Investigation

Describe the solids with these plans and elevations

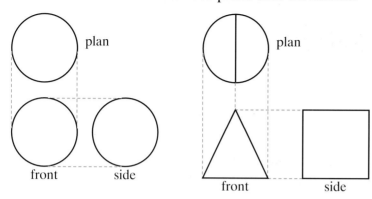

Draw your own set of three shapes. Try to describe a solid which has your shapes as its plan, front elevation and side elevation.

Summary of key points

1 A flat surface is called a plane.

2 Surfaces of 3-dimensional shapes are called faces.

3 The line where two faces meet is called an edge.

4 The point where three edges meet is called a vertex.

5 A prism is a shape whose cross section is the same all the way through.

6 Nets are the 2-D shapes that can be made into a 3-D shape.

7 • The plan of a solid is the view when seen from above.

   • The front elevation is the view when seen from the front.

   • The side elevation is the view when seen from the side.

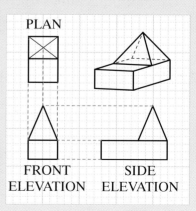

12A Using and applying mathematics

In your GCSE course you will be using and applying mathematics all the time.

For your GCSE exam, using and applying mathematics is sometimes tested through coursework and sometimes through an examination. Whichever way it is tested the basic ideas are the same.

This unit uses an **investigative task** to show you how to tackle this type of work. Here is the task:

Diagonals

The diagram on the right shows a regular hexagon.

A diagonal of any shape is a line which joins any two of the vertices (corners) of the shape but which is not an edge of the shape.

1   Show that the regular hexagon has a total of 9 diagonals.

2   Work out the number of diagonals for a regular heptagon (seven-sided shape).

3   Draw some regular shapes of your own.
    For each shape work out the total number of diagonals.
    Investigate regular shapes to find the relationship between the number of diagonals and the number of vertices of any shape.

How investigative tasks are marked

Your work will be marked on three qualities which can be described as:

Deciding and doing   **Showing**   **Explaining**

Deciding and doing means deciding what to do and then doing it to carry out the investigation.

Showing means writing down your results so that others can understand what you have done. Tables and diagrams are good ways of showing your results.

Explaining means giving reasons for your answers.

It is important that your work contains all three qualities.

What to do in an investigation

Here is a list of things you should try to do during an investigation *and* things you should provide evidence of in your final report.

Don't worry if you don't understand it all yet. The rest of this unit will help.

If you are doing an investigative task as coursework you can refer to this list:

- Make sure you understand the problem.
- Check to see if you have worked on a similar problem. If you have, try to make use of this experience.
- Try some simple and special cases.
- Plan your work in an ordered way. Have a strategy.
- Record what you are trying to do.
- Record your observations.
- Use appropriate diagrams and forms of communication.
- Record and tabulate any findings and results.
- Predict what you think may happen and test it. This is called testing a conjecture.
- Try to find and make use of any counter examples.
- Generalise, especially in symbols, if you can.
- Comment on any generalisations.
- Explain or justify your generalisations.
- Try to prove any generalisations.

The rest of this unit shows you how the task called **Diagonals** could be investigated.

Understanding the problem

You need to understand the main idea of the task:

You are being asked to investigate the number of diagonals there are in regular shapes in two dimensions.

An answer to question **1** about the regular hexagon will help to show that you understand the basic idea of the problem.

> **Remember:**
> A diagonal is a line from one corner to another that is not a side of the shape.

> **1** Show that the regular hexagon has a total of 9 diagonals.

Here are the 9 diagonals drawn on the hexagon:

If you label the six vertices of the regular hexagon as:

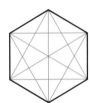

then you can record the 9 diagonals as:

$①→③$ $①→④$ $①→⑤$
$②→④$ $②→⑤$ $②→⑥$
$③→⑤$ $③→⑥$
$④→⑥$

This way of recording the results for the regular hexagon may help at a later stage.

Have you worked on a similar problem?

Only you can answer this but it is quite likely that you have done similar work before.

Compare the way the diagonals of the regular hexagon are recorded:

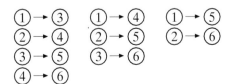

with an array of dots like this:

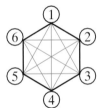

You may have seen similar dot patterns when dealing with number patterns such as the triangular numbers or the square numbers. This may help later.

Question **2** helps to develop your understanding of the problem:

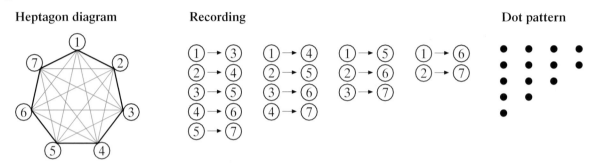

2   Work out the number of diagonals for a regular heptagon (seven-sided shape).

Here are the diagonals, the recording of them and the dot pattern:

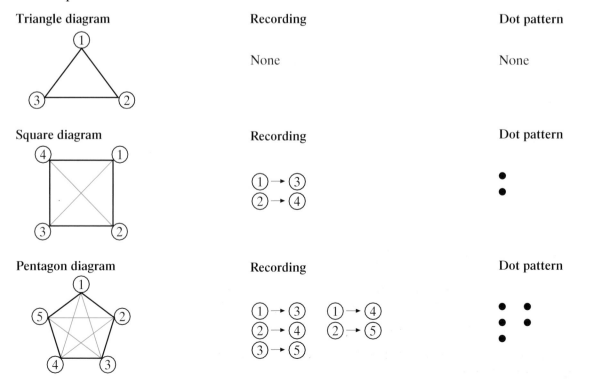

Try some simple and special cases

You could just keep exploring different shapes but it is more sensible to choose an order in which to investigate them. For example: start with a triangle, square and pentagon, increasing the number of sides by one each time.

Here are the results in diagrams, the recording of them and the dot patterns:

Strategies

It is always worth writing a sentence or two to explain what you have done and how you have done it.

The strategy used so far has been:

Step 1 Draw the shape.

Step 2 Number the vertices.

Step 3 Join up the diagonals starting with those that go from 1; then those that go from 2 and so on.

Step 4 Make a record of the diagonals.

Step 5 Draw the dot pattern.

Once ① → ③ is drawn you do not need to draw ③ → ① in the opposite direction. This is an important part of the strategy.

Recording and tabulating observations and results

You are now at a stage where you can record the results of the investigation so far in a table:

Shape	Number of vertices	Number of diagonals
Triangle	3	0
Square	4	2
Pentagon	5	5
Hexagon	6	9
Heptagon	7	14

Forms of communication

Several forms of **communication** have been used so far: diagrams, the ① ② ③ labelling system, dot patterns and numbers and the table of results.

Now you can look for any pattern in the number of diagonals of the regular shapes.

Observations and predictions

Without the shape names the table of results looks like this:

Vertices	3	4	5	6	7
Diagonals	0	2	5	9	14

You can make many comments or **observations** about this sequence of results.

At a very basic level you might observe that the larger the number of vertices the larger the number of diagonals.

At a higher level you might observe that the numbers of diagonals follow a pattern of:

> even, even, odd, odd, even

You might **predict** that for 8 vertices there would be an even number of diagonals.

Remember:
Write down any observations you try out even if they don't work.

Using differences to find patterns

One of the best ways of finding patterns about which you can make observations is the method of **differences**.

Here is the table of results so far. Find the difference between each pair of results in the sequence. Find the difference between each pair of differences.

Vertices	Diagonals		1st difference		2nd difference
3	0				
4	2	2 – 0	2	3 – 2	1
5	5	5 – 2	3	4 – 3	1
6	9	9 – 5	4	5 – 4	1
7	14	14 – 9	5		

You might now notice that the numbers under the 1st difference column follow a very simple pattern. The numbers under the second difference column are all the same and equal to 1. This is very useful information as it *suggests* that the pattern of results can be continued as:

Vertices	Diagonals	1st difference	2nd difference
3	0		
4	2	2	1
5	5	3	1
6	9	4	1
7	14	5	1
8	20	6 14 + 6	1 5 + 1
9	27	7 20 + 7	6 + 1

Making and testing predictions

The method of differences allows you to **predict** (or **conjecture**) that the number of diagonals for a regular shape with 8 vertices is 20 and for a regular shape with 9 sides is 27.

It also tells you that if this is correct you may have found a general pattern.

You can **test** your predictions or conjectures using the diagrams, and the recording and dot pattern method used on page 194:

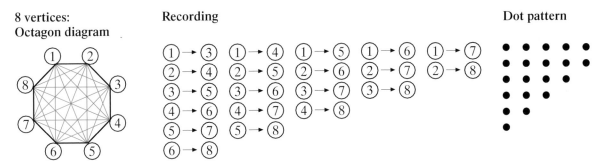

8 vertices:
Octagon diagram

Recording

Dot pattern

Counting the diagonals confirms the result of 20.

9 vertices:
Nonagon diagram

Recording

Dot pattern

Counting the diagonals confirms the result of 27.

Making and testing generalisations

When you can explain a pattern with a **general rule** for finding other numbers in the pattern you are **generalising** or **making a generalisation**.

Look at the results and the first differences again:

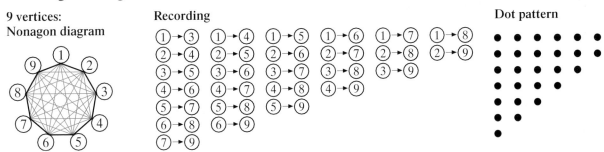

Vertices	3	4	5	6	7	8	9		10	11	12
Diagonals	0	2	5	9	14	20	27		35	44	54
1st difference		2	3	4	5	6	7		8	9	10

You can see that the 1st differences follow the sequence of **natural numbers** (apart from 1): 2, 3, 4, 5, 6, 7 and so on. This is a generalisation. You can use it to extend the table of results like this:

Generalising in symbols

The next stage in developing this investigation is to try to put the generalisation in **symbolic** (or **algebraic**) **form**. The correct use of **symbols** is a very important step on the road to achieving a good GCSE grade.

Sometimes it is quite easy to spot a general rule, but this one is a bit harder. One of the keys here is to see that we have:

Vertices	Diagonals
3	$0 = 3 \times 0$
5	$5 = 5 \times 1$
7	$14 = 7 \times 2$
9	$27 = 9 \times 3$

which suggests that with 11 vertices the number of diagonals $= 11 \times 4 = 44$. This does work – it is a result we have already obtained.

Now we can look again at the full set of results:

Vertices	Diagonals
3	$0 = 3 \times 0 = 3 \times \frac{0}{2}$
4	$2 = 4 \times \frac{1}{2} = 4 \times \frac{1}{2}$
5	$5 = 5 \times 1 = 5 \times \frac{2}{2}$
6	$9 = 6 \times 1\frac{1}{2} = 6 \times \frac{3}{2}$
7	$14 = 7 \times 2 = 7 \times \frac{4}{2}$
8	$20 = 8 \times 2\frac{1}{2} = 8 \times \frac{5}{2}$
9	$27 = 9 \times 3 = 9 \times \frac{6}{2}$
10	$35 = 10 \times 3\frac{1}{2} = 10 \times \frac{7}{2}$
11	$44 = 11 \times 4 = 11 \times \frac{8}{2}$

This suggests that the **general rule** is:

Diagonals = Vertices × ((Vertices − 3) ÷ 2)

Writing D for the number of diagonals and V for the number of vertices the **symbolic** or **algebraic form** of the general rule is:

$$D = V \times \frac{(V - 3)}{2}$$

or: $D = \frac{1}{2}V(V - 3)$

Testing the generalisation

You can test the general rule in one of two ways:

One way is to use a value of V and D for which you know the result, for example, choose $V = 6$, $D = 9$. Putting this in the rule gives:

$$D = \frac{1}{2}V(V - 3)$$
$$D = \frac{1}{2}6(6 - 3)$$
$$= 3(3) = 9$$

So the general rule works: when $V = 6$ it gives $D = 9$.

Another way is to test the rule for some new value of V and check the result by drawing, recording and using the dot pattern, or by extending the results table. Choose $V = 12$, for instance:

$$D = \tfrac{1}{2} (12) (12 - 3)$$
$$= 6 \times 9$$
$$= 54$$

You could try either extending the table of results or using the drawing, recording and dot pattern method. If you do you will find that when $V = 12$ then $D = 54$.

This investigation does not have to stop here. You may have ideas for further work. In the Intermediate and Higher courses the same problem is developed to include the last five items on the more advanced work list at the start of this unit.

Hint:
Even if a test doesn't work, still record it and try to say why it failed.

12B Handling data

You can use mathematics to analyse real life information, or data.

For your GCSE you will need to complete a Handling Data coursework project. This chapter will work through a sample project, showing you how to approach your Handling Data project. Some of the techniques used in this chapter are covered in Units 20 and 25.

In 2001 a national census was used to collect data about everyone in the country.

Jordan Hill County High School

Jordan Hill County High School is a school for students aged 11 to 16. It is a growing school so the number of students in each year varies. Although fictional, the data is based on a real school.

Data has been provided about all the students at Jordan Hill County High School. Data that has been gathered for you is called **secondary data**. Data you gather yourself is called **primary data**.

Secondary data suitable for use in the Handling Data project is available to download from Heinemann's website at www.mathsmatrix.com.

The data

This table shows how many boys and girls there are in each year group at Jordan Hill:

Year	Boys	Girls	Total
7	150	150	300
8	145	125	270
9	120	140	260
10	100	100	200
11	84	86	170

Data is provided for each pupil in a number of categories including:

Name, age, year group, IQ, height, hair colour, eye colour, number of brothers/sisters, distance travelled between home and school, gender (boy or girl).

Lines of enquiry

With so much information it is important to know exactly what you want to find out about Jordan Hill. You will have to choose a **line of enquiry** to investigate. Here are some possibilities:

the hair colour of students

the range of distances travelled to school

the relationship between shoe size and weight or height

the relationship between IQ and Key Stage 3 English results

the height to weight ratio for boys and girls

You need to choose a line of enquiry that will give you plenty to say without being too difficult. It should also make sense. There wouldn't be any value in investigating the relationship between hair colour and weight, for example. You should ask your teacher whether your line of enquiry is suitable before beginning your project.

The project in this chapter will investigate the relationship between shoe size and height for the students at Jordan Hill.

Collecting data

You should begin by stating what you plan to investigate and what data you need to collect to do it. In this project we will take a **random sample** of 30 students from the school register, and recording their shoe sizes and heights. We will then look at the average shoe size and the average height for our sample.

There are different ways to take a random sample of size 30. You could write the names of each of the 1200 students on pieces of paper, put them in a hat, and choose 30 without looking. This would take a very long time. One alternative would be to give each student a number from 1 to 1200 and use the **random number button** on your calculator to choose your sample:

Press **SHIFT** **RAN#** . Your calculator will display a random number between 0 and 1.

You need a number between 1 and 1200 so multiply the number by 1200. Round your answer to the nearest whole number.

Repeat this process 30 times to select a random sample of 30 students.

If the calculator selects a number you have already chosen, ignore it and try again.

How big should my sample be?

The bigger a sample the more useful the data will be. If you select a lot of people, your results will be closer to the actual results for the whole school. However, if you choose too many people the data becomes difficult to analyse. A sample of size 25 is an adequate minimum. 30 is a sensible size for a sample because it is bigger than 25 and because it divides 360 exactly. This makes it easier to draw pie charts from your data.

Generate a random number between 0 and 1.

> *0.252*

Multiply by 1200:

> *302.4*

Round to the nearest whole number:

$$302.4 \approx 302$$

You have selected the 302nd student on the school register.

A sample of 30 sudents is shown below. They have been renumbered 1 to 30 to make the data easier to represent and analyse.

Number	Gender	Shoe size	Height (m)
1	B	5	1.60
2	G	4	1.58
3	G	2	1.37
4	G	6	1.63
5	B	11	1.88
6	G	3	1.57
7	B	8	1.77
8	B	5	1.58
9	B	2	1.35
10	G	4	1.58
11	B	12	1.97
12	B	4	1.56
13	G	7	1.72
14	G	5	1.61
15	G	4	1.63
16	B	7	1.68
17	B	7	1.71
18	G	2	1.41
19	B	5	1.59
20	B	9	1.80
21	B	6	1.58
22	G	1	1.32
23	B	6	1.66
24	G	6	1.69
25	B	3	1.55
26	G	4	1.60
27	B	2	1.40
28	B	8	1.78
29	G	3	1.52
30	B	7	1.75

Representing your data

The table above is not a very useful way of representing the data. You can represent the shoe sizes and heights on tally charts. The values for shoe size are **discrete data** so you can list the shoe sizes in your tally chart. Because the values for height are **continuous data** you have to use class intervals. Class intervals of width 10 cm are sensible.

There is more about discrete and continuous data in Section 10.6.

Shoe size	Tally	Frequency
1	|	1
2	||||	4
3	|||	3
4	|||||	5
5	||||	4
6	||||	4
7	||||	4
8	||	2
9	|	1
10		0
11	|	1
12	|	1

Height, h (cm)	Tally	Frequency
$130 \leqslant h < 140$	|||	3
$140 \leqslant h < 150$	||	2
$150 \leqslant h < 160$	|||| ||||	9
$160 \leqslant h < 170$	|||| |||	8
$170 \leqslant h < 180$	|||||	5
$180 \leqslant h < 190$	||	2
$190 \leqslant h < 200$	|	1

In the height column, $130 \leqslant h < 140$ means '130 up to but not including 140'. Any value greater than or equal to 130 but less than 140 would go in this class interval. 140 would go in this class interval.

You need to record your results on a diagram as well. The bar chart and histogram below illustrate the data from this sample.

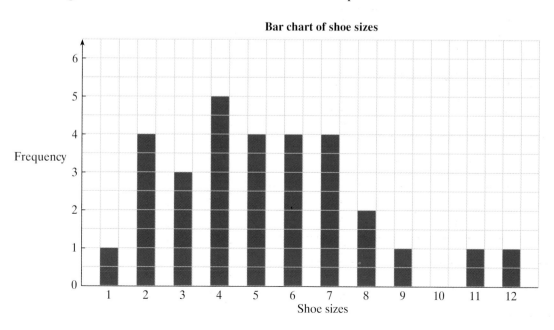

Bar chart of shoe sizes

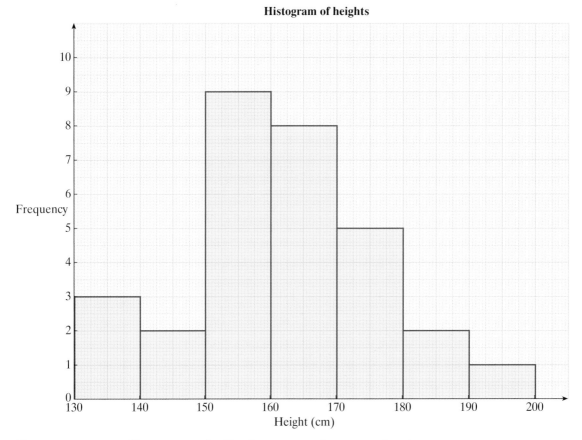

Histogram of heights

You can use your diagrams and tally charts to make some simple statements about your sample:

The modal shoe size for the sample is 4.
There are very few people in the sample who are taller than 180 cm.

These are statements about your sample. You should try to make **generalizations** about *all* the students at Jordan Hill:

The evidence from our sample suggests that at Jordan Hill High School there are likely to be fewer people with shoe size 1, 9, 10, 11 or 12 than with other shoe sizes.

This sample suggests that the modal class interval for height at Jordan Hill is likely to be 150–159.

> You should make it clear in your generalizations that you are using evidence to make a statement about what is *likely* to be true for all the values.

Extending your investigation

You are now ready to look at your sample in a bit more detail. You should extend your line of enquiry and give yourself a **hypothesis** to test. A hypothesis is a statement that could be true or false. You can test a hypothesis by looking at data.

We will extend our investigation by looking at the difference in height and shoe size between boys and girls. We will also test the following hypothesis:

In general the larger a person's shoe size, the taller that person is likely to be.

A new sample

To investigate the difference between boys and girls you will need to take a new sample. You need to make sure you have an equal number of girls and boys. There are 599 boys at Jordan Hill and 601 girls. We will take a sample of 30 boys and a sample of 30 girls, using the random number button on a calculator.

> To take separate random samples of boys and girls you need to create two lists – one of all the boys and one of all the girls.

Here is our sample of 30 boys and 30 girls:

Boys		Girls	
Shoe size	**Height (m)**	**Shoe size**	**Height (m)**
6	1.63	11	1.90
8	1.73	5	1.63
7	1.71	7	1.68
2	1.32	8	1.77
5	1.65	1	1.28
13	1.98	2	1.31
9	1.83	4	1.60
7	1.70	4	1.62
8	1.72	6	1.66
3	1.40	5	1.67
7	1.73	2	1.29
10	1.89	3	1.38
7	1.74	7	1.68
11	1.79	8	1.91
6	1.66	5	1.65
7	1.72	5	1.66
4	1.51	5	1.64
6	1.68	4	1.61
12	1.81	9	1.72
7	1.70	6	1.67
7	1.73	4	1.59
6	1.69	3	1.48
8	1.77	4	1.55
8	1.76	5	1.68
7	1.74	5	1.69
7	1.80	7	1.71
5	1.58	6	1.65
5	1.65	5	1.66
9	1.83	6	1.72
10	1.84	5	1.80

Once again, you need a more useful representation of this data.
Here are the frequency tables for shoe size and height separated
into boys and girls:

Boys

Shoe size	Tally	Frequency
1		0
2			1
3			1
4			1
5					3
6						4
7	ⅢⅠ ||||	9
8						4
9				2
10				2
11			1
12			1
13			1

Height, h (cm)	Tally	Frequency
$120 \leqslant h < 130$		0
$130 \leqslant h < 140$			1
$140 \leqslant h < 150$			1
$150 \leqslant h < 160$	||	2
$160 \leqslant h < 170$	ⅢⅠ		6
$170 \leqslant h < 180$	ⅢⅠ ⅢⅠ |||	13
$180 \leqslant h < 190$	ⅢⅠ		6
$190 \leqslant h < 200$			1

Girls

Shoe size	Tally	Frequency
1			1
2				2
3				2
4	ⅢⅠ	5
5	ⅢⅠ ||||	9
6						4
7					3
8				2
9			1
10		0
11			1
12		0
13		0

Height, h (cm)	Tally	Frequency
$120 \leqslant h < 130$	||	2
$130 \leqslant h < 140$	||	2
$140 \leqslant h < 150$			1
$150 \leqslant h < 160$	||	2
$160 \leqslant h < 170$	ⅢⅠ ⅢⅠ ⅢⅠ		16
$170 \leqslant h < 180$	||||	4
$180 \leqslant h < 190$			1
$190 \leqslant h < 200$	||	2

Shoe sizes

You are now ready to record your results in a diagram. We will begin by analysing the data about shoe sizes, using bar charts to compare the results for boys and girls.

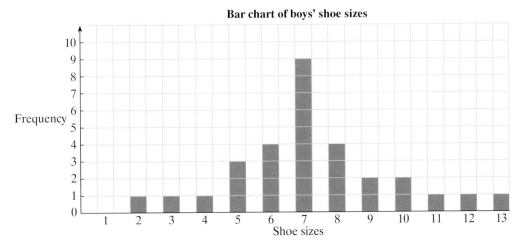

If you want to compare two sets of discrete data you can record them on a **dual bar chart**.

You can use a dual bar chart because there are the same number of boys and girls in your sample. If there were different numbers you could use pie charts to compare the data. In a pie chart you can see the **proportion of 360°** for each shoe size, so you are always comparing like with like.

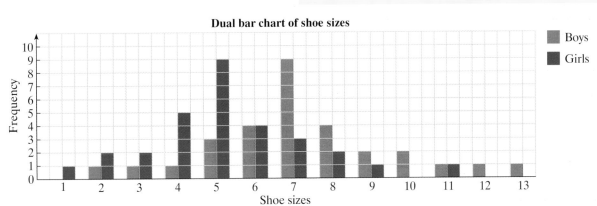

You can now make some simple statements to compare the shoe sizes of boys and girls:

The modal shoe size for the boys in my sample was higher than the modal shoe size for girls.

The evidence from the sample suggests that there will be fewer boys with shoe sizes 1 to 6 than girls.

You need to gather more information to support your statements. Comparing the **mean, median, mode** and **range** of shoe sizes for boys and girls will give you more evidence.

Mean shoe sizes

You can calculate the mean easily from the frequency tables. To calculate the mean shoe size for boys you work out the sum of the shoe sizes divided by the total number of boys:

$$\frac{(1 \times 2) + (1 \times 3) + (1 \times 4) + (3 \times 5) + (4 \times 6) + (9 \times 7) + (4 \times 8) + (2 \times 9) + (2 \times 10) + (1 \times 11) + (1 \times 12) + (1 \times 13)}{30}$$

You can calculate the mean shoe size for girls in the same way.

Mean shoe size for boys = 7.23

Mean shoe size for girls = 5.23

There is more about finding the mean from frequency tables on page 318.

Modal shoe sizes

You can read the modes of the shoe sizes for boys and girls straight off your bar charts or frequency tables:

Modal shoe size for boys = 7

Modal shoe size for girls = 5

Median shoe sizes

There are 30 people in each sample, so the median will be half way between the 15th and 16th values:

Median shoe size for boys = 7

Median shoe size for girls = 5

Range of shoe sizes

The range of shoe sizes will show you how spread out your data is:

Range of shoe sizes for boys = 11
Range of shoe sizes for girls = 10

You can summarise these results in a table:

Shoe sizes	Mean	Mode	Median	Range
Boys	7.23	7	7	10
Girls	5.23	5	5	11

You now have more evidence to describe the difference in shoe size between boys and girls:

All three measures of average (mean, median and mode) are greater for boys than for girls. The range of shoe sizes for boys is slightly greater for boys than for girls. In conclusion, although there are a small number of boys with small shoe sizes and girls with large shoe sizes, the evidence suggests that, in general, the shoe sizes for the boys are greater than the shoe sizes for the girls.

You can also use your data to make specific comments:

Evidence from the sample suggests that 17 out of 30, or 57% of boys have a shoe size of either 6, 7 or 8, and that 18 out of 30, or 60% of girls have a shoe size of either 4, 5 or 6.

Height

You can analyse the data about height in exactly the same way. Because height is continuous you need to record it on a histogram.

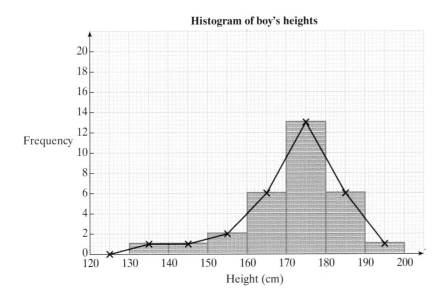

Histogram of boy's heights

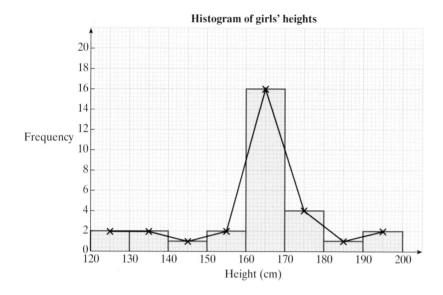

You can compare continuous data by drawing the frequency polygons on the same graph.

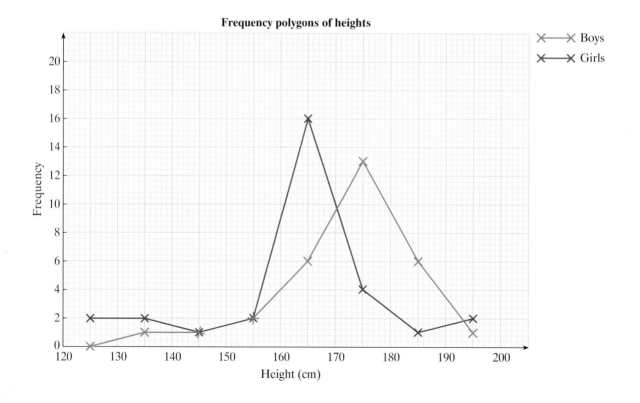

Since the data is grouped into class intervals, it also makes sense to record it in a stem and leaf diagram. This will make it easier to read off the median values.

Boys

Stem	Leaf	Frequency
120		0
130	2	1
140	0	1
150	1, 8	2
160	3, 5, 5, 6, 8, 9	6
170	0, 0, 1, 2, 2, 3, 3, 3, 4, 4, 6, 7, 9	13
180	0, 1, 3, 3, 4, 9	6
190	8	1

Girls

Stem	Leaf	Frequency
120	8, 9	2
130	1, 8	2
140	8	1
150	5, 9	2
160	0, 1, 2, 3, 4, 5, 5, 6, 6, 6, 7, 7, 8, 8, 8, 9	16
170	1, 2, 2, 7	4
180	0	1
190	0, 1	2

Averages

You can also record the mean, median and range for the data. Because the data is continuous it makes more sense to find the **modal class interval** rather than the mode. This is the class interval that contains the most values. The values for the mean and median have been rounded to two decimal places.

Heights (cm)	Mean	Modal class interval	Median	Range
Boys	171	$170 \leqslant h < 180$	173	66
Girls	163	$160 \leqslant h < 170$	166	63

You now make comments about the heights of students at Jordan Hill:

> *All three measures of average in the sample were higher for boys than for girls, though the sample for boys was more spread out, with a range of 0.66 m compared to 0.63 m for the girls. The evidence from the sample suggests that 13 out of 30, or 43% of the boys have a height between 170 and 180 cm, whilst 16 out of 30, or 53% of the girls have a height between 160 and 170 cm. The frequency polygons show that there are fewer boys with heights below 140 cm than girls.*

Always support your statements with evidence from your analysis. You should give actual numbers wherever possible.

You need to remember that all your comments are based on a *sample* of the whole school:

> *These conclusions are based on a sample of only 30 girls and 30 boys. I could extend the sample or repeat the whole exercise to confirm my results.*

You shouldn't actually repeat your test, or use a larger sample, but it is very important that you show that you know how to find more evidence to support your statements.

Comparing shoe size and height

When we extended the investigation we made this hypothesis:

> *In general the larger a person's shoe size, the taller that person is likely to be.*

To test this hypothesis we need a new random sample of 30 students.

We can use the sample we chose at the beginning of the chapter:

Shoe size	5	4	2	6	11	3	8	5	2	4	12	4	7	5	4
Height (cm)	160	158	137	163	188	157	177	158	135	158	197	156	172	161	163

Shoe size	7	7	2	5	9	6	1	6	6	3	4	2	8	3	7
Height (cm)	168	171	141	159	180	158	132	166	169	155	160	140	178	152	175

The most sensible way to compare this data is to draw a scatter diagram:

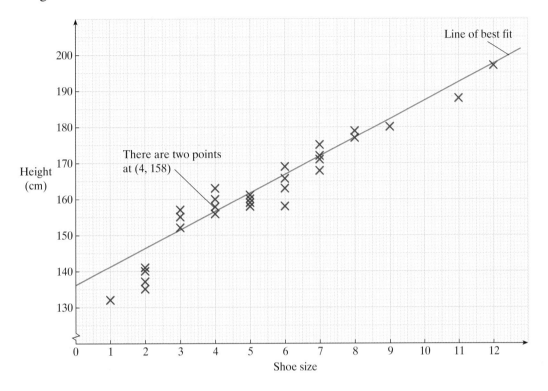

You can make simple comments based on your scatter diagram:

There is a positive correlation between shoe size and height. This suggests that the larger a person's shoe size, the taller they will be.

You can use a **line of best fit** to make predictions from your results:

The line of best fit suggests that somebody with shoe size 10 will be 187 cm tall.

Summarising your results

It is very important to analyse your data, interpret the outcomes and discuss your results and findings. Here is a summary of some of the findings from this investigation – this shows what has been achieved by the investigation, and might produce ideas for further lines of enquiry. You will need to refer back to all of your graphs and comments when you are summarising your results.

- There is a positive correlation between height and shoe size. In general taller people will have larger shoes than shorter people.

- The scatter graphs can be used to give reasonable estimates of shoe size and height. This can be done by using lines of best fit.

- The median height for boys is higher than the median height for girls.

You should also make some statements about the limitations of your analysis:

- We could have had greater confidence in our results if we had taken larger samples or given some consideration to the ages of the students in the sample.

- Our predictions are based on general trends observed in the data. There will probably be exceptional individuals whose results fall outside the general trend.

Summary of key points

When completing your Data Handling coursework you should:

- decide on the line of enquiry for your project

- plan out your line of enquiry, outline its aims and include any hypotheses you will make

- decide on the type of data and how much data you will require

- explain how you intend to collect the data

- record all the data

- use appropriate statistical techniques and diagrams

- explain what you intend to do with any very unusual datum points which seem to be exceptions to the rule

- make appropriate observations and comments on your data

- in cases when you make comparisons remember that a valid comparison usually makes use of a measure of central tendency (or average) such as mean, mode or median and a measure of dispersion (or spread) such as range

- provide a summary of your line of enquiry; this should be related to the aims and any hypotheses

- make a final conclusion based on the evidence you have obtained

13 Measure 2

13.1 Changing metric units

Before you tackle this section you will need to be able to multiply and divide whole numbers and decimals by 10, 100 and 1000. Check pages 106–108 if you need to remind yourself.

■ **You need to know that:**

Length	Weight	Capacity
10 mm = 1 cm 100 cm = 1 m 1000 mm = 1 m 1000 m = 1 km	1000 mg = 1 g 1000 g = 1 kg 1000 kg = 1 tonne	100 cl = 1 litre 1000 ml = 1 litre 1000 l = 1 cubic metre

You need to remember:

■ **When you change from small units to large units you divide.**

■ **When you change from large units to small units you multiply.**

Example 1

(a) Change 2 kilometres to metres.

(b) Change 250 mm to cm.

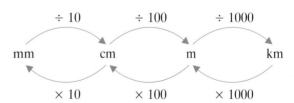

(a) Kilometres are larger units than metres so you multiply by the number of metres in a kilometre which is 1000.

$2 \times 1000 = 2000$ m

(b) Millimetres are smaller units than centimetres so you divide by the number of millimetres in a centimetres which is 10.

$250 \div 10 = 25$ cm

Exercise 13A

1 Change these lengths to centimetres.
   (a) 3 m   (b) 30 mm   (c) 6 m
   (d) 12 m   (e) 100 mm

2 Change these lengths to millimetres.
   (a) 2 cm   (b) 5 cm   (c) 12 cm
   (d) 20 cm   (e) 100 cm

3 Change these lengths to metres.
   (a) 5 km   (b) 300 cm   (c) 10 km
   (d) 2000 cm   (e) 60 km

4 Change these weights to grams.
   (a) 5 kg   (b) 40 kg   (c) 100 kg
   (d) 250 kg   (e) 1000 kg

5 Change these volumes to litres.
   (a) 3000 ml   (b) 8000 ml   (c) 50 000 ml
   (d) 75 000 ml

6 Change these volumes to millilitres.
   (a) 6 l   (b) 40 l   (c) 100 l
   (d) 350 l   (e) 25 l

7 Change these lengths to kilometres.
   (a) 3000 m   (b) 7000 m   (c) 40 000 m
   (d) 45 000 m

8 Change these weights to tonnes.
   (a) 4000 kg   (b) 7000 kg   (c) 30 000 kg
   (d) 55 000 kg

9 Change these weights to kilograms.
   (a) 2000 g   (b) 3 tonnes   (c) 50 000 g
   (d) 12 tonnes   (e) 100 000 g

10 Work out the number of millimetres in 1 metre.

11 Jim's lorry weighs 8 tonnes. How many grams is this?

12 Jeremy walks 10 kilometres. How many centimetres is this?

13 Work out the number of centimetres in 1 kilometre.

14 Gerry has an average pace length of 75 cm. When she
   walks 3 km how many paces does she take?

15 Work out the number of millimetres in 1 kilometre.

Exercise 13B

1 Put these lines in order, smallest first.

A —————————— E F

B ——

C ————

D ————————————

2 Put these weights in order, largest first.

250 g   25 g   2 kg   250 kg   3000 g

3 Put these lengths in order, smallest first.

3 m   5 mm   20 cm   3 km   50 mm

4 Put these lengths in order, smallest first.

75 cm   3000 mm   2 m   4000 m   4 cm

5 Put these volumes in order, smallest first.

200 ml   5 l   600 ml   2000 ml   1 l

So far all of the answers in this section have been whole numbers. In the following exercises you will find it easier if you use your calculator *but* don't forget to **check if your answer is sensible**.

Example 2

(a) Change 450 g to kg.

(b) Change 2.4 l to ml.

(a) 450 ÷ 1000 = 0.45 kg

(b) 2.4 × 1000 = 2400 ml

Changing small units to larger units so divide. 1000 g in 1 kg so ÷ 1000

Changing large units to smaller units so multiply. 1000 ml in 1 l so × 1000

Exercise 13C

1 Change these lengths to metres.

   (a) 250 cm   **(b)** 50 cm   **(c)** 3.6 km
   (d) 75 cm   **(e)** 0.005 km   **(f)** 35 cm
   (g) 475 cm   **(h)** 0.6 km   **(i)** 0.04 km
   (j) 5 mm

2 Change these weights to grams.

   (a) 4.5 kg   **(b)** 0.4 kg   **(c)** 10.3 kg
   (d) 0.03 kg   **(e)** 0.005 kg

3 Change these lengths to centimetres.
    (a) 350 mm   **(b)** 2.5 m   **(c)** 5.4 m
    (d) 5 mm   **(e)** 0.08 m   **(f)** 0.8 m
    (g) 35 mm   **(h)** 50 mm   **(i)** 85 mm
    (j) 275 mm

4 Change these volumes to millilitres.
    (a) 3.5 *l*   **(b)** 0.5 *l*   **(c)** 15.4 *l*
    (d) 0.05 *l*   **(e)** 0.003 *l*

5 Change these lengths to millimetres.
    (a) 3.5 cm   **(b)** 0.7 cm   **(c)** 0.08 cm
    (d) 12.5 cm   **(e)** 0.005 m

6 Write these lengths in kilometres.
    (a) 300 m   **(b)** 50 m   **(c)** 1250 m
    (d) 75 m   **(e)** 375 m

7 Change these volumes to litres.
    (a) 250 m*l*   **(b)** 100 m*l*   **(c)** 50 m*l*
    (d) 3500 m*l*   **(e)** 1 m*l*

8 Change these weights to kilograms.
    (a) 500 g   **(b)** 0.3 tonnes   **(c)** 50 g
    (d) 5.5 tonnes   **(e)** 0.006 tonnes

9 Write these weights in tonnes.
    (a) 3500 kg   **(b)** 450 kg   **(c)** 50 kg
    (d) 3000 g   **(e)** 75 kg

10 How many 75 m*l* glasses can be filled from a bottle holding 1.5 *l* of cola?

11 How many 50 mm pieces of wood can be cut from a piece of wood of length 3 m assuming there is no waste?

12 It takes 150 g of flour to make a batch of rock cakes. How many batches of rock cakes can be made from 1.5 kg of flour?

Exercise 13D

1 Write these lengths in order, smallest first.
    (a) 25 mm, 3 cm, 2.4 cm, 50 mm, 6 cm, 57 mm
    (b) 30 cm, 0.4 m, 270 mm, 1.2 m, 500 mm, 45 cm
    (c) 2 m, 340 cm, 4000 mm, 4 m, 370 cm, 3500 mm
    (d) 5 cm, 45 mm, 36 cm, 0.3 m, 55 mm, 0.2 cm, 4 mm
    (e) 50 cm, 0.4 m, 560 mm, 0.45 m, 34 cm

2 Write these weights in order, smallest first.
   (a) 250 g, 0.3 kg, 500 g, 0.05 kg
   (b) 500 g, 350 g, 0.4 kg, 0.52 kg
   (c) 5000 g, 3000 g, 4 kg, 4.5 kg, 0.5 tonnes, 400 kg

3 Write these volumes in order, smallest first.
   (a) 300 ml, 0.4 l, 500 ml, 250 ml, 0.3 l
   (b) 500 ml, 450 ml, 0.4 l, 360 ml, 0.05 l, 45 ml

13.2 Metric and imperial conversions

You may need to change from metric units to imperial units (the old style units) and vice versa. To do this it helps to memorise these facts.

Metric	Imperial
8 km	5 miles
1 kg	2.2 pounds
25 g	1 ounce
1 l	$1\frac{3}{4}$ pints
4.5 l	1 gallon
1 m	39 inches
30 cm	1 foot
2.5 cm	1 inch

Most petrol pumps have a conversion table to show how many litres or gallons you are buying.

Example 3

Change 10 km to miles.

8 km is approximately 5 miles
so 1 km = 5 ÷ 8 = 0.625 miles
then 10 km = 0.625 × 10 = 6.25 miles

Exercise 13E

The Evans family are going on holiday to Scotland. The family consist of Mr and Mrs Evans and their three children, Glenys, Eira and Gareth.

1 Mr Evans works out the distance from their home in London to Scotland. He makes it 400 miles. What is this distance in kilometres?

2 Mrs Evans packs a 3 litre bottle of cola for the trip.
How many pints is 3 litres?

3 The petrol tank of the family's car holds 15 gallons.
How many litres is this?

4 Glenys estimates the weight of all the luggage as 100 kg.
How many pounds is that?

5 Gareth puts 1 pint of water in the radiator.
How many litres is that?

6 Mr Evans puts 30 litres of petrol in the car.
How many gallons is that?

7 Mrs Evans puts 0.5 l of oil in the engine.
How many pints is that?

8 The family stop at a service station 150 km from home.
How many miles is that?

9 Eira buys an 800 g bar of chocolate.
How many pounds is that?

10 When they get to Scotland there is half a bottle of cola left.
How many pints is that?

Exercise 13F

Class 11E decide to raise money for charity by holding a cola morning.

1 Claire cooks some rock cakes. She makes 72 cakes and uses 24 ounces of fat, 40 ounces of flour and 32 ounces of dried fruit. She only has a metric set of scales.
How many grams of each ingredient should she use?

2 Sybil buys ten 3 l bottles of cola.
How many gallons is this?

3 Henri buys 20 packets of biscuits. Each pack weighs 1 pound.
How many kilograms is this altogether?

4 Jonathan makes plates that are 4 inches across.
How many centimetres is this?

5 The trays the class use are 24 inches long and 15 inches across.
Change these measurements to centimetres.

6 The tables the class use are circular with a diameter of 48 inches. Will tablecloths with a 1.2 m diameter fit the tables?

7 Nilmini makes sandwiches and uses 5 one-kilogram loaves and 500 g of spread.
   Change these weights to pounds.

8 At the end of the cola morning there were 5 pints of cola left.
   How many litres is this?

9 In the 'guess the weight of the cake' competition the correct answer was 5 pounds. Robin said the weight was 5.1 pounds and Hazel said 2.3 kilograms.
   Which of these two answers was nearer the correct one?

10 Claire sold 3.5 kilograms of cakes.
   How many pounds is that?

13.3 Calculating time

■ **You need to know that:**
   60 seconds = 1 minute
   60 minutes = 1 hour
   24 hours = 1 day
   365 days = 1 year
   366 days = 1 leap year
   3 months = 1 quarter
   12 months = 1 year

Many people make mistakes when they are calculating with time because they forget that there are 60 minutes in an hour and not 100.

Example 4

(a) How many hours are there in 135 minutes?

(b) How many minutes are there in 3 hours?

(a) $135 \div 60 = 2.25$ hours
     Many people write this as 2 hours 25 minutes but they would be *wrong*. To change the 0.25 hours to minutes you must multiply 0.25 by 60.
        $0.25 \times 60 = 15$
     so 135 minutes is 2 hours 15 minutes.

(b) $3 \times 60 = 180$ minutes

Why does 30 minutes and 45 minutes give 1 hour and 15 minutes?

You can't use an ordinary calculator to add time.

Exercise 13G

1   Change these times into minutes.
    (a) 2 hours   **(b)** 5 hours
    (c) 2 hours 30 minutes   **(d)** $5\frac{1}{2}$ hours
    (e) $6\frac{1}{4}$ hours   **(f)** 5 hours 15 minutes

2   Change these times into hours.
    (a) 180 minutes   **(b)** 240 minutes   **(c)** 75 minutes
    (d) 260 minutes   **(e)** 325 minutes   **(f)** 90 minutes
    (g) 3 days   **(h)** $5\frac{1}{2}$ days   **(i)** 500 minutes

3   How many seconds are there in 1 hour?

4   How many minutes are there in 1 day?

5   How many seconds are there in:
    (a) a year
    (b) a leap year?

When you come to make calculations in time you have to be careful when it comes to dealing with the carry digit.

Example 5

(a) Add $2\frac{1}{2}$ hours to the time of 10:40

(b) Take 3 hours 15 minutes away from 11:10

> **Don't forget:**
> 60 minutes make 1 hour

(a)   10:40
         2:30
        13:10   Not 70 because 70 minutes make 1 hour 10 minutes
         1

(b)   11:10   You have to carry 60 minutes so 70 – 15 gives 55.
    – 3:15
      7:55

    11:10 could be written as 10:70.

    11:10 $\overline{10:70}$
   – 3:15   is   – 3:15
      7:55   7:55

Exercise 13H

1   Add 15 minutes to each of these times.
    (a) 10:30   **(b)** 09:45   **(c)** 11:40   **(d)** 09:55

2   Add 50 minutes to each of these times.
    (a) 09:00   **(b)** 10:30   **(c)** 11:40   **(d)** 08:05

3   Add 2 hours 40 minutes to each of these times.
    (a) 09:40   **(b)** 10:45   **(c)** 11:50   **(d)** 06:10

4   Add 12 hours 45 minutes to each of these times.
    (a) 02:30   **(b)** 07:15   **(c)** 12:50   **(d)** 16:45

5   Subtract 15 minutes from each of these times.
    (a) 09:55   **(b)** 11:40   **(c)** 08:10   **(d)** 09:05

6   Subtract 50 minutes from each of these times.
    (a) 08:55   **(b)** 11:40   **(c)** 10:30   **(d)** 09:00

7   Subtract 2 hours 30 minutes from each of these times.
    (a) 09:55   **(b)** 11:40   **(c)** 08:10   **(d)** 09:05

8   Subtract 12 hours 45 minutes from each of these times.
    (a) 14:50   **(b)** 17:30   **(c)** 12:00   **(d)** 08:30

13.4 Dealing with dates

You will often need to add days onto dates. This old rhyme can help:

■   **30 days hath September**
    April June and November.
    All the rest have 31
    except for February alone
    which has just 28 days clear
    and 29 in each leap year.

Example 6

Jane agreed to go out with John in 10 days time.
Today is Tuesday 23rd April. When is their date?

	April					May			
Monday	1	8	15	22	29	6	13	20	27
Tuesday	2	9	16	23	30	7	14	21	28
Wednesday	3	10	17	24		1	8	15	22	29
Thursday	4	11	18	25		2	9	16	23	30
Friday	5	12	19	26		3	10	17	24	31
Saturday	6	13	20	27		4	11	18	25
Sunday	7	14	21	28		5	12	19	26

If you start at 23 and count 10 days on you get to Friday 3rd of
May. You must of course remember that April only has 30 days.

Exercise 13I

1   Count on 10 days from the following dates.
   (a) 1st Jan   **(b)** 2nd March   **(c)** 3rd June
   (d) 5th July   **(e)** 10th Sept   **(f)** 20th May
   (g) 25th June   **(h)** 27th Aug

2   Count on 14 days from the following dates.
   (a) 2nd Feb   **(b)** 3rd March   **(c)** 3rd April
   (d) 7th Dec   **(e)** 15th Nov   **(f)** 18th Sept
   (g) 23rd March   **(h)** 25th Nov

3   Count on 30 days from the following dates.
   (a) 5th April   **(b)** 6th June   **(c)** 17th May
   (d) 1st Sept   **(e)** 7th June   **(f)** 20th May
   (g) 30th Nov   **(h)** 5th Dec

   Use the calendar in Example 6 to answer the following questions.

4   What day of the week and date is 5 days from the 5th April?

5   Which day and date is 5 days after the 27th April?

6   Which day and date is 7 days before 20th May?

7   What is the day and date 10 days before the 5th May?

8   Write down the day and date two weeks after the 19th April.

13.5 Timetables

Bus and train timetables are often used to test your knowledge of time in the GCSE examination.

On this timetable each train's times starts at the top and then go down the page. The time it should leave a stopping place can then be read off opposite the place name.

Example 7

Find how long it takes for the 8:15 train from Swindon to get to London (Paddington).

You first have to find the 08:15 train from Swindon. It is in the third column along. Follow that column down to the bottom and that gives the arrival time in London of 09:10.

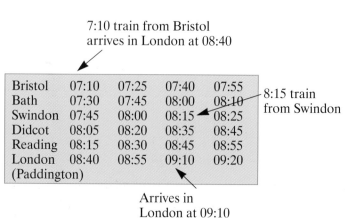

7:10 train from Bristol arrives in London at 08:40

8:15 train from Swindon

Arrives in London at 09:10

You could subtract times by doing a subtraction sum and carrying a 60. Here is another way of dealing with the problem that you may find easier:

15 minutes

08:15 to 08:30 is 15 minutes
08:30 to 09:00 is 30 minutes
09:00 to 09:10 is 10 minutes

30 minutes

Total time 15 + 30 + 10 = 55 minutes

10 minutes

Exercise 13J

Bus timetable

Coate	07:05	07:35	08:05
Piper's Way	07:10	07:40	08:10
Old Town	07:20	07:50	08:20
Drove Road	07:25	07:55	08:25
New Town	07:30	08:00	08:30
Bus Station	07:35	08:05	08:35

Train timetable

Bristol	07:10	07:25	07:40	07:55
Bath	07:30	07:45	08:00	08:15
Swindon	07:45	08:00	08:15	08:30
Didcot	08:05	08:20	08:35	08:50
Reading	08:15	08:30	08:45	09:00
London (Paddington)	08:40	08:55	09:10	09:25

1 At what time should the 07:35 bus from Coate arrive at Drove Road?

2 At what time should the 07:40 train from Bristol arrive at Didcot?

3 At what time should the 08:30 train from Reading start from Bristol?

4 At what time should the 08:00 bus from New Town start at Piper's Way?

5 Buses from Coate depart every half an hour.
   Continue the bus timetable for the next three buses. You can assume that each bus takes the same amount of time between stops as the previous ones.

6 Trains from Bristol depart every quarter of an hour.
   Continue the train timetable for the next three trains. You can assume that each train takes the same amount of time between stops as the previous ones.

7 Rashmi arrives at the train station in Bath at 07:35.
   What time is the next train he could catch to London?

8 Claude arrives at the train station in Didcot at 08:25.
   What time is the next train he could catch to London?

9 Cecille arrives at the train station in Swindon at 09:10.
   What time is the next train she could catch to Reading?

10 Sophia arrives at the train station in Bristol at 08:00.
   What time is the next train she could catch to Bath?

11 How long should it take to travel by bus from Coate to the Bus Station?

12 How long should it take to travel by bus from Pipers Way to New Town?

13 How long should it take to travel by train from Bristol to London (Paddington)?

14 How long should it take to travel by train from Swindon to Reading?

15 How long should it take to travel by train from Bath to Didcot?

16 The bus timetable is for a bus route in Swindon. It takes five minutes to walk from the bus station to the train platform. Use both timetables to work out which bus and train:

(a) Gareth catches at Pipers Way to be in London (Paddington) at 09:00

(b) Susan catches if she needs to be in Reading at 08:30 and she travels by bus from Coate

(c) Mario catches if he travels by bus from Old Town and needs to be in Didcot at 08:40

(d) Claudette catches at Drove Road to be in London Paddington by 10:00

(e) Bridgette catches if she needs to be in Reading at 09:30 and she travels by bus from Old Town

(f) Katrina caught if she travelled from Drove Road and arrived in Reading at 08:45.

Exercise 13K

1

This signpost is on the road from Paris to Dijon.
(a) Work out the distance, in kilometres, from Paris to Dijon along this road.
(b) Work out the approximate distance, in miles, from the signpost to Paris.

[E]

2 Here is a timetable for Amina's school bus.

Bus stop	Time
Bus Station	08:00
Station Road	08:15
Grange Drive	08:20
Holley Ave.	08:30
Kings Road	08:40
School Road	08:45

Amina catches the bus at Grange Drive.
(a) At what time should the bus be at Grange Drive?
(b) How long should the bus take to get from Grange Drive to School Road?

[E]

3 Chippy the carpenter marks a 3 metre length of wood into three pieces.

One piece is 1.40 metres long.
Another piece is 84 centimetres long.

How long is the third piece of wood?

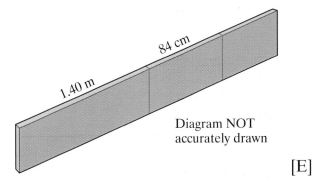

Diagram NOT accurately drawn

[E]

Summary of key points

1

Length	Weight	Capacity
10 mm = 1 cm 100 cm = 1 m 1000 mm = 1 m 1000 m = 1 km	1000 mg = 1 g 1000 g = 1 kg 1000 kg = 1 tonne	100 c*l* = 1 litre 1000 m*l* = 1 litre 1000 *l* = 1 cubic metre

2 When you change from small units to large units you divide.

3 When you change from large units to small units you multiply.

4
60 seconds	=	1 minute
60 minutes	=	1 hour
24 hours	=	1 day
365 days	=	1 year

366 days	=	1 leap year
3 months	=	1 quarter
12 months	=	1 year

5 30 days hath September,
April, June and November.
All the rest have 31
except for February
alone which has just 28 days clear
and 29 in each leap year.

14 Percentages

14.1 Understanding percentages

■ $\left.\begin{array}{l}\textbf{Percentage}\\ \textbf{\%}\\ \textbf{pc}\end{array}\right\}$ means 'number of parts per hundred'.

Look at the large square below. It has been divided into 100 equal small squares.

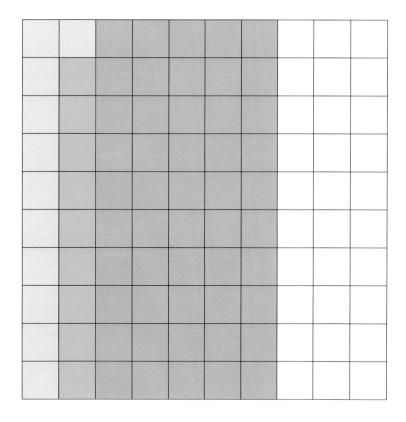

11 of the 100 small squares are shaded blue.
So 11% of the large square is shaded blue.

9 of the 100 small squares are shaded red.
So 9% of the large square is shaded red.

50 of the 100 small squares are shaded green.
So 50% of the large square is shaded green.

30 of the 100 small squares are unshaded.
So 30% of the large square is unshaded.

Exercise 14A

1 Look at the large square below which is divided into 100 equal small squares.

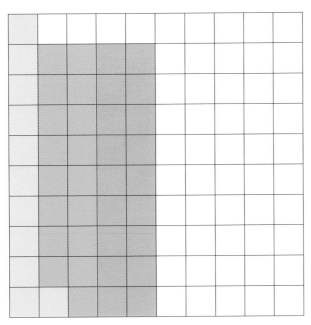

   (a) What percentage of the large square is shaded:
      (i) blue   **(ii)** red   **(iii)** green?

   (b) What percentage of the large square is unshaded?

   (c) What fraction of the large square is unshaded?

2 Look at the large rectangle below which is divided into 100 equal small rectangles.

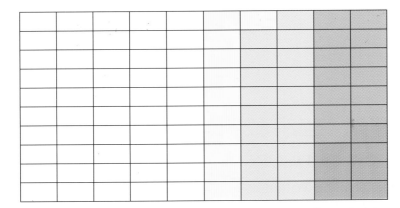

   (a) What percentage of the large rectangle is shaded.
      (i) yellow   **(ii)** blue   **(iii)** red   **(iv)** green?

   (b) What percentage of the large rectangle is unshaded?

   (c) What fraction of the large rectangle is unshaded?

14.2 Percentages, fractions and decimals

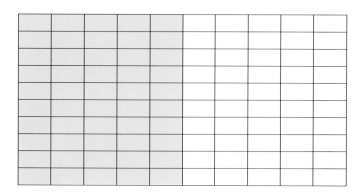

In the large rectangle above, 50% is shaded blue and 50% is unshaded.

50% means 50 in a hundred, which can be written as $\frac{50}{100}$.

$\frac{50}{100}$ simplifies to $\frac{1}{2}$

So 50% is the same as $\frac{1}{2}$. You can write $50\% = \frac{1}{2} = 0.5$.

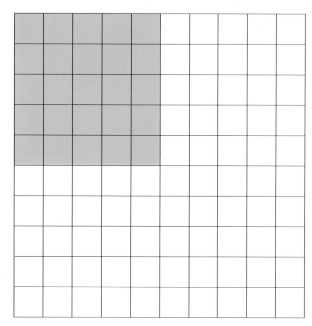

In the large square above, 25% is shaded red and 75% is unshaded.

25% means 25 in a hundred, which can be written as $\frac{25}{100}$.

$\frac{25}{100}$ simplifies to $\frac{1}{4}$

So 25% is the same as $\frac{1}{4}$. You can write $25\% = \frac{1}{4} = 0.25$.

Similarly, $75\% = \frac{75}{100} = \frac{3}{4} = 0.75$.

■ **To write a percentage as a fraction, always use the denominator 100.**

Example 1

Write 19% as a fraction.

$$19\% = \frac{19}{100}$$

Example 2

Write 85% as a fraction.

$$85\% = \frac{85}{100}$$

Simplify $\frac{85}{100}$ $\xrightarrow{85 \div 5}$ $\frac{17}{20}$

$$100 \div 5$$

So $85\% = \frac{17}{20}$

Example 3

Write $3\frac{1}{2}\%$ as a fraction.

$$3\frac{1}{2}\% = \frac{3\frac{1}{2}}{100}$$

Simplify $\frac{3\frac{1}{2}}{100}$ $\xrightarrow{3\frac{1}{2} \times 2}$ $\frac{7}{200}$

$$100 \times 2$$

So $3\frac{1}{2}\% = \frac{7}{200}$

■ **To write a percentage as a decimal:**
  ● **write the percentage as a fraction**
  ● **convert the fraction to a decimal.**

Example 4

Write 63% as a decimal.

$$63\% = \frac{63}{100} = 63 \div 100 = 0.63$$

Example 5

Write 15% as a decimal.

$$15\% = \frac{15}{100} = 15 \div 100 = 0.15$$

Exercise 14B

1

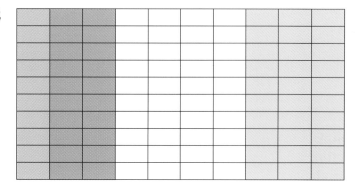

For the large rectangle above, state:

(a) (i) what percentage is shaded blue
      (ii) what fraction is shaded blue

(b) (i) what percentage is shaded red
      (ii) what fraction is shaded red

(c) (i) what percentage is shaded green
      (ii) what fraction is shaded green

(d) (i) what percentage is unshaded
      (ii) what fraction is unshaded.

2

For the large rectangle above, state:

(a) (i) what percentage is shaded red
      (ii) what fraction is shaded red

(b) (i) what percentage is shaded green
      (ii) what fraction is shaded green

(c) (i) what percentage is shaded blue
      (ii) what fraction is shaded blue

(d) (i) what percentage is unshaded
      (ii) what fraction is unshaded.

3 Write these percentages as fractions in their simplest form:

(**a**) 17%   (**b**) 99%   (**c**) 41%   (**d**) 3%

(**e**) 60%   (**f**) 80%   (**g**) 90%   (**h**) 30%

(**i**) 10%   (**j**) 70%   (**k**) 22%   (**l**) 6%

(**m**) 64%   (**n**) 96%   (**o**) 15%   (**p**) 65%

4 Write these percentages as decimals:

(**a**) 37%   (**b**) 49%   (**c**) 87%   (**d**) 7%   (**e**) 40%

(**f**) 15%   (**g**) 8%   (**h**) 28%   (**i**) 36%   (**j**) 95%

(**k**) 45%   (**l**) 3%   (**m**) $3\frac{1}{2}$%   (**n**) $6\frac{1}{2}$%   (**o**) $12\frac{1}{2}$%

■ **To change a decimal to a percentage, multiply the decimal by 100.**

Example 6

Change to a percentage:

(a) 0.47   (b) 0.075

(a) $0.47 \times 100 = 47\%$

(b) $0.075 \times 100 = 7.5\%$

■ **To write a fraction as a percentage:**
  - **change the fraction to a decimal**
  - **multiply the decimal by 100.**

Example 7

Change to a percentage:

(a) $\frac{7}{10}$ (b) $\frac{17}{40}$

(a) $\frac{7}{10} = 7 \div 10 = 0.7$

$0.7 \times 100 = 70$

So $\frac{7}{10} = 70\%$

(b) $\frac{17}{40} = 17 \div 40 = 0.425$

$0.425 \times 100 = 42.5$

So $\frac{17}{40} = 42.5\%$

Exercise 14C

1 Change these decimals to percentages:

(a) 0.37   (b) 0.59   (c) 0.11   (d) 0.1   (e) 0.36

(f) 0.7   (g) 0.03   (h) 0.771   (i) 0.09   (j) 0.055

(k) 0.83   (l) 0.56   (m) 0.075   (n) 0.125   (o) 0.675

2 Write these fractions as percentages:

(a) $\frac{1}{2}$ (b) $\frac{3}{4}$ (c) $\frac{2}{5}$ (d) $\frac{4}{5}$ (e) $\frac{9}{10}$

(f) $\frac{7}{20}$ (g) $\frac{8}{25}$ (h) $\frac{19}{25}$ (i) $\frac{3}{20}$ (j) $\frac{3}{8}$

(k) $\frac{5}{8}$ (l) $\frac{3}{16}$ (m) $\frac{7}{50}$ (n) $\frac{9}{100}$ (o) $\frac{13}{1000}$

3

Percentage	Decimal	Fraction
40%	0.4	$\frac{2}{5}$
	0.61	
		$\frac{7}{10}$
35%		
$8\frac{1}{2}\%$		
	0.15	
		$\frac{3}{25}$
	0.07	
$1\frac{1}{4}\%$		
		$\frac{2}{3}$

14.3 Comparing different proportions using percentages

Example 8

Write in order of size, smallest first:

$65\%, \frac{3}{5}, 0.66, \frac{5}{8}$

First, change them to percentages:

$\frac{3}{5} = 3 \div 5 = 0.6 = 60\%$

$0.66 \qquad\qquad = 66\%$

$\frac{5}{8} = 5 \div 8 = 0.625 = 62.5\%$

Then reorder $\frac{3}{5}, \frac{5}{8}, 65\%, 0.66$

Example 9

In three class tests, Robin scored 8 out of 10 in English, 17 out of 20 in science and 42 out of 50 in history. Which is his best subject?

First, change the marks to percentages:

English: 8 out of 10 = $\frac{8}{10}$ = 8 ÷ 10 = 0.8 = 80%

Science: 17 out of 20 = $\frac{17}{20}$ = 17 ÷ 20 = 0.85 = 85%

History: 42 out of 50 = $\frac{42}{50}$ = 42 ÷ 50 = 0.84 = 84%

So the best subject is science.

Exercise 14D

1 Rearrange in order of size, smallest first:
   (a) 52%, 0.53, $\frac{9}{15}$
   (b) 72%, $\frac{7}{10}$, 0.71
   (c) 0.07, $\frac{1}{10}$, 8%
   (d) $\frac{3}{8}$, 30%, 0.36

2 In end of term tests, Sheila's results were:

English: 16 out of 24
Maths: 27 out of 40
History: 31 out of 50
Geography: 58 out of 100
Science: 11 out of 20
Technology: 33 out of 60

   (a) Which was her best result?
   (b) Which was her worst result?

14.4 Working out a percentage of an amount

There are two methods of working out a percentage of an amount.

Example 10

Work out 15% of 40 kg.

Method 1

Change the percentage to a fraction.

Multiply the amount by the fraction:

$$15\% = \frac{15}{100}$$

$$\frac{15}{100} \times 40\,\text{kg} = \frac{15 \times 40\,\text{kg}}{100} = \frac{600\,\text{kg}}{100} = 6\,\text{kg}$$

So 15% of 40 kg = 6 kg.

Your calculator may have a short way of doing this with the **%** key.

Method 2

Change the percentage to a decimal.

Multiply the amount by the decimal:

$$15\% = \frac{15}{100} = 15 \div 100 = 0.15$$

$$0.15 \times 40\,\text{kg} = 6\,\text{kg}$$

So 15% of 40 kg = 6 kg.

Example 11

Work out 8.57% of £685.

Method 1

$$8.57\% = \frac{8.57}{100}$$

$$\frac{8.57}{100} \times £685 = \frac{8.57 \times £685}{100} = \frac{£5870.45}{100} = £58.7045$$

So 8.57% of £685 = £58.70.

Remember: when you work out a percentage of an amount of **money** make sure that your answer always has two figures after the decimal point.

Method 2

$$8.57\% = 0.0857$$

$$0.0857 \times £685 = £58.7045$$

So 8.57% of £685 = £58.70.

Exercise 14E

Work out:

1 60% of 165	**2** 45% of 920 kg
3 17% of £7000	**4** 90% of 80
5 6% of £420	**6** 10% of £16.80
7 17.5% of £164	**8** 30% of £264
9 5% of £31 240	**10** 0.3% of 250 tonnes
11 7.2% of £600	**12** 96% of 32 000
13 25% of 36 km	**14** 25% of 14.2 km
15 6.8% of £9840	**16** 4.8% of 3.6 litres

14.5 Using percentages

Percentages are very often used in commerce and industry for reckoning profits, discounts, wage rises and so on.

Take off...30% of the cost of airfares!

Cotherston Building Society yesterday launched two new fixed-rate mortgages. It is offering 8.75pc fixed for one year and 9.45pc fixed for three years.

Save 10% now. Vehicle Rescue from £25

8.57% TAX-FREE INCOME

8pc pay rise bid for nurses

Example 12

Satinder earns £160 per week working at the supermarket. She is awarded a 5% pay rise.

(a) Work out the amount of the rise.

(b) Work out her new weekly wage.

(a) **Either:**

Method 1 $5\% = \dfrac{5}{100}$

$$\dfrac{5}{100} \times £160 = \dfrac{5 \times £160}{100} = \dfrac{£800}{100} = £8$$

Rise = £8

Method 2 $5\% = 0.05$

$0.05 \times £160 = £8$

Rise $= £8$

(b) **Either:** $£160 + £8 = £168$

or New wage is $100\% + 5\% = 105\%$ of old wage

So new wage $= 105\%$ of £160   **or** 105% of £160

$\qquad\qquad = \dfrac{105}{100} \times £160$ $\qquad 105\% = 1.05$

$\qquad\qquad\qquad\qquad\qquad\qquad 1.05 \times £160 = £168$

$\qquad\qquad = \dfrac{105 \times £160}{100}$

$\qquad\qquad = \dfrac{£16\,800}{100}$

$\qquad\qquad = £168$

Example 13

At the beginning of a year the number of jobless people in a city was 5500. This number fell by 1% during the year.

(a) Work out the fall in the number of jobless.

(b) Work out the number of jobless remaining.

(a) Fall in the number of jobless:

Either $1\% = \dfrac{1}{100}$

$\dfrac{1}{100} \times 5500 = \dfrac{1 \times 5500}{100} = \dfrac{5500}{100} = 55$

or $1\% = 0.01$

$0.01 \times 5500 = 55$

(b) Number of jobless remaining:

Either $5500 - 55 = 5445$

or new number is $100\% - 1\% = 99\%$ of old number

So the number remaining:

$\qquad = 99\%$ of 5500 $\qquad$ **or** $99\% = 0.99$

$\qquad = \dfrac{99}{100} \times 5500$ $\qquad\qquad 0.99 \times 5500 = 5445$

$\qquad = \dfrac{99 \times 5500}{100}$

$\qquad = \dfrac{544\,500}{100}$

$\qquad = 5445$

Index numbers are used to mark changes in retail prices. The cost of food, tobacco, fuel, clothing, etc is calculated by working out the amount an average family uses. The overall rise or fall in the amount of money spent is given as a percentage of the price in a certain year.

This table shows the retail prices for January to April 1999.

January	February	March	April
100	104.5	103	106.8

This means there was a rise of 4.5% from January to February and although in March there was a fall of 1.5% compared with February, it was still 3% above the January figure. April shows a rise of 6.8% on the January figure.

Exercise 14F

1 Good Price Record Shop reduced the prices of all CDs and cassettes by 5%.
Calculate: **(a)** the reduction in price   **(b)** the new price of a CD usually costing £12.

2 A furniture store reduced all normal prices by 15% for the Spring Sale.
Work out **(i)** the reduction in price   **(ii)** the Spring Sale price of:

(a) a table normally priced at £80

(b) a settee normally priced at £740

(c) a TV table normally priced at £40.60.

3 A school entered 150 candidates for the GCSE maths exam. The table shows the percentage of candidates gaining each grade.

Grade	A*	A	B	C	D	E	F	G
Percentage	2%	6%	12%	42%	20%	6%	8%	4%

How many pupils gained each grade?

4 John wanted to buy a CD player and speakers costing £376. The shop keeper asked for a deposit of 30% of the price. How much deposit did John pay?

5 A new car, bought for £12 600, had lost 60% of its value after 3 years.

(a) What was the total loss?

(b) What was the value of the car after 3 years?

6 Janice left £450 in her building society account for one year. The building society paid interest of $6\frac{1}{2}\%$ per year. How much interest did Janice's money earn in the year?

7 Value Added Tax (VAT) is charged at the rate of $17\frac{1}{2}\%$. How much VAT is there to pay on:

(a) a telephone bill of £52.40 before VAT

(b) a restaurant bill of £28 before VAT

(c) a wedding reception costing £1980 before VAT

(d) a builder's bill for repairs of £158.80 before VAT.

8 A travel company advertises a discount of 15% off all holiday prices.

How much would it cost for:

(a) a fly drive holiday priced at £360

(b) an activity holiday priced at £168?

9 A sales representative is paid $6\frac{1}{2}\%$ commission on the value of all sales made. How much commission is the representative paid on sales valued at:

(a) £1000   (b) £3600   (c) £8040?

10 An estate agency charges $2\frac{1}{2}\%$ commission on the value of houses that the agency sells. How much commission does the agency charge for selling these houses:

11 A market trader buys knitted jumpers for £8.60 each and sports shirts for £4.80 each. She sells them for 65% more than she bought them for.

Work out the selling price of each item.

12 In 1996, a skiing holiday in Italy cost £276 made up of Fares £80, Hotel £120, Skiing Instruction £30 and Hire of Equipment £46.

By 1997, the cost of Fares had risen by 12%, Hotel costs had risen by 8%, the cost of instruction had increased by 50% and Hire of Equipment by 10%.

Work out the total cost of the holiday in 1997.

14.6 Writing one quantity as a percentage of another quantity

Example 14

The top mark in a test is 34 out of 40. What is this as a percentage?

To find one quantity as a percentage of another follow these steps:

Step 1 Write the two amounts as a fraction: $\frac{34}{40}$

Step 2 Convert the fraction to a decimal: $\frac{34}{40} = 34 \div 40 = 0.85$

Step 3 Multiply the decimal by 100: $\quad 0.85 \times 100 = 85$

The top mark is 85%.

Example 15

A jacket is reduced in price from £80 to £62.

What is the percentage reduction?

The actual reduction is   £80 – £62 = £18

Step 1 Write as a fraction: $\quad\quad\quad \frac{18}{80}$

Step 2 Convert to a decimal: $\quad 18 \div 80 = 0.225$

Step 3 Multiply by 100: $\quad\quad 0.225 \times 100 = 22.5\%$

The percentage reduction is 22.5%.

■ **To write an amount as a percentage of another:**
  1 write the amounts as a fraction
  2 convert the fraction to a decimal
  3 multiply the decimal by 100.

Why the method works

These three numbers are **equivalent**. They represent the same quantity written in three different forms:

percentage		fraction		decimal
85%	=	$\frac{85}{100}$	=	0.85

To convert from the decimal form to percentage form multiply by 100: $0.85 \times 100 = 85$ or 85%

Exercise 14G

1 (a) What percentage of £100 is £5?
   (b) What percentage of 5 kg is 600 g?
   (c) What percentage of £160 is £24?
   (d) What percentage of 2 *l* is 150 m*l*?
   (e) What percentage of 3 hours is 1 hour 15 mins?
   (f) What percentage of £4.20 is 35p?
   (g) What percentage of £16 000 is £480?
   (h) What percentage of 20 tonnes is 200 kg?
   (i) What percentage of 3.6 m is 180 cm?

2 (a) Write £15 as a percentage of £200.
   (b) Write £600 as a percentage of £3000.
   (c) Write 15 minutes as a percentage of 1 hour 15 mins.
   (d) Write 6.3 m as a percentage of 157.5 m.
   (e) Write £5.60 as a percentage of £140.
   (f) Write 36 kg as a percentage of 900 kg.
   (g) Write 20 tonnes as a percentage of 320 tonnes.
   (h) Write 4.5 mm as a percentage of 90 cm.
   (i) Write £850 as a percentage of £4000.
   (j) Write £1575 as a percentage of £63 000.

3 Shoes in a sale are reduced in price from £60 to £42.
   What is the percentage reduction?

4 A factory employing 300 people made 18 people redundant.
   What percentage of the employees were made redundant?

5 Peter bought a new motor bike for £840 and sold it two
   years later for £378.
   By what percentage had its value fallen in the two years?

6 A stereo system, normally sold for £215, is offered for sale
   at £180.60.
   Calculate the percentage reduction in the normal price.

7 A shopworker received a wage increase from £160 per
   week to £165 per week.
   What was the percentage increase?

8 Shona deposited £450 in a bank savings account. Interest
   paid after a year increased the amount in her account to
   £478.80.
   What percentage rate of interest did the account pay?

9 Ali paid a deposit of £108 on a second-hand car costing £1350.
What percentage of the price was the deposit?

10 The number of pupils in a school grew from 1050 to 1500.
What was the percentage increase in the number of pupils?

Exercise 14H

1 The same sort of bike is for sale in two shops.
Work out the price you would have to pay in each shop.

[E]

2

Mega Ace Games System

Normal Price   £320

Sale Price   £272

Find the percentage reduction on the Mega Ace Games System in the sale.

[E]

3 Janet invests £50 in a building society for one year.
The interest rate is 6% per year.
(a) How much interest, in pounds, does Janet get?

Nisha invests £60 in a different building society. She gets £3 interest after one year.
(b) Work out the percentage interest rate that Nisha gets.

[E]

4 Nigel works in a service station and earns £8500 per year.
He is given a pay rise of 15%.
Ryan is a football player earning £75 000 per year. He is given a pay rise of 1.5%.
Who ends up with the biggest rise in pay?

5 A shopkeeper increases the prices of goods in the shop in line with the rate of inflation.
When the rate of inflation is 2.1%, calculate:

(a) the increase in the price of a table marked at £160

(b) the new price of a chair marked at £85

(c) the increase in the price of a bed marked at £249

(d) the new price of a settee marked at £1350.

6 In an election, 6600 people were eligible to vote. 45% voted for Edwards, 28% voted for Philips and 6% voted for Fortescue. The remainder did not vote.

(a) What percentage of voters did not vote?

(b) How many votes did each candidate receive?

7 House prices in a town increased on average by 8% during the year 2000. Calculate:

(a) the price increase of a house previously valued at £84 000

(b) the new price of a house previously valued at £102 250.

8 An insurance agent is paid 2% commission on the value of any policies she sells. How much commission is she paid when she sells a policy worth:

(a) £4500   (b) £6550?

Summary of key points

1 A percentage can be written as a fraction with the denominator (bottom) 100.

2 To convert a percentage to a decimal divide by 100.

3 Remember these percentages and their equivalent fractions:

$$50\% = \tfrac{1}{2} \quad 25\% = \tfrac{1}{4} \quad 75\% = \tfrac{3}{4} \quad 33\tfrac{1}{3}\% = \tfrac{1}{3} \quad 66\tfrac{2}{3}\% = \tfrac{2}{3}.$$

4 To work out a percentage of an amount write the percentage as a fraction with a denominator of 100. Then multiply the fraction by the amount.

5 To write an amount as a percentage of another:
- write the amounts as a fraction
- convert the fraction to a decimal
- multiply the decimal by 100.

15 Algebra 3

15.1 Simple equations

■ **In algebra letters are often used to represent numbers.**

$a + 3 = 7$ The letter a must equal 4 because 4 add 3 equals 7.

     $a = 4$

$a - 3 = 2$ a must equal 5 because 5 take away 3 equals 2.

     $a = 5$ These are examples of equations.

Exercise 15A

Find the value of the letter in these equations:

1 $a + 2 = 5$	**2** $b + 1 = 4$	**3** $c + 2 = 9$	**4** $w + 5 = 7$
5 $m + 3 = 4$	**6** $y + 5 = 5$	**7** $x - 2 = 3$	**8** $k - 4 = 1$
9 $n - 3 = 3$	**10** $h - 5 = 3$	**11** $g - 2 = 2$	**12** $f - 5 = 4$
13 $d + 2 = 7$	**14** $2 + e = 5$	**15** $4 + y = 7$	**16** $10 + x = 14$
17 $7 - m = 3$	**18** $5 - d = 2$	**19** $k + 6 = 15$	**20** $12 = y + 2$
21 $15 - t = 5$	**22** $z - 2 = 2$	**23** $z + 2 = 2$	**24** $n - 5 = 0$

In Exercise 15A you probably spotted the answers and then wrote them down. There is another way of looking at equations.

An equation is a balancing act!

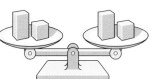

Here is a way of working out the value of the letter in an equation that does not rely on guessing the answer.

Example 1

To find a from: $a + 6 = 9$

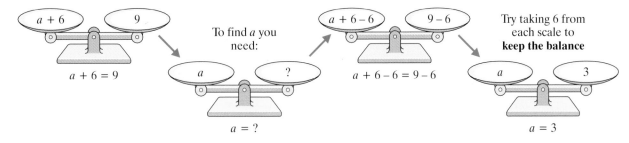

Example 2

To find k from:

$k - 7 = 13$

$k - 7 + 7 = 13 + 7$ (keeping the balance)

$k = 20$

Exercise 15B

Find the value of the letter in these equations. Use the balancing method to help you.

1 $a + 6 = 7$	**2** $y + 3 = 5$	**3** $h + 2 = 9$
4 $p - 5 = 4$	**5** $q - 3 = 7$	**6** $d - 6 = 2$
7 $x + 3 = 3$	**8** $t - 4 = 0$	**9** $r + 7 = 10$
10 $k + 2 = 3$	**11** $n + 1 = 2$	**12** $x - 2 = 3$
13 $m + 7 = 12$	**14** $y - 7 = 9$	**15** $w + 5 = 5$
16 $q - 10 = 2$	**17** $5 + p = 7$	**18** $6 + t = 6$
19 $a + 19 = 31$	**20** $21 + x = 21$	**21** $p - 15 = 23$
22 $7 = a + 3$	**23** $6 = b + 5$	**24** $10 = y + 10$

Exercise 15C

Find the value of the letter in each of these equations:

1 $a + 5 = 10$	**2** $p - 4 = 7$	**3** $q - 3 = 5$
4 $x + 7 = 15$	**5** $y + 4 = 17$	**6** $s + 12 = 15$
7 $x - 7 = 15$	**8** $y - 4 = 17$	**9** $s - 12 = 15$
10 $a + 5 = 6$	**11** $p - 5 = 6$	**12** $c + 17 = 21$
13 $5 + a = 6$	**14** $11 + p = 16$	**15** $12 + q = 12$
16 $4 = a + 2$	**17** $5 = b + 3$	**18** $12 = c - 3$
19 $10 = p + 5$	**20** $11 = y - 10$	**21** $15 = t + 10$
22 $12 = p + 12$	**23** $12 = p - 12$	**24** $p + 12 = 12$

In this next section you will learn to find the value of the letter when the letter is multiplied or divided by a number.

Example 3

$$3a = 12$$

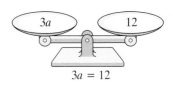

$3a = 12$

This means $a \times 3 = 12$

To work out $a = ?$ you need to keep the balance in the equation.

To make the equation balance you must do the same to each side.

Try: $3a \div 3 = 12 \div 3$
$a = 4$

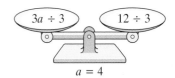

$3a \div 3$ $12 \div 3$

$a = 4$

■ **To make an equation balance you must do the same to each side.**

Exercise 15D

Find the value of the letter in these equations:

1 $3a = 6$	**2** $4p = 8$	**3** $5p = 15$	**4** $6s = 18$	**5** $2k = 10$
6 $7u = 28$	**7** $2g = 14$	**8** $5k = 35$	**9** $6j = 12$	**10** $8f = 32$
11 $3r = 27$	**12** $5v = 45$	**13** $2t = 42$	**14** $4d = 48$	**15** $7t = 63$

Example 4

$$\frac{w}{3} = 5 \quad (\text{means } w \div 3 = 5)$$

Look for the opposite process.
The opposite process of $\div 3$ is $\times 3$
Multiply each side of the equation by 3

$$\frac{w}{3} \times 3 = 5 \times 3$$

$$w = 15$$

Exercise 15E

Find the value of the letter in these equations:

1 $\frac{a}{2} = 5$	**2** $\frac{b}{5} = 4$	**3** $\frac{s}{4} = 3$	**4** $\frac{c}{6} = 5$	**5** $\frac{t}{4} = 6$
6 $\frac{s}{8} = 9$	**7** $\frac{h}{6} = 12$	**8** $\frac{f}{4} = 7$	**9** $\frac{d}{3} = 15$	**10** $\frac{a}{3} = 15$
11 $\frac{b}{5} = 8$	**12** $\frac{r}{4} = 13$	**13** $\frac{a}{12} = 5$	**14** $\frac{b}{2} = 16$	**15** $\frac{k}{3} = 16$

Exercise 15F Mixed questions

1 $a + 4 = 5$	**2** $b + 3 = 6$	**3** $c + 4 = 9$
4 $p - 3 = 6$	**5** $q - 2 = 2$	**6** $d - 6 = 2$
7 $2p = 6$	**8** $4r = 8$	**9** $5t = 20$
10 $\dfrac{a}{2} = 6$	**11** $\dfrac{b}{5} = 12$	**12** $\dfrac{s}{4} = 5$
13 $4 + r = 7$	**14** $6 + e = 7$	**15** $7 + p = 7$

15.2 Equations combining operations

You have dealt with equations where you added or subtracted numbers and where you multiplied or divided numbers. You are now going to look at what happens when these are combined into one equation.

Example 5

$$
\begin{aligned}
2a + 3 &= 11 \\
2a + 3 - 3 &= 11 - 3 \quad &\text{Take 3 from each side.} \\
2a &= 8 \\
2a \div 2 &= 8 \div 2 \quad &\text{Divide each side by 2} \\
a &= 4
\end{aligned}
$$

Example 6

$$
\begin{aligned}
5p - 3 &= 7 \\
5p - 3 + 3 &= 7 + 3 \quad &\text{Add 3 to each side.} \\
5p &= 10 \\
5p \div 5 &= 10 \div 5 \quad &\text{Divide each side by 5} \\
p &= 2
\end{aligned}
$$

Example 7

$$
\begin{aligned}
\frac{m}{4} + 3 &= 8 \\
\frac{m}{4} + 3 - 3 &= 8 - 3 \quad &\text{Take 3 from each side} \\
\frac{m}{4} &= 5 \\
\frac{m}{4} \times 4 &= 5 \times 4 \quad &\text{Multiply each side by 4} \\
m &= 20
\end{aligned}
$$

Exercise 15G

Find the value of the letter in these equations:

1 $2a + 1 = 5$ **2** $2a - 1 = 5$ **3** $3a + 2 = 8$ **4** $3a - 5 = 4$

5 $3p + 7 = 7$ **6** $3p + 7 = 13$ **7** $4q + 5 = 17$ **8** $5r - 6 = 4$

9 $6t - 12 = 18$ **10** $7f - 12 = 9$ **11** $2r - 11 = 15$ **12** $10a - 5 = 5$

13 $10a + 5 = 5$ **14** $4d + 7 = 19$ **15** $5c - 2 = 18$ **16** $\dfrac{a}{3} + 2 = 3$

17 $\dfrac{z}{5} + 1 = 2$ **18** $\dfrac{r}{6} + 4 = 7$ **19** $\dfrac{s}{4} + 6 = -9$ **20** $\dfrac{b}{3} + 7 = -3$

21 $\dfrac{c}{4} - 2 = 4$ **22** $\dfrac{f}{3} - 6 = 3$ **23** $\dfrac{h}{2} - 4 = -2$ **24** $\dfrac{x}{5} - 1 = -2$

All the equations dealt with so far have answers that are whole numbers. Look at the following example. You can see that answers can be fractions or decimals as well.

Example 8

$$
\begin{aligned}
4p + 7 &= 16 \\
4p + 7 - 7 &= 16 - 7 \qquad \text{Take 7 from each side.} \\
4p &= 9 \\
4p \div 4 &= 9 \div 4 \qquad \text{Divide each side by 4} \\
p &= 2\tfrac{1}{4}
\end{aligned}
$$

Example 9

$$
\begin{aligned}
5q - 8 &= 3 \\
5q - 8 + 8 &= 3 + 8 \qquad \text{Add 8 to each side.} \\
5q &= 11 \\
5q \div 5 &= 11 \div 5 \qquad \text{Divide each side by 5} \\
q &= 2.2
\end{aligned}
$$

Exercise 15H

1 $2a + 3 = 6$ **2** $2a - 4 = 3$ **3** $3a + 7 = 15$ **4** $3a - 6 = 7$ **5** $5p + 7 = 15$

6 $5p - 7 = 15$ **7** $5e + 3 = 3$ **8** $4t + 3 = 9$ **9** $8j - 7 = 5$ **10** $7c - 4 = 7$

11 $8k + 3 = 5$ **12** $3d - 7 = 3$ **13** $9u + 7 = 9$ **14** $4q - 4 = 5$ **15** $7y + 6 = 15$

So far all the answers to the equations you have looked at have been positive. This is how to deal with equations when the answers are negative.

Example 10

$$2a + 7 = 1$$
$$2a + 7 - 7 = 1 - 7 \qquad \text{Take 7 from each side.}$$
$$2a = -6$$
$$2a \div 2 = -6 \div 2 \qquad \text{Divide each side by 2}$$
$$a = -3$$

Exercise 15I

Find the value of the letter in these equations.

1 $2a + 3 = 1$	**2** $2a + 5 = 1$	**3** $2a + 9 = 1$	**4** $3a + 8 = 5$
5 $3a + 7 = 1$	**6** $5p + 12 = 2$	**7** $2s + 7 = -3$	**8** $5p - 2 = -12$
9 $4k - 5 = -9$	**10** $8h + 10 = 2$	**11** $4y + 12 = -8$	**12** $3e + 47 = 20$
13 $6t - 12 = -12$	**14** $3w + 4 = 1$	**15** $2c + 15 = 11$	**16** $13a + 9 = 9$

Exercise 15J Mixed questions

Find the value of the letter in these equations:

1 $2s + 4 = 10$	**2** $5d + 3 = 18$	**3** $8m - 7 = 33$	**4** $4h - 2 = 14$
5 $4k + 7 = 43$	**6** $3y + 7 = 13$	**7** $5p + 2 = 9$	**8** $4f + 4 = 17$
9 $3s - 6 = 5$	**10** $-7g - 4 = 12$	**11** $4f - 5 = 12$	**12** $5k - 12 = 6$
13 $-3s - 15 = 2$	**14** $6j - 3 = 19$	**15** $9b + 7 = 2$	**16** $-2r + 12 = 5$
17 $5t + 15 = -12$	**18** $7y - 15 = -21$	**19** $3e - 5 = -6$	**20** $-4f - 7 = -2$
21 $5g + 17 = 15$	**22** $4h + 4 = 0$	**23** $-3c - 5 = 0$	**24** $8s + 9 = 4$
25 $\dfrac{z}{2} + 2 = 4$	**26** $\dfrac{x}{5} - 3 = 2$	**27** $\dfrac{p}{2} - 5 = -3$	**28** $\dfrac{c}{3} + 4 = -2$
29 $\dfrac{a}{8} - 1 = 5$	**30** $-\dfrac{e}{3} + 2 = 10$		

15.3 Equations with brackets

Algebra 1 dealt with quite complicated algebraic expressions so
you can use what you have learnt to solve quite complicated equations:

Example 11

$$3(2p + 3) = 5$$
$$3 \times 2p + 3 \times 3 = 5$$
$$6p + 9 = 5$$
$$6p + 9 - 9 = 5 - 9$$
$$6p = -4$$
$$6p \div 6 = -4 \div 6$$
$$p = -\tfrac{4}{6} \text{ or } -\tfrac{2}{3}$$

$$3(2p + 3)$$

$$3 \times 2p + 3 \times 3$$

Remember to deal with brackets first and multiply the inside by 3.

■ **In an equation with brackets remove the brackets first.**

Exercise 15K

Find the value of the letter in these equations:

1 $2(p + 4) = 10$ **2** $3(d - 2) = 9$ **3** $2(c + 5) = 16$

4 $3(b - 2) = 1$ **5** $3(g + 2) = 15$ **6** $5(g - 2) = 15$

7 $2(v + 3) = 2$ **8** $4(4 + s) = 20$ **9** $2(3d + 3) = 4$

10 $3(t - 5) = 2$ **11** $4(h - 3) = 0$ **12** $2(3h - 7) = 10$

13 $2(2s + 5) = 22$ **14** $4(2y - 3) = 16$ **15** $4(4r - 12) = 32$

Exercise 15L

Solve these equations:

1 $2(a + 2) = 6$ **2** $3(h - 4) = 12$ **3** $4(g + 5) = 8$

4 $6(f - 3) = 18$ **5** $5(q + 7) = 35$ **6** $9(k - 2) = 18$

7 $5(4 + g) = 25$ **8** $4(5 + h) = 12$ **9** $3(d + 2) = 3$

10 $2(v + 7) = 3$ **11** $3(s + 7) = 4$ **12** $5(2n - 3) = 20$

13 $4(4f + 5) = 6$ **14** $6(7d - 12) = 30$ **15** $2(5m + 11) = 0$

If there are letters on both sides of an equation you have to deal with the problem slightly differently.

It is a good idea to always keep the letter on the side with the most; in this case the side with $5p$.

Example 12

$$5p - 2 = 3p + 6$$
$$5p - 3p - 2 = 3p - 3p + 6 \qquad \text{Take } 3p \text{ from both sides.}$$
$$2p - 2 = 6$$
$$2p = 6 + 2 \qquad \text{Add 2 to both sides.}$$
$$2p = 8$$
$$p = 4 \qquad \text{Divide by 2}$$

Exercise 15M

Find the value of the letter in these equations:

1 $2k - 3 = k + 2$ **2** $5s - 4 = 3s + 3$

3 $4p + 2 = 3p + 6$ **4** $5g + 4 = 3g + 2$

5 $7t - 4 = 4t + 7$ **6** $2k + 6 = 3k - 3$

7 $4d + 9 = 5d + 2$ **8** $5c + 8 = 7c + 2$

9 $5z + 6 = 3z + 4$ **10** $7b + 12 = 3b - 6$

11 $9p + 8 = 3p - 2$ **12** $4g + 9 = 3g + 17$

Exercise 15N

Solve these equations:

1 $5h - 5 = 4h + 7$

2 $7t + 11 = 6t + 3$

3 $5d + 3 = 3d + 1$

4 $6f + 9 = 4f + 3$

5 $3s + 5 = 4s - 2$

6 $4d + 13 = 5d + 7$

7 $2a + 7 = 4a - 2$

8 $3q + 9 = 6q - 3$

9 $4y + 4 = 7y + 6$

10 $2e + 6 = 5e + 9$

11 $12s + 6 = 6s - 4$

12 $5u + 9 = 3u - 2$

13 $5t - 5 = 2t - 9$

14 $6s + 9 = 2s + 2$

15 $3q + 6 = 7q - 5$

16 $2w + 3 = 7w + 9$

17 $8h + 4 = 3h - 4$

18 $3s + 4 = 2s - 3$

19 $r + 2 = 5r + 6$

20 $5a - 7 = 2a + 4$

The most complicated equations you will be asked to solve will have letters on both sides and perhaps brackets as well. Here is an example.

Example 13

$$4(2x - 3) = 2(x + 3)$$
$$8x - 12 = 2x + 6$$
$$8x - 2x - 12 = 2x - 2x + 6$$
$$6x - 12 = 6$$
$$6x - 12 + 12 = 6 + 12$$
$$6x = 18$$
$$x = 3$$

The first step is to get rid of the brackets then sort out the equation in the usual way.

Exercise 15O

Find the value of the letters in these equations:

1 $4(2p + 3) = 2(p + 8)$

2 $5(2h - 9) = 3(3h + 7)$

3 $6(5r - 7) = 4(3r + 7)$

4 $7(2t + 6) = 3(5t + 7)$

5 $4(3g + 5) = 2(5g + 7)$

6 $5(4d + 9) = 6(3d + 5)$

7 $8(2k - 6) = 5(3k - 7)$

8 $7(2m + 3) = 4(5m - 3)$

9 $7(9d - 5) = 12(5d - 6)$

10 $5(3j + 7) = 4(4j + 3)$

11 $4(8y + 3) = 6(7y + 5)$

12 $3(6t + 7) = 5(4t + 7)$

Exercise 15P Mixed questions

1 $5t + 7 = 3t + 10$ **2** $4g + 7 = 3g + 9$

3 $6s - 6 = 4s + 2$ **4** $5q - 5 = 3q + 7$

5 $2d + 4 = 5d - 6$ **6** $6k - 3 = 2k + 7$

7 $2(a + 3) = 7$ **8** $5(2k - 4) = 15$

9 $5 = 2(2d + 7)$ **10** $8 = 4(7p - 3)$

11 $2(3p + 2) = 5p - 7$ **12** $3(5r + 2) = 12r - 7$

13 $2(6t + 2) = 3(5t - 6)$ **14** $6(g + 7) = 3(4g + 2)$

15 $5(2a + 1) + 3(3a - 4) = 4(3a - 6)$

Summary of key points

1 In algebra letters are often used to represent numbers like $a = 5$.

2 To make an equation balance you must do the same to each side:

$a + 4 = 7 \;\;\rightarrow\;\; a + 4 - 4 = 7 - 4 \;\;\rightarrow\;\; a = 3$

3 In an equation with brackets remove the brackets first:

$3(x + 1) = 4 \;\;\rightarrow\;\; 3x + 3 = 4$

16 Pie charts

■ **A pie chart is a good way of displaying data when you want to show how something is shared or divided:**

... how Wayne spent the last 24 hours:

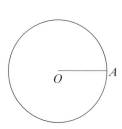

... how the land is used in the UK:

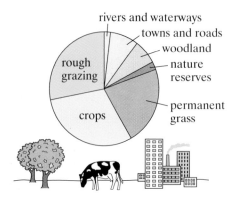

... how much market share different cat food brands have:

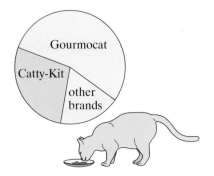

This unit shows you how to represent data on a pie chart.

16.1 Drawing a pie chart

Each slice of a pie chart is called a sector.

When you know the angle of each sector at the centre of a pie chart, here is how to draw it.

Example 1

Draw a pie chart whose angles at the centre are: 108°, 90°, 72°, 60° and 30°.

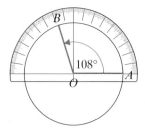

Draw a circle. Draw a line *OA* from its centre to its circumference.

Use your protractor to measure the angle 108°. Mark it and draw the line *OB*.

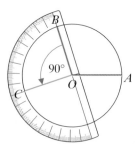

Place your protractor along *OB*. Measure the angle 90°, mark it and draw *OC*.

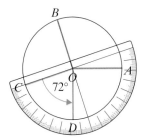

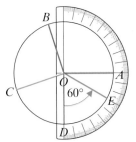

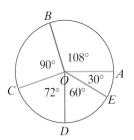

Place your protractor along *OC*. Measure the angle 72°, mark it and draw *OD*.

Place your protractor along *OD*. Measure the angle 60°, mark it and draw *OE*.

Check that the angle remaining is 30° and mark it.

16.2 Calculating the angles

Twenty students were asked on which day they would help paint the scenery for the school play. The replies were: Monday 5, Tuesday 4, Wednesday 8 and Thursday 3. Here is the data shown on a pie chart.

The circle represents all 20 students and each section represents one of the days.

Separating the sections shows that to fit together again the shaded angles must add up to 360°: the total sum of the angles at the centre of a circle.

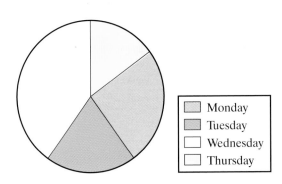

☐ Monday
☐ Tuesday
☐ Wednesday
☐ Thursday

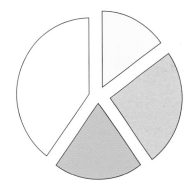

■ **The angles at the centre of a pie chart add up to 360°.**

Here is how to calculate the angles of each sector:

20 students are represented by 360°

1 student is represented by $\dfrac{360}{20} = 18°$

Monday:	5 students so the angle is	5 × 18°	= 90°
Tuesday:	4 students so the angle is	4 × 18°	= 72°
Wednesday:	8 students so the angle is	8 × 18°	= 144°
Thursday:	3 students so the angle is	3 × 18°	= 54°
Check:	20 students so the angle is	20 × 18°	= 360°

Another way of calculating the angles is to find what fraction of 360° the students represent:

Monday: $\quad\dfrac{5}{20} \times 360° = 90°$

Tuesday: $\quad\dfrac{4}{20} \times 360° = 72°$

Wednesday: $\dfrac{8}{20} \times 360° = 144°$

Thursday: $\quad\dfrac{3}{20} \times 360° = 54°$

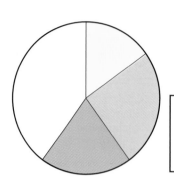

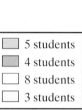

�	5 students
	4 students
	8 students
	3 students

16.3 Calculating using pie charts

The same twenty students agreed to help the next week. Here is the new pie chart:

Here is how to find how many students worked each day:

20 students are represented by 360°

1 student is represented by

$\dfrac{360°}{20} = 18°$

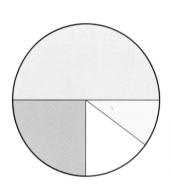

	Monday
	Tuesday
	Wednesday
	Thursday

Monday: $\qquad$ angle is 180° so the number of students is $\dfrac{180}{18} = 10$

Tuesday: $\qquad$ angle is 90° so the number of students is $\dfrac{90}{18} = 5$

Wednesday: $\quad$ angle is 54° so the number of students is $\dfrac{54}{18} = 3$

Thursday: $\qquad$ angle is 36° so the number of students is $\dfrac{36}{18} = 2$

Another way of finding out how many students worked each day is to find what fraction of the twenty students each angle represents.

Monday: $\quad\dfrac{180°}{360°} \times 20 = 10$ $\qquad$ Tuesday: $\qquad\dfrac{90°}{360°} \times 20 = 5$

Wednesday: $\dfrac{54°}{360°} \times 20 = 3$ $\qquad$ Thursday: $\qquad\dfrac{36°}{360°} \times 20 = 2$

Exercise 16A

1   Donna wanted to draw pie charts to show examination
entries for three classes and she calculated these angles:

	English	Maths	Science	History	Art	French
Class IIC	180	60	45	45	30	–
Class IIB	108	72	60	50	40	30
Class IIA	120	72	60	54	36	18

Draw a pie chart for each class.

2   The number of pens and pencils five students had with
them is shown in the table.

Name	Number	Angle needed
Gwyneth	3	
Peter	8	
Wes	4	
Mario	2	
Nesta	7	

   (a) Copy and complete the table by calculating the angle
   needed to draw a pie chart.
   (b) Draw the pie chart.

3   Draw pie charts to represent the following data:
   (a) the number of A grades gained by students in Class IIB

Subject	English	Maths	French	Science	Other
Number	5	6	2	4	3

   (b) the number of pets kept by students in Class IIB

Pets	Dogs	Hamsters	Cats	Birds	Other	No pets
Number	10	5	6	4	3	8

   (c) vehicles passing while Haji waited for the bus

Type	Car	Bus	Lorry	Van	M/Cycle
Number	24	12	10	6	8

4 A class of 30 students listed their favourite sports:

Sport	Swimming	Tennis	Football	Hockey	Athletics
Number	10	4	8	6	2

Draw a pie chart to illustrate this data.

Worked examination question 1

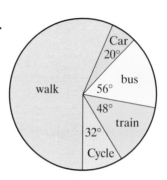

720 students were asked how they travelled to school.

The pie chart shows the results of this survey.

Work out:

(a) how many of the students travelled to school by bus
(b) how many students walked to school.

(a) How many travelled by bus?

360° represents 720 students

$$1° \text{ represents } \frac{720}{360} = 2 \text{ students}$$

The angle of the bus sector of the pie chart is 56°, so the number of students is:

$$56 \times 2 = 112$$

Here is another way of doing this:

The bus sector is 56° and the whole pie chart is 360°, so the fraction travelling by bus is: $\frac{56}{360}$

There are 720 students so the number of students travelling by bus is:

$$\frac{56}{360} \times 720 = 112$$

(b) First, find the angle of the 'walk' sector:

$$360 - (20 + 56 + 48 + 32) = 360 - 156$$
$$= 204°$$

Either each 1° represents 2 students   **or** $\frac{204}{360} \times 720$

$$204 \times 2 = 408 \text{ students} \qquad\qquad = 408 \text{ students}$$

Exercise 16B

1 In a pilot survey Henry asked the question, 'Which main course did you choose?' His results were:

Main course	Tally	Frequency
Fish	IIII	5
Sausages	II	2
Salad	III	3
Cheese roll	II	2

(a) How many students' results did Henry record?
(b) Henry wants to represent this information on a pie chart. How many degrees represent one person on the pie chart?
(c) Copy and complete the table:

Column 1	Column 2	Column 3	Column 4
Main course	**Frequency tally**	**Degrees per student**	**Sector angle** (Column 2 × Column 3)
Fish Sausages Salad Cheese roll	5		
Total			

(d) Use your answers to column 4 to draw a pie chart.
(e) Suggest a title for your pie chart.

2 The table shows the sweets served:

Rice pudding	Apple tart	Ice cream	Trifle	Fruit
2	6	16	12	4

Draw a pie chart to illustrate this data.

3 Emily asked staff what drink they had at break. She made this table.

Drink	Tea	Coffee	Milk	Water	Chocolate
Number	12	11	7	4	6

(a) Illustrate this data by drawing:
    (i) a pictogram
    (ii) a bar chart
    (iii) a pie chart.
(b) Comment on each of your charts.

4 The pie chart shows the results of another question
'Which vegetables did you have?'

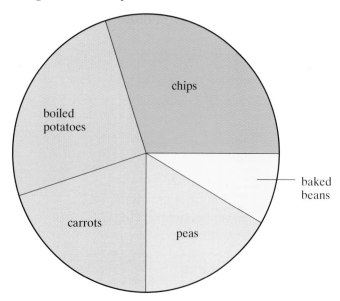

(a) Measure each angle, copy and complete the table:

Column 1	Column 2	Column 3	Column 4
Vegetable	**Sector angle**	**Degrees per vegetable**	**Number of pupils** (Column 2 ÷ Column 3)
Chips		6	
Boiled potatoes		6	
Carrots		6	
Peas		6	10
Baked beans		6	
Total			

(b) How many portions of vegetables were served?

5 Kesh takes home £120 per week. He allocates
his money like this:
Rent £30   Travel £9   Clothes £14
Food £40   Savings £20   Spare £7
Draw a pie chart to show how he allocates his money.

6 Kesh had a £40 rise and the changes he made
are shown in the pie chart.
(a) Calculate how much he spent on rent,
travel, clothes, food, savings and how
much he had spare.
(b) Which items did not change?

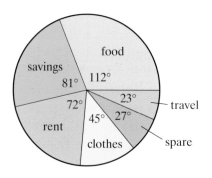

7 Thirty pupils were asked to name their favourite ice cream. The result is shown in this pie chart.

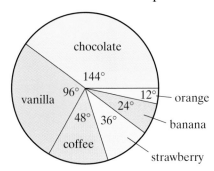

(a) What does the whole circle represent?

(b) Which ice cream does the largest sector represent?

(c) What does the smallest sector represent?

(d) Use the given angles to calculate the number of pupils who liked strawberry ice cream.

(e) How could you select pupils so that this was a random sample?

8 Sita spent £90 as in the table.

(a) Calculate the angle of sector.

(b) Draw a pie chart to show her spending.

Items	Amount spent	Angle of sector
Bus fares	£12	
Going out	£25	
Clothes	£30	
Records	£15	
Other	£8	
Total spending	£90	**Total angles** 360°

[E]

Summary of key points

1 You can use a pie chart to display data when you want to show how something is shared or divided.

2 The angles at the centre of a pie chart add up to 360°.

17 Ratio and proportion

17.1 What is a ratio?

■ **A ratio is a way of showing the relationship between two numbers.**

Ratios can be used to compare costs, weights and sizes . . .

On the deck there are 2 women and 1 man, 5 cars and 2 bicycles. You can say that:

the ratio of cars to bicycles is 5 to 2. This is written as 5 : 2
the ratio of bicycles to cars is 2 to 5, written 2 : 5
the ratio of men to women is 1 to 2, or 1 : 2
the ratio of women to men is 2 to 1, or 2 : 1

Ratios make it easier to work out the exact quantities needed in mixtures. Chemists making up medicines, manufacturers making chocolate biscuits, cooks baking cakes and builders mixing cement all need to be able to make exact mixtures.

Example 1

Frank is making pastry for 5 apple pies. How much of each ingredient does he need?

flour 4 × 5 = 20 ounces
fat   2 × 5 = 10 ounces

This can be written as a ratio:

the ratio of flour to fat is 4 : 2

I always use 4oz flour and 2oz fat to make pastry for my apple pies

Example 2

Vijay needs 4 mixers full of cement. How
much sand, gravel and cement will he need?

I always use
15 shovels of sand
9 shovels of gravel
6 shovels of cement to
make a mixer full of
cement

    sand $15 \times 4 = 60$ shovels
    gravel $9 \times 4 = 36$ shovels
    cement $6 \times 4 = 24$ shovels

This can be written as a ratio:

    the ratio of sand to gravel to cement is
    $15 : 9 : 6$

Example 3

Find the ratio of white tiles to blue tiles:

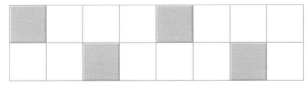

There are 12 white tiles and 4 blue tiles.

    The ratio of white to blue is $12 : 4$

This can be simplified to $3 : 1$. (There is more about
simplifying in the next section.)

Exercise 17A

1   Here are some tile patterns. For each one write down:
- what fraction of the pattern is red
- the ratio of red to blue in the pattern.

(a)

(b)

(c)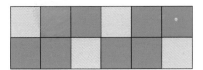

(d)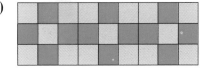

2   A recipe for 6 rock cakes needs 40 g of margarine and 100 g
of flour. How much margarine and flour are needed to make:
   (a) 12 rock cakes **(b)** 18 rock cakes **(c)** 30 rock cakes
   (d) 3 rock cakes   **(e)** 15 rock cakes?

3 A recipe for a rice cake weighing 1200 g is:
    200 g of butter   400 g of ground rice
    400 g of sugar   4 eggs (eggs weigh 50 g each)
How much of each ingredient would you use to make a rice cake weighing:

(a) 2400 g   (b) 600 g   (c) 1800 g   (d) 3000 g?

4 A builder uses 5 buckets of sand and 2 buckets of cement to make concrete. How many buckets of cement will she need if she uses:

(a) 15 buckets of sand   (b) 35 buckets of sand?

5 The builder prepares 120 kg of mortar by mixing 20 kg of cement with 10 kg of lime and 90 kg of sand. How much cement, lime and sand does she use to prepare:

(a) 60 kg of mortar   (b) 180 kg of mortar
(c) 12 kg of mortar?

6 Brass is an alloy (mixture) of zinc and copper in the ratio 3 : 17
How much copper would you expect to be in a brass cross which contained 120 g of zinc?

This hot rolling mill makes mild steel – an alloy of iron and carbon.

7 To make 25 kg of bronze you mix 6 kg of tin with 19 kg of copper.
How much tin and copper is needed to make:

(a) 50 kg of bronze   (b) 250 kg of bronze
(c) 5 kg of bronze?

17.2 Simplifying ratios

These ratios below are **equivalent** – the relationship between each pair of numbers is the same:

    10 : 20
     3 : 6
     2 : 4
     1 : 2   This ratio is a **simpler form** of the ratio 10 : 20

■ **You can simplify a ratio if you can divide both its numbers by a common factor.**

Example 4

Simplify the ratio 30 : 100

10 is a common factor of 30 and 100. Dividing both numbers by 10 gives:

$$\ldots 3 : 10 \ldots$$
$$30 \div 10 \qquad 100 \div 10$$

■ **When a ratio cannot be simplified it is said to be in its lowest terms. Ratios are usually written in their lowest terms.**

Example 5

Write 40p to £1 as a ratio.

First, you make the units the same so you are comparing pennies with pennies: 40p to 100p

The ratio is:

40 : 100

Dividing both numbers by 20 gives:

2 : 5

The ratio of 40p to £1 is 2 : 5 in its lowest terms.

Example 6

Patrick and Colleen share £35 in the ratio 3 : 4
How much does each person get?

The total number of shares is 3 + 4 = 7

Each share is worth $\dfrac{£35}{7} = £5$

Patrick gets $3 \times £5 = £15$

Colleen gets $4 \times £5 = £20$

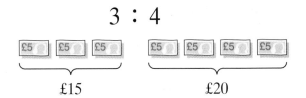

$$3 : 4$$

£15   £20

Exercise 17B

1 Write these ratios in their lowest terms:
   (a) 2 : 4   **(b)** 3 : 9   **(c)** 3 : 18   **(d)** 6 : 24
   (e) 8 : 24   **(f)** 16 : 24   **(g)** 10 : 2   **(h)** 28 : 4
   (i) 32 : 48   **(j)** 15 : 9

2 Write these ratios in their lowest terms. Remember that the units must be the same before you write a ratio.
   (a) 20 cm : 100 cm   **(b)** 20 cm : 1 m
   (c) 25p : £2   **(d)** £3 : 60p

3 Write these as ratios in their lowest terms:
 (a) a small loaf costing 49p and a large loaf costing 63p
 (b) Jeanette weighs 40 kg and Pauline weighs 44 kg
 (c) Spey school has 660 boys and 750 girls
 (d) Mr Johnson takes 1 hour to get to work by train but 2 hours by car
 (e) a salesman made 27 successful calls and 21 unsuccessful calls
 (f) a factory employs 66 craft workers, 18 clerical staff and 12 sales staff.

4 £360 is divided between Sally and Nadir in the ratio 5 : 4
 How much should each person receive?

5 Hermit, Mark and Gavin share £480 in the ratio 4 : 5 : 3
 How much should each person receive?

6 Share 40 sweets in the ratio 2 : 5 : 1

7 Henry, Sue and Rebecca agree to look after the cake stall in the ratio 2 : 4 : 3   Rebecca looked after it for 1 hour. How long did:
 (a) Henry   **(b)** Sue
 spend looking after the stall?

8 Copy and complete this table. The first one is done for you.

Quantity	Divided in the ratio		
	4 :	3 :	2
£27	£12	£9	£6
9 lbs			
36 km			
63 miles			
81 tonnes			
£144			

9 A kilo of cherries costs £1.20 and a kilo of apples cost 72p. Write these prices as a ratio in its lowest terms.

10

Mortar is made by mixing 5 parts by weight of sand with 1 part by weight of cement. How much sand is needed to make 8400 kg of mortar?   [E]

17.3 Proportion

■ **Two quantities are in proportion if their ratio stays the same as the quantities get larger or smaller.**

For example, if the ratio of teachers to students is 1 : 30 (or one per class of 30) then three classes will need a ratio of 3 : 90

Example 7

A factory employs office staff and production staff in the ratio 2 : 7   The number of office staff is increased to 10.

If the number of production staff is kept in proportion to the number of office staff, how many production staff will there be?

Here are the ratios before and after increasing the staff:

  ratio before increase: 2 : 7
  ratio after increase: 10 : ?

You need to find the number that makes these ratios **equivalent**.

Multiplying both numbers in the 'before' ratio by the same number will give an equivalent ratio. Try 5:

$$\begin{array}{c} 2 : 7 \\ \times 5 \qquad \times 5 \\ 10 : 35 \end{array}$$

2 : 7 and 10 : 35 are equivalent ratios.

The number of production staff after the increase is 35.

Exercise 17C

1 Calculate the missing number in these ratios:
  (a) $3 : 5 = 12 : ?$ **(b)** $4 : 7 = 16 : ?$ **(c)** $6 : 5 = 3 : ?$
  (d) $4 : 5 = ? : 35$ **(e)** $8 : 3 = ? : 15$ **(f)** $7 : ? = 49 : 63$

2 A telegraph pole 60 feet high casts a shadow 12 feet long. At the same time of day, how long is a shadow cast by:
(a) a 90 foot pole **(b)** a 40 foot pole **(c)** a 25 foot pole?

3 The numbers of telephone boxes made in a factory in the morning and in the afternoon are in the ratio 4 : 9 They are always completed in this ratio.
(a) How many boxes are made in the afternoon if 60 are made in the morning?
(b) How many are made in the morning if 189 are made in the afternoon?

4 The ratio of students going home for lunch to students staying at school for lunch is 3 : 5 If 273 students go home for lunch how many stay at school?

5 A machine can produce 1120 plastic mugs in 8 hours. At that rate:
(a) how many plastic mugs will it produce in 10 hours
(b) how long will it take to produce 840 plastic mugs?

6 The ratio of the length of a room to its width is 5 : 4
If the length is 6 metres, what is the width?

7 A car travels 126 miles on 18 litres of petrol.
(a) How far will it travel on 40 litres?
(b) How many litres will be needed to travel 540 miles?

8 Mario is paid £21.60 for working 6 hours in the supermarket. At that rate:
(a) how much will he be paid for working 8 hours
(b) how long would it take him to earn £54?

17.4 Solving ratio and proportion problems by the unitary method

Another way of solving ratio problems is to first find the value of **one unit** of the quantity. This is called the unitary method.

Example 8

If 6 similar CDs cost £30, how much will 8 CDs cost?

6 CDs cost £30

Find the cost of one CD.
It costs less.

1 CD costs $\dfrac{£30}{6}$

8 CDs cost $\dfrac{£30}{6} \times 8 = £40$

Eight CDs cost 8 × the cost of one CD.

Example 9

If 6 men can build a shed in 3 days, how long will it take 4 men working at the same rate?

6 men take   3 days

1 man takes 3×6 days

Find how long it would take one man.
It takes longer.

4 men take $\dfrac{3 \times 6}{4} = 4\frac{1}{2}$

Four men take $\frac{1}{4}$ the time taken by one man.

$\dfrac{3 \times 6}{4}$ is the same as $3 \times 6 \times \frac{1}{4}$.

Example 10

A factory employs office staff and production staff in the ratio 2 : 7   How many production staff are employed if there 10 office staff?

2 office staff are needed for 7 production staff

Find how many production staff are needed for one office staff.

1 office staff is needed for $\dfrac{7}{2}$

10 times as many production staff are needed for 10 office staff.

10 office staff are needed for $\dfrac{7}{2} \times 10$

The number of production staff needed is 35.

Example 11

Zoe paid £3.20 for 8 mince pies. How much would 12 mince pies cost?

8 mince pies cost   320 pence

1 mince pie costs $\dfrac{320}{8}$

Find how much one mince pie costs.

12 mince pies cost $\dfrac{320}{8} \times 12$

12 pies cost 12 × the cost of one mince pie.

12 mince pies cost 480 pence or £4.80

Exercise 17D

1   Ten leisure club tickets cost £48. What would 25 tickets cost?

2   Twenty daffodil bulbs cost £2.50. What would 36 bulbs cost?

3 Paul paid £7.20 for 24 Christmas cards. How much would he have to pay for 36 similar cards?

4 Camilla paid £40 for 15 CDs. How much would she have to pay for 24 CDs at the same price?

5 Vijay bought 18 postcards for £2.16 How much would he pay if he bought 27?

6 Bronwen bought 15 roses for £9. How many roses could she have bought for £12.60?

7 A train travels at 80 miles per hour. How long will it take to travel:

(**a**) 140 miles   (**b**) 440 miles   (**c**) 640 miles?

8 A cyclist travels at an average speed of 16 km per hour. At the same rate:

(**a**) how far would he travel in 4 hours

(**b**) how long would it take him to cycle 100 miles?

9 A wall took 6 hours for 4 men to build. At the same rate:

(**a**) how long would it have taken 10 men

(**b**) how many men would have been needed to build it in 12 hours?

10 Eight men can build a chalet in 18 days. Working at the same rate how long would it take:

(**a**) 12 men   (**b**) 5 men?

(**c**) How many men would be needed to build it in 3 days?

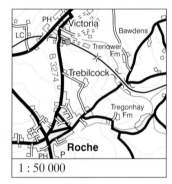

1 : 50 000

17.5 Scales in maps and diagrams

Ratios called **scales** are used to show the relationship between distances on the ground and distances on a map. A common scale is 1 : 50 000, this means that 1 cm on the map represents 50 000 cm on the ground.

This is part of an OS 1:50 000 scale map. 1 cm on the map represents 500 m or half a kilometre.

Example 12

Two towns are 5.2 cm apart on a map whose scale is 1 : 50 000. How far apart are the towns on the ground?

    1 cm   represents   50 000 cm
  5.2 cm   represents   50 000 cm × 5.2 = 260 000 cm

The towns are 2.6 kilometres apart.

To convert from cm to m ÷ by 100
260 000 cm ÷ 100
= 2600 m

To convert from m to km ÷ by 1000
2600 m ÷ 1000
= 2.6 km

Other uses for scales

Model makers also use scales to make copies of objects, for example model aeroplanes, cars, trains, dresses or pieces of furniture. Architects use scales to draw plans of houses.

Example 13

A dolls house has a scale of 1 : 40   If the height of a room in the dolls house is 7.5 cm what would this represent in real life?

    1 cm   represents   40 cm
    7.5 cm   represents   40 × 7.5 cm = 300 cm or 3 metres

In real life the room would be 3 metres high.

Exercise 17E

1   A map has a scale of 1 : 50 000   What is the distance on the ground if the distance on the map is:

    (a) 2.5 cm   **(b)** 3.6 cm   **(c)** 5.2 cm   **(d)** 6.2 cm?

2   What is the distance on the map from question **1** if the distance on the ground is:

    (a) 6 km   **(b)** 5.2 km   **(c)** 8.4 km   **(d)** 25.6 km?

3   A town map has a scale of 1 : 6000

    (a) The town hall is 1500 metres from the station. How far would it be on the map?

    (b) On the map the hotel is 4.6 cm from the harbour. How far is this on the ground?

    (c) Adrian plans a walk round the town. He measures it on the map to be 36.5 cm. How far is he planning to walk?

4   A gardener draws this plan of a patio using a scale of 1 : 125

    (a) Measure the sides.

    (b) Copy the diagram and put on the actual measurements the patio will have. Give your answer in metres.

Patio

Scale 1 : 125

5   A model aircraft has a scale of 1 cm represents 1.8 metres

    (a) Write this scale as a ratio.

    (b) What is the length of the model if the length of the aircraft is 42 metres?

Summary of key points

1   A ratio such as 3 : 2 is a way of showing the relationship between two numbers.

2   You can simplify a ratio if you can divide both its numbers by a common factor. For example:

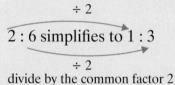

divide by the common factor 2

2 : 6   and   1 : 3   are equivalent ratios.

3   When a ratio cannot be simplified it is said to be in its lowest terms. Ratios are usually written in their lowest terms.

4   Two quantities are in proportion if their ratios stay the same as the quantities get larger or smaller.

5   The unitary method is a way of solving ratio and proportion problems by finding the value of one unit of a quantity first. For example, to find the cost of 12 pies when you know the cost of 8 pies:

8 pies cost   320 pence

1 pie costs $\frac{320}{8}$ ————— find the cost of one

So 12 pies cost $\frac{320}{8} \times 12$ ——— 12 × the cost of one pie

6   A ratio called a scale is used to show the relationship between a distance on a map and the distance on the ground. On a 1 : 50 000 scale map 1 cm represents 50 000 cm on the ground.

18 Symmetry

18.1 Reflective symmetry in 2-D shapes

The two-dimensional (2-D) picture of a butterfly below is **symmetrical**. If you could fold it in half along the dotted line each half would fit exactly on top of the other.

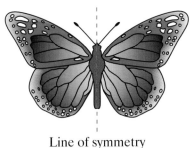

Line of symmetry

The dotted line is called a **line of symmetry**. One half of the shape is a mirror image of the other half.

Some shapes have more than one line of symmetry. The flags below have more than one line of symmetry.

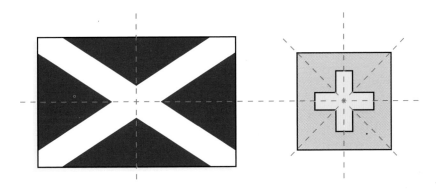

A good way to find the lines of symmetry of a shape is to draw it on tracing paper. Then you can actually fold the shape and check that each half fits exactly on top of the other. Another method is to use a mirror.

- **A line of symmetry is sometimes called a mirror line.**

- **A 2-D shape has a line of symmetry if the line divides the shape into two halves and one half is the mirror image of the other half.**

Example 1

Half of a symmetrical shape is shown here. The dotted line is a line of symmetry. Copy and complete the shape.

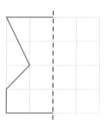

Mark the mirror images of the points (or vertices) in first...

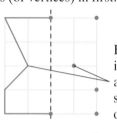

... then complete the lines.

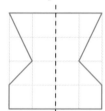

Each point has a mirror image the same distance away from the line of symmetry, but on the other side.

Exercise 18A

1 This question is about the following tiles.

A

B

C

D

E

F

G

H

Which shape(s) have:
(a) only one line of symmetry
(b) no lines of symmetry
(c) exactly three lines of symmetry
(d) exactly four lines of symmetry
(e) exactly two lines of symmetry?

2 Copy these shapes and draw all the lines of symmetry on each flag.

(a)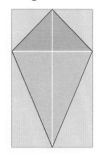

(b)

(c)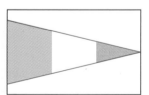

3 In each of these shapes the dotted line is a line of symmetry. Copy and complete each shape.

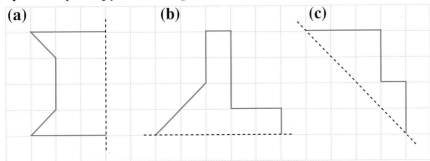

(a) **(b)** **(c)**

4 Use the two lines of symmetry to copy and complete each of these shapes. Hint: use just one line of symmetry at a time.

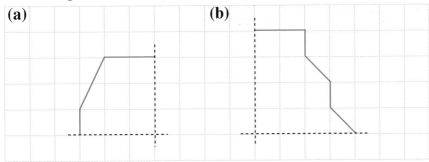

(a) **(b)**

5 Write out the capital letters of the alphabet which have:
   (a) one line of symmetry
   (b) more than one line of symmetry
   (c) no lines of symmetry.

18.2 Reflective symmetry in pictures and patterns

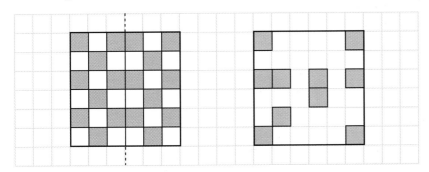

This pattern is **symmetrical**. It can be folded in half so all the lines and colours in each half fit exactly on top of each other.

This pattern is **asymmetrical** (not symmetrical). When it is folded in half some of the lines and colours do not match.

Exercise 18B

You may want to use a mirror or tracing paper in this exercise.

1 Look at these patterns.

A 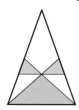 B C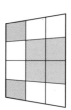

D 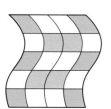 E 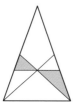 F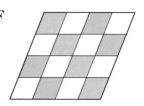

(a) Which patterns have only one line of symmetry?

(b) Which of these patterns are asymmetrical?

(c) Which of these patterns have two lines of symmetry?

> **Remember:**
> Asymmetrical shapes have no lines of symmetry.

2 Copy these patterns. Draw in all the lines of symmetry on each one.

(a)  **(b)** **(c)**

(d) 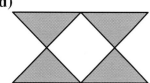 **(e)** **(f)**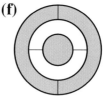

3 In how many different ways can you arrange six squares (which must touch) to form a symmetrical pattern?

Draw your patterns. Two are done for you.

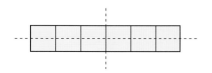 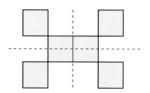

18.3 Rotational symmetry

When a square is rotated or turned through 360° it looks exactly as it did at the start on four different occasions during the rotation. Any shape that looks exactly as it did at the start at least twice during a full turn has **rotational symmetry**.

The number of times a shape looks the same during a full turn is its **order of rotational symmetry**. A square has order of rotational symmetry 4 as it looks the same 4 times during one complete turn.

- ■ **A 2-D shape with rotational symmetry repeats the appearance of its starting position two or more times during a full turn.**

- ■ **The order of rotational symmetry is the number of times the original appearance is repeated in a full turn.**

A square looks the same four times during a full turn:

Starting point

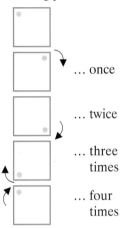

... once

... twice

... three times

... four times

Looking for rotational symmetry

Sometimes rotational symmetry can be hard to spot, so you may want to try one of the following ideas.

Either actually turn the book or paper the shape is drawn on, **or** trace the shape onto tracing paper and turn it on top of the shape in the book. Then you will be able to see when the shapes match.

Example 2

Does a kite have rotational symmetry?

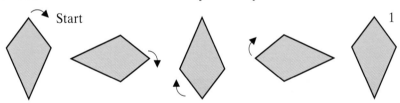

A kite only looks the same once during a complete turn, so it does not have rotational symmetry.

Example 3

Write down the order of rotational symmetry of this flag.

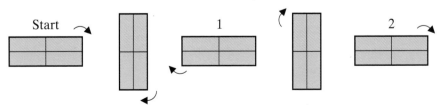

When the flag is turned through 360° it looks the same twice.
The order of rotational symmetry is 2.

Exercise 18C

You may want to use tracing paper in this exercise.

1 Write down whether or not these shapes have rotational symmetry.

(a) 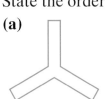 (b) 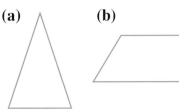 (c) 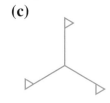 (d) 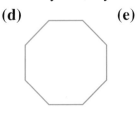 (e)

2 State the order of rotational symmetry of these shapes.

(a) 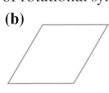 (b) 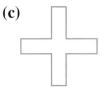 (c)  (d)   (e)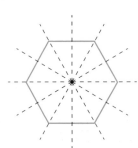

18.4 The symmetry of regular polygons

A regular hexagon has 6 lines of symmetry and rotational symmetry of order 6.

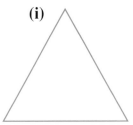

Exercise 18D

1 For each of these regular polygons, find:
  (a) the number of lines of symmetry
  (b) the order of rotational symmetry.

(i)   (ii)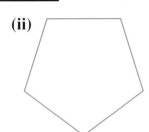

■ **Regular polygons have the same number of lines of symmetry as they have sides.**

■ **The order of rotational symmetry of a regular polygon is the same as the number of sides.**

18.5 Symmetry in 3-D shapes

■ **A plane of symmetry separates a 3-D shape into two halves which are mirror images of each other.**

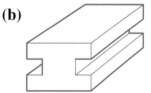

plane of symmetry

Example 4

Draw diagrams to show the planes of symmetry of a cuboid.

A cuboid has 3 planes of symmetry. To show the planes of symmetry clearly, draw a diagram for each plane.

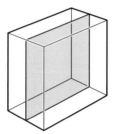 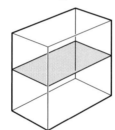 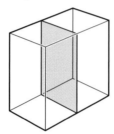

Exercise 18E

You may want to use tracing paper in this exercise.

1   Write down the number of planes of symmetry for each of the following shapes. (Some shapes may not have any.)

(a)

(b)

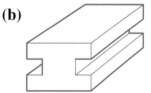

(c)

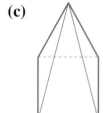

(d)

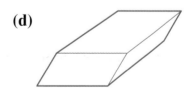

2   Copy or trace the following shapes and clearly mark any planes of symmetry on your drawings.

(a)    **(b)**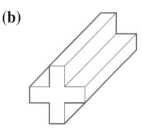

(c)    **(d)**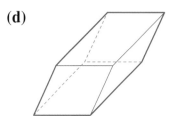

3   The drawings below show half a 3-D solid. Copy and complete each solid so that the shaded face forms a plane of symmetry.

(a) 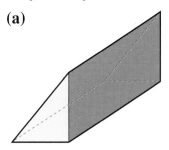   **(b)**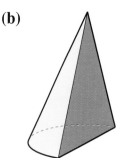

Exercise 18F Mixed questions

1   How many lines of symmetry does a kite have?

2   Draw a rectangle showing all the lines of symmetry.

3   Which regular polygon has five lines of symmetry?

4   Sketch a regular octagon showing its lines of symmetry.

5   What is the order of rotational symmetry of a regular octagon?

6   Which special triangle has rotational symmetry of order 3?

7 Copy or trace the shapes below. Draw all the lines of symmetry and write down the order of rotational symmetry for each shape.

(a) **(b)** 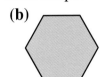 **(c)** 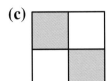 **(d)** 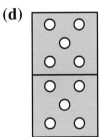 **(e)**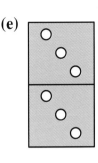

8 Copy and complete the shapes below using the lines of symmetry given.

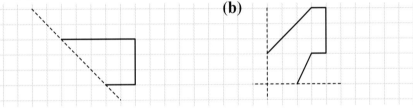

(a) **(b)**

9 Using the lines of symmetry shown copy and complete the missing parts of the patterns.

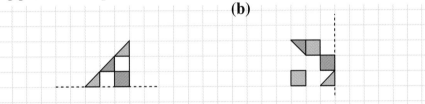

(a) **(b)**

10 How many planes of symmetry does each of the following shapes have?

(a) **(b)** **(c)**

11 Copy or trace the shapes below. Show the plane or planes of symmetry on your diagrams.

(a) **(b)**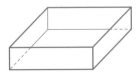

Exercise 18G

1   On a copy of each of these shapes, draw all of its lines of
    symmetry, if it has any.

                                                                    [E]

2   These shapes each have at least one line of symmetry.
    (a) Copy these shapes and show all the lines of symmetry.
    (b) Explain how you could check whether or not a line of
        symmetry was correct.

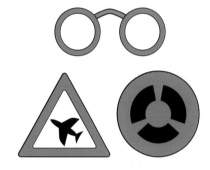

                                                                    [E]

3

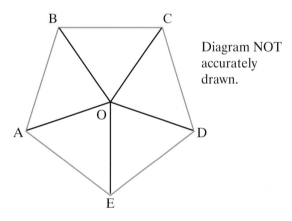

Diagram NOT
accurately
drawn.

ABCDE is a regular pentagon.
O is the centre of the pentagon.
    (a) Write down the order of rotational symmetry of the
        regular pentagon.
    (b) Write down the number of lines of symmetry of
        triangle OCD.

                                                                    [E]

4 Make a list of the designs below which have line symmetry.

(a)

Floor tile

(b)

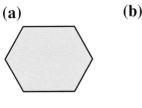

Asian carpet design

(c)

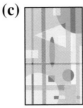

Contemporary art

(d)

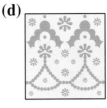

Wallpaper pattern

(e)

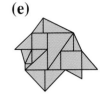

Tile design

[E]

5

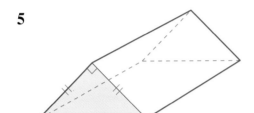

The diagram represents a prism.
The cross section (shaded region) of
the prism is a right-angled isosceles triangle.
Draw one plane of symmetry of the prism. [E]

Summary of key points

1 A line of symmetry is sometimes called a mirror line.

2 A 2-D shape has a line of symmetry if the line divides the shape into two halves and one half is the mirror image of the other half.

3 A 2-D shape with rotational symmetry repeats the appearance of its starting position two or more times in a full turn.

4 The order of rotational symmetry is the number of times the original appearance is repeated in a full turn.

5 Regular polygons have the same number of lines of symmetry as they have sides.

6 The order of rotational symmetry of a regular polygon is the same as the number of sides.

7 A plane of symmetry separates a 3-D shape into two halves which are mirror images of each other.

19 Measure 3

19.1 Perimeter

■ **The perimeter of a shape is the distance around the edge of the shape.**

Example 1

Find the perimeter of this rectangle.

The rectangle has two sides of length 5 cm and two sides of length 3 cm. The perimeter is:

$$5 + 3 + 5 + 3 = 16 \text{ cm}$$

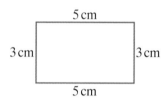

Exercise 19A

1 Work out the perimeters of these shapes.

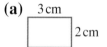 (a) 3 cm, 2 cm

(b) 4 cm, 6 cm

 (c) 2 cm, 3 cm, 4 cm

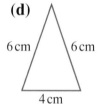 (d) 6 cm, 6 cm, 4 cm

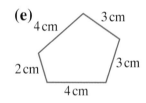 (e) 4 cm, 3 cm, 3 cm, 4 cm, 2 cm

2 Work out the perimeters of these shapes:
   (a) a square with side 5 centimetres
   (b) a rectangle with sides 4 centimetres and 2 centimetres
   (c) an equilateral triangle with all sides 6 centimetres
   (d) an isosceles triangle with two sides of 5 centimetres and one of 6 centimetres.

Example 2

Find the perimeter of this shape.

The shape is made up from a rectangle and a triangle. The rectangle is 9 cm by 6 cm and the triangle has sides 8 cm, 6 cm and 10 cm. The perimeter is:

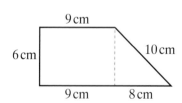

$$6 + 9 + 10 + 8 + 9 = 42 \text{ cm}$$

Remember that the perimeter is the distance around the outside edge of the shape. It does not include lines inside the shape.

Exercise 19B

Find the perimeters of these shapes.

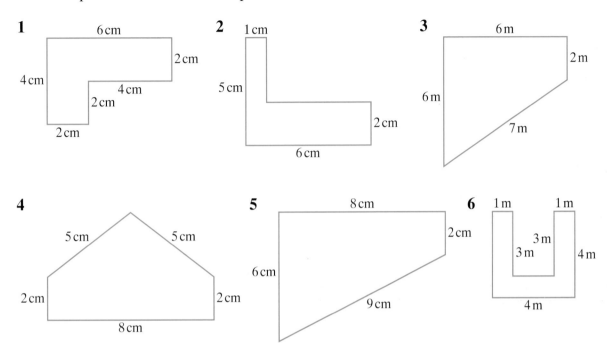

1 6 cm 2 cm 4 cm 4 cm 2 cm 2 cm

2 1 cm 5 cm 2 cm 6 cm

3 6 m 2 m 6 m 7 m

4 5 cm 5 cm 2 cm 2 cm 8 cm

5 8 cm 2 cm 6 cm 9 cm

6 1 m 1 m 3 m 3 m 4 m 4 m

Exercise 19C

Measure these shapes and then find their perimeters.

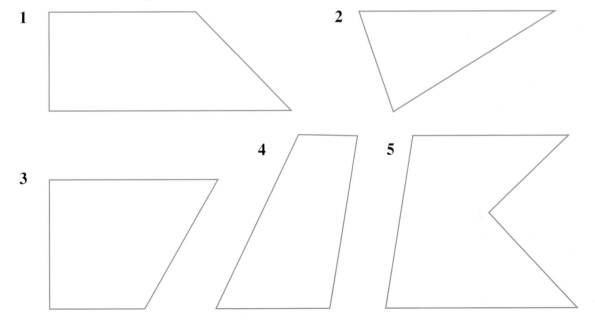

1

2

3

4

5

Circumference of a circle

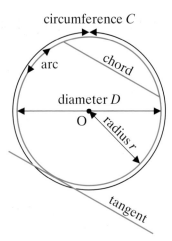

circumference C

■ **The perimeter of a circle is called the circumference. Part of the circumference is called an arc.**

There is a simple formula that you need to know to work out the circumference of a circle it is:

■ $C = \pi \times D$ or $C = 2 \times \pi \times r$

where C is the circumference, D is the diameter and r is the radius of the circle. The diameter of a circle is **twice** the length of the radius.

To find the perimeter or circumference of a circle you have to multiply the diameter of the circle by a number called pi or π. This number is approximately 3.14 but you may be able to display a more accurate value on your calculator.

Example 3

Find the circumference of a circle with diameter of 5 cm.

$C = \pi \times D$
$C = 3.14 \times 5 = 15.7 \text{ cm}$

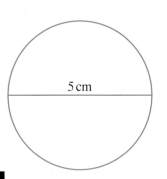

5 cm

Exercise 19D

1 Work out the circumferences of the circles with these diameters.

(a) 2 cm   (b) 3 cm   (c) 4 cm
(d) 10 cm   (e) 8 cm   (f) 12 cm

2 Calculate the circumferences of the circles with these diameters.

(a) 20 cm   (b) 5 m   (c) 50 cm
(d) 2.5 cm   (e) 3.6 cm   (f) 8.25 cm

Example 4

Find the circumference of a circle with radius 4 cm.

If the radius is 4 then the diameter is $2 \times 4 = 8$ and the circumference is:

$C = \pi \times D$
$C = 3.14 \times 8 = 25.1 \text{ cm}$

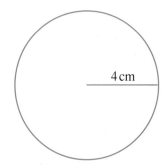

4 cm

Exercise 19E

Find the circumference of a circle with radius:

1 2 cm	**2** 3 cm	**3** 5 cm	**4** 10 cm
5 8 cm	**6** 12 cm	**7** 20 cm	**8** 5 m
9 50 cm	**10** 2.5 cm	**11** 3.6 cm	**12** 8.25 cm

Exercise 19F

Work out the circumference in each of the following cases:

1 a circle with a radius of 9 cm

2 a circle with a diameter of 9 cm

3 a circular table with a diameter of 90 cm

4 a circular fish pond whose radius is 60 cm

5 a circular paving slab of radius 15 cm

6 a circular flower bed of diameter 2.5 m

7 a circular candle of diameter 1.5 cm

8 a circular flower pot of diameter 6 inches

9 a pencil whose radius is 0.3 cm

10 a pen whose radius is 0.5 cm

Finding the diameter when you know the circumference
Sometimes you know the circumference of the circle and you have to calculate the radius or diameter.

Example 5

Susie's bike wheel has a circumference of 62.8 cm. Work out the diameter of the wheel.

$$C = \pi \times D$$
$$\text{so} \quad 62.8 = 3.14 \times D$$

Dividing both sides by 3.14 gives:

$$D = 62.8 \div 3.14$$
$$D = 20 \text{ cm}$$

To find the radius divide the diameter by 2. The radius is 10 cm.

Exercise 19G

Copy this table and work out the diameter and radius of each
circular shape.

	Shape	Circumference	Diameter	Radius
1	Circle	9.42 cm		
2	Circle	12.56 cm		
3	Circular table	3.14 m		
4	Circular fish pond	10 m		
5	Circular paving slab	90 cm		
6	Circular flower bed	7.5 m		
7	Circular candle	5 cm		
8	Circular flower pot	6 inches		
9	Pencil	1 cm		
10	Pen	2.5 cm		

19.2 Area

■ **The area of a shape is a measure of the amount of space it
covers. Typical units of area are square centimetres (cm^2),
square meters (m^2) and square kilometres (km^2).**

You can use a cm^2 grid to help you estimate the area of a
shape.

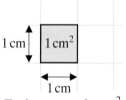

Each square of a cm^2
grid has an area of one
square centimetre (1 cm^2).

Example 6

Find the area of this rectangle.

This rectangle covers up 6 squares.

Each square is 1 cm by 1 cm and has
an area of 1 square centimetre or 1 cm^2.

The rectangle has an area of 6 cm^2.

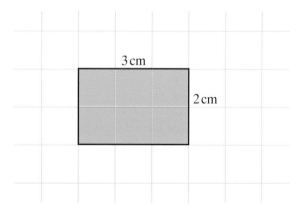

Exercise 19H

By counting squares find the areas of these shapes in cm².

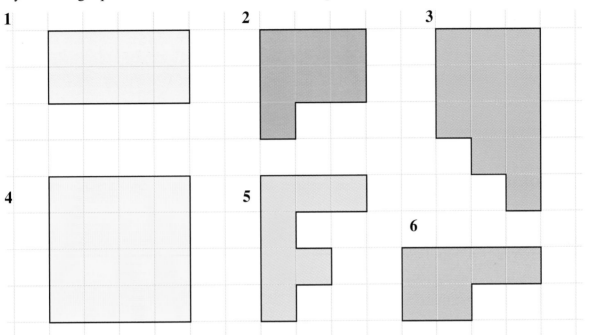

1   **2**   **3**

4   **5**

6

Estimating the areas of irregular shapes

Sometimes shapes do not fit exactly into whole squares. In these cases try and match up part squares to make whole ones.

It is a good idea to number the squares as you count to make sure that you don't miss a square or part of a square.

This shape has a total area of $14\frac{1}{2}$ square centimetres.

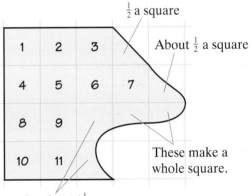

$\frac{1}{2}$ a square

About $\frac{1}{2}$ a square

These make a whole square.

These make about $1\frac{1}{2}$ squares.

Exercise 19I

Estimate the areas of these shapes in cm² by counting squares.

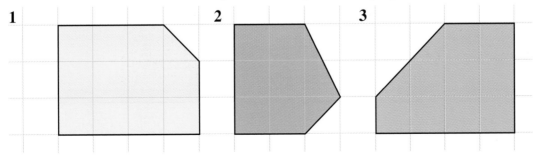

1   **2**   **3**

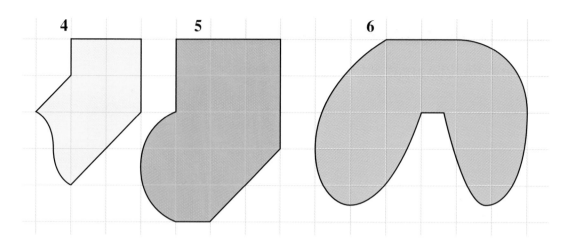

4 **5** **6**

19.3 Volume

■ **The volume of a 3-D shape is a measure of the amount of space it occupies. Typical units of volume are cubic centimetres (cm^3) and cubic metres (m^3).**

You can estimate an area using a grid of squares with area 1 cm^2. To estimate a volume you can use cubes of volume 1 cm^3.

This cuboid is made up of 24 identical small cubes.

Each small cube is 1 cm by 1 cm by 1 cm and has a volume of 1 cm^3. The cuboid has a volume of 24 cm^3.

1 cm
1 cm 1 cm^3
1 cm

This cube has a volume of one cubic centimetre (1 cm^3).

Exercise 19J

Find the volumes of these shapes in cm^3 by counting cubes.

1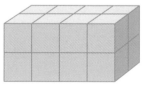

2

3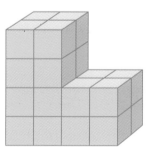

4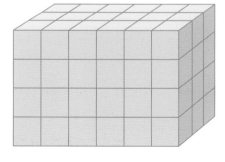

5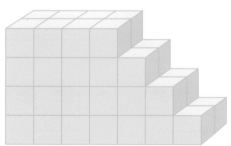

19.4 Finding areas from formulae

You can find the areas of some shapes using formulae:

■ **Area of a rectangle = length × width**
   $$= l \times w$$

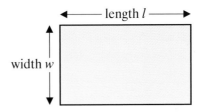

Since a square is a special type of rectangle then:

■ **Area of a square = length × length**
   $$= l \times l$$
   $$= l^2$$

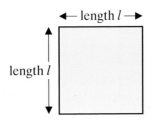

The area of a triangle is half the area of a rectangle that encloses it so:

■ **Area of a triangle = $\frac{1}{2}$ × base × vertical height**
   $$= \frac{1}{2} \times b \times h$$

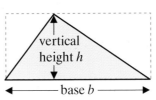

You can cut a corner off a rectangle and replace it on the other side to make a parallelogram so:

■ **Area of a parallelogram = base × vertical height**
   $$= b \times h$$

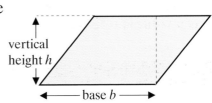

You will need to learn these formulae for your exam.

Example 7

Find the areas of these shapes.

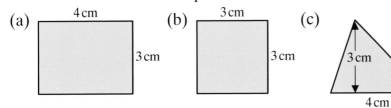

(a) Area of rectangle $= l \times w$
     $$= 4 \times 3 = 12 \, \text{cm}^2$$

(b) Area of square $= l \times l$
     $$= 3 \times 3 = 9 \, \text{cm}^2$$

(c) Area of triangle $= \frac{1}{2} \times b \times h$
     $$= \frac{1}{2} \times 4 \times 3$$
     $$= \frac{1}{2} \times 12 = 6 \, \text{cm}^2$$

Exercise 19K

Find the areas of these shapes.

1 **(a)**

4 cm
2 cm

(b)

5 cm
3 cm

(c)

3 cm
8 cm

(d)

5 cm
6 cm

2 **(a)**

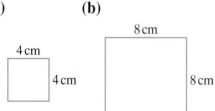
4 cm
4 cm

(b)

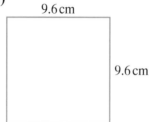
8 cm
8 cm

(c)

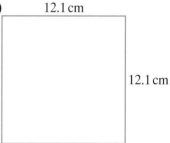
9.6 cm
9.6 cm

(d)

12.1 cm
12.1 cm

3 **(a)**

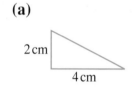
2 cm
4 cm

(b)

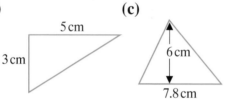
5 cm
3 cm

(c)

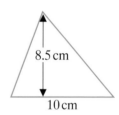
6 cm
7.8 cm

(d)

8.5 cm
10 cm

4 **(a)**

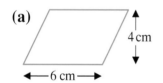
4 cm
6 cm

(b)

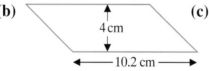
4 cm
10.2 cm

(c)

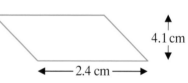
4.1 cm
2.4 cm

Area of a composite shape

You can find the area of a more complicated shape by splitting it up into simple shapes.

Example 8

Find the area of this shape.

Area of rectangle $= l \times w$
$= 8 \times 6$
$= 48\,\text{cm}^2$

Area of triangle $= \frac{1}{2} \times b \times h$
$= \frac{1}{2} \times 5 \times 8$
$= 20\,\text{cm}^2$

Total area $= 48 + 20 = 68\,\text{cm}^2$

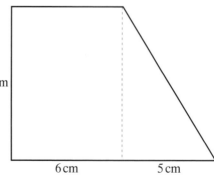

8 cm

6 cm   5 cm

You can split this shape into a rectangle and a triangle.

You also need to be able to work out the surface area of simple solid shapes.

Example 9

Work out the surface area of this cuboid.

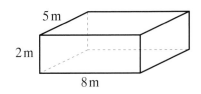

The top and bottom faces are 5 m × 8 m rectangles. They have area 40 m².
The sides are 2 m × 5 m rectangles. They have area 10 m².

The front and back faces are 2 m × 8 m rectangles. They have area 16 m².

The total surface area of the cuboid is:

top + bottom + right side + left side + front + back = 40 + 40 + 10 + 10 + 8 + 8
= 116 m²

Exercise 19L

Find the areas of these shapes.

1

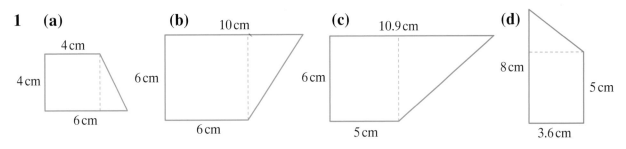

(a) 4 cm, 4 cm, 6 cm

(b) 10 cm, 6 cm, 6 cm

(c) 10.9 cm, 6 cm, 5 cm

(d) 8 cm, 5 cm, 3.6 cm

2

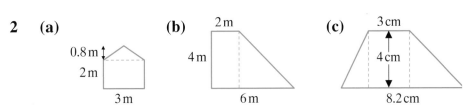

(a) 0.8 m, 2 m, 3 m

(b) 2 m, 4 m, 6 m

(c) 3 cm, 4 cm, 8.2 cm

3 Copy this table and complete the columns.

	Shape	Length	Width	Area
(a)	Rectangle	5 cm	6 cm	
(b)	Rectangle	4 cm		20 cm²
(c)	Rectangle	2 cm		30 cm²
(d)	Rectangle		5 cm	40 cm²
(e)	Rectangle		12 cm	60 cm²

4 Copy this table and complete the columns.

	Shape	Base	Vertical height	Area
(a)	Triangle	10 cm	5 cm	
(b)	Triangle		12 cm	60 cm²
(c)	Triangle		5 cm	40 cm²
(d)	Triangle	4 cm		32 cm²
(e)	Triangle	16 cm		64 cm²

5 Work out the surface area of a cube with sides of length 4 cm.

Hint:
The cube has six square faces.

6 Work out the surface area of these shapes. Remember to consider the sides that are hidden.

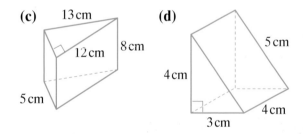

(a) 3 cm, 3 cm, 6 cm

(b) 9 m, 3 m, 2 m

(c) 13 cm, 12 cm, 8 cm, 5 cm

(d) 5 cm, 4 cm, 4 cm, 3 cm

Area of a circle

There is a simple formula that you need to know to work out the area of a circle:

■ $A = \pi \times r^2$ or $\pi \times r \times r$

Where A is the area and r is the radius of the circle.

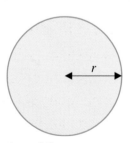

Area $(A) = \pi \times r \times r$

Example 10

Find the area of a circle with a radius of 5 cm.

$A = \pi \times r \times r$
$A = 3.14 \times 5 \times 5 = 78.5 \text{ cm}^2$

Remember:
The value of pi (π) is approximately 3.14. You may have a more accurate value on your calculator.

Example 11

Find the area of a circle with diameter 4 cm.

If the diameter is 4 cm then the radius is 4 ÷ 2 = 2 cm and the area is:

$$A = \pi \times r \times r$$
$$A = 3.14 \times 2 \times 2 = 12.56 \, \text{cm}^2$$

Exercise 19M

Round answers to 3 s.f.

1 Find the areas of the circles with these radii:
   (a) 2 cm   **(b)** 3 cm   **(c)** 4 cm
   (d) 10 cm   **(e)** 8 cm   **(f)** 12 cm

2 Work out the areas of the circles with these radii:
   (a) 20 cm   **(b)** 5 m   **(c)** 50 cm
   (d) 2.5 cm   **(e)** 3.6 cm   **(f)** 8.25 cm

3 Find the areas of the circles with these diameters:
   (a) 2 cm   **(b)** 3 cm   **(c)** 5 cm
   (d) 10 cm   **(e)** 8 cm   **(f)** 12 cm

4 Work out the areas of circles with these diameters:
   (a) 20 cm   **(b)** 5 m   **(c)** 50 cm
   (d) 2.5 cm   **(e)** 3.6 cm   **(f)** 8.25 cm

Exercise 19N

Work out the area, rounding off to 3 s.f., of each of the following:

1 a circle of radius 9 cm
2 a circle of diameter 9 cm
3 a circular table of diameter 90 cm
4 a circular fish pond of radius 60 cm
5 a circular paving slab of radius 15 cm
6 a circular flower bed of diameter 2.5 m
7 a circular candle of diameter 1.5 cm
8 a circular flower pot of diameter 6 inches
9 a pencil of radius 0.3 cm
10 a pen of radius 0.5 cm

Finding the radius when you know the area

Sometimes you will be given the area of a circle and asked to calculate its radius or diameter.

Example 12

A circle has an area of $2826\,\text{cm}^2$. Work out its radius.

$$\pi \times r \times r \quad = A$$
$$3.14 \times r \times r = 2826$$

Dividing both sides by 3.14 gives:

$$r \times r = 2826 \div 3.14$$
$$r \times r = 900$$

To find r when you know $r \times r$ take the square root of each side:

$$\sqrt{r \times r} = \sqrt{900}\ \text{cm}$$
$$r = \sqrt{900} = 30\ \text{cm}$$

If you need to find the diameter then you multiply the radius by 2. The diameter is 60 cm.

Exercise 19O

1 A circle has an area of $28.26\,\text{cm}^2$. Find the diameter.

2 A circle has an area of $12.56\,\text{cm}^2$. Find the radius.

3 A circular table has an area of $3.14\,\text{m}^2$. Work out the diameter.

4 A circular fish pond has an area of $10\,\text{m}^2$. Work out the diameter.

5 A circular paving slab has an area of $90\,\text{cm}^2$. What is the radius?

6 A circular flower bed has an area of $7.5\,\text{m}^2$. What is the diameter?

7 A circular candle has an area of $5\,\text{cm}^2$. Work out the radius.

8 A circular flower pot has an area of 6 square inches. What is the diameter?

9 A pencil has an area of $1\,\text{cm}^2$. Work out the diameter.

10 A pen has an area of $2.5\,\text{cm}^2$. Work out the radius.

19.5 Finding volumes of cuboids from formulae

Section 19.3 shows how to find the volumes of cuboids by counting cubes. There is also a formula for finding the volume of such shapes.

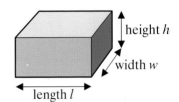

- **Volume of a cuboid = length × width × height**
  $$= l \times w \times h$$

If the shape is a cube then the length, width and height are all the same size so the formula is:

- **Volume of a cube = length × length × length**
  $$= l^3$$

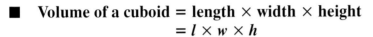

Example 13

Find the volume of a cuboid with length 8 cm, width 6 cm and height 4 cm.

Volume = length × width × height
$$= 8 \times 6 \times 4 = 192 \, \text{cm}^3$$

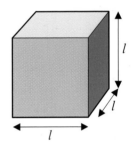

Example 14

This shape is made from 2 cm cubes.
Find its volume.

For each cube:

Volume = length × length × length
$$= 2 \times 2 \times 2 = 8 \, \text{cm}^3$$

There are 7 cubes in the shape so the total volume is:

$$7 \times 8 = 56 \, \text{cm}^3$$

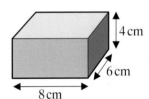

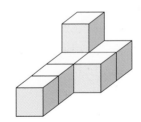

Exercise 19P

Find the volumes of these shapes.

1 6 cm, 2 cm, 2 cm

2 4 cm, 3 cm, 3 cm

3 6 cm, 6 cm, 3 cm

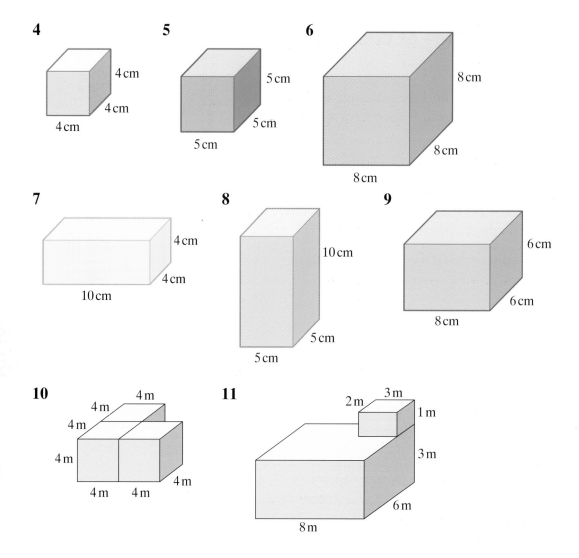

4

4 cm
4 cm
4 cm

5

5 cm
5 cm
5 cm

6

8 cm
8 cm
8 cm

7

4 cm
4 cm
10 cm

8

10 cm
5 cm
5 cm

9

6 cm
6 cm
8 cm

10

4 m
4 m
4 m
4 m
4 m
4 m
4 m
4 m

11

2 m
3 m
1 m
3 m
6 m
8 m

Exercise 19Q

Keith and Mary have bought an old house and are having some work done on it.

1 The lounge is a rectangle 4 m by 3 m and the carpet they buy covers it completely. The carpet costs £7.95 per square metre.
How much does the carpet cost them?

2 The kitchen is rectangular and measures 3.4 m by 2.7 m. They have to buy a whole number of square metres of tiles to cover the floor. The tiles cost £12.50 per square metre. How much will it cost to tile the floor?

3 The bathroom walls need to be tiled to a height of 1.5 m.
The tiles are all 15 cm square. The bathroom measures
3 m by 3 m. The tiles cost 65p each.
 (a) How many tiles are needed for the height?
 (b) How many tiles are needed for the length of one wall?
 (c) How many tiles are needed for all 4 walls? You can
 ignore the space taken up by the door.
 (d) Work out the cost of all the tiles.

4 The main bedroom measures 4 m by 3 m. Strips of
wallpaper are 50 cm wide and they can get four strips out
of one roll of wallpaper. Work out how many rolls of
wallpaper they need to buy, ignoring doors and windows.

5 The small bedroom measures 3 m by 2 m and the walls are
2.5 m high. The walls of this room are going to be painted
with emulsion paint. Each tin of paint will cover 15 m^2.
How many tins will they need?

6 The central heating system runs on oil. The oil tank is a
cuboid with length 2.5 m, width 1 m and height 1.5 m.
How many litres of oil will the tank contain? (Remember
that $1\,l = 1000\,\text{m}l$ or $1000\,\text{cm}^3$).

7 There are 13 doors in the house. Each door measures 2 m
by 0.9 m and has to be painted on both sides. All the doors
are going to be painted white with two coats of paint.
Each litre of paint covers 8 m^2.
How many litres of paint will be needed?

8 All the tiles on the roof need to be replaced. Each tile
measures 30 cm by 30 cm. Each of the two sides of the roof
measures 10 m by 8 m. There needs to be an allowance of
an extra 50% for overlaps.
How many tiles are needed for the roof?

9 The lawn needs replacing and they decide to replace the
worn-out grass with turves that measure 2 m by 0.3 m. The
new lawn is to measure 20 m by 12 m.
Work out how many turves they need to buy.

10 Keith decides to dig out the old drive which measures
12 m by 8 m by 30 cm.
 (a) Work out the volume of rubble that Keith needs to
 dig out.
 (b) How many 4 m^3 skips will be needed to carry away the
 rubbish?

19.6 Converting units of area and volume

You may be asked in the examination to convert square centimetres (cm²) to square metres (m²) or m² to cm².

Here are two pictures of the same square. The only difference is that one is measured in metres and the other in centimetres.

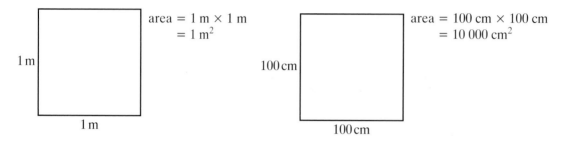

area = 1 m × 1 m
    = 1 m²

1 m

1 m

area = 100 cm × 100 cm
    = 10 000 cm²

100 cm

100 cm

■ **1 m² = 100 × 100 cm² = 10 000 cm²**

Similarly you may be asked to convert cubic centimetres (cm³) to cubic metres (m³).

Here are two pictures of the same cube. The only difference is that one is measured in metres and the other in centimetres.

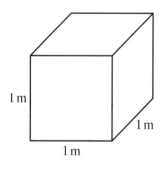

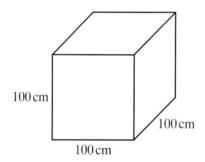

1 m

1 m

1 m

volume = 1 m × 1 m × 1 m
     = 1 m³

100 cm

100 cm

100 cm

volume = 100 cm × 100 cm × 100 cm
     = 1 000 000 cm³

■ **1 m³ = 100 × 100 × 100 cm² = 1 000 000 cm³**

Example 15

A square has an area of 1.2 m².
Write down its area in cm².

$$1 \text{ m}^2 = 10\,000 \text{ cm}^2$$
$$\text{So } 1.2 \text{ m}^2 = 1.2 \times 10\,000 \text{ cm}^2$$
$$= 12\,000 \text{ cm}^2$$

Example 16

The volume of an oil drum is 2 340 000 cm^3.
Write down the volume of the oil drum in m^3.

$$1 \text{ m}^3 = 1\ 000\ 000 \text{ cm}^3$$

$$\text{So } 2\ 340\ 000 \text{ cm}^3 = \frac{2\ 340\ 000}{1\ 000\ 000} \text{ m}^3$$

$$= 2.34 \text{ m}^3$$

Exercise 19R

1 Work out the number of:
  (a) cm^2 in 2 m^2 **(b)** cm^2 in 13 m^2
  (c) cm^2 in 2.4 m^2 **(d)** cm^2 in 15.2 m^2
  (e) m^2 in 120 000 cm^2 **(f)** m^2 in 23 000 cm^2
  (g) m^2 in 164 300 cm^2 **(h)** cm^2 in 0.42 m^2
  (i) cm^2 in 0.03 m^2 **(j)** m^2 in 3000 cm^2
  (k) m^2 in 100 cm^2

2 Work out the number of:
  (a) cm^3 in 7 m^3 **(b)** cm^3 in 15 m^3
  (c) cm^3 in 3.5 m^3 **(d)** cm^3 in 4.78 m^3
  (e) m^3 in 4 000 000 cm^3 **(f)** m^3 in 3 780 000 cm^3
  (g) m^3 in 14 789 000 cm^3 **(h)** cm^3 in 0.8 m^3
  (i) cm^3 in 0.002 m^3 **(j)** cm^3 in 0.000 024 m^3
  (k) m^3 in 37 800 cm^3 **(l)** m^3 in 142 000 cm^3
  (m) m^3 in 3000 cm^3

3 The area of a pane of glass is 12 000 cm^2.
  Write down the area of the pane of glass in m^2.

4 The diagram shows a door.
  (a) Work out the area of the surface
  of the door in m^2.
  (b) Write down your answer to
  part **(a)** in cm^2.

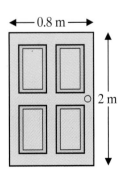
0.8 m

2 m

5 The volume of a large packing case is 2.3 m^3.
  Write down the volume of the packing case in cm^3.

6 The dimensions of a washing machine are shown in the diagram.

(a) Work out the volume of the washing machine in cm³.

(b) Write down your answer to part (a) in m³.

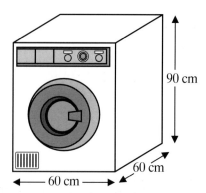

19.7 Compound measures: speed

Sometimes you need to work with two units at the same time. For example, the speed of a car can be measured in miles per hour – a measurement involving a unit of length and a unit of time.

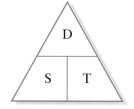

You can use this triangle to help you remember the formulae.

■ **For an object moving at a constant speed:**

$$\textbf{speed} = \frac{\textbf{distance}}{\textbf{time}} \qquad \textbf{time} = \frac{\textbf{distance}}{\textbf{speed}}$$

distance = speed × time

Typical units are miles per hour, and metres per second.

Usually the speed of a car is not constant for the whole journey so you use the average speed:

■ **average speed** $= \dfrac{\textbf{total distance travelled}}{\textbf{total time taken}}$

Example 17

What speed does my car average if I travel 90 miles in 3 hours?

$$\text{average speed} = \frac{\text{total distance travelled}}{\text{total time taken}}$$

$$= 90 \div 3 = 30 \text{ miles per hour}$$

Example 18

How long does it take to travel 400 miles at a constant speed of 50 miles per hour?

$$\text{time} = \frac{\text{distance}}{\text{speed}}$$

$$= 400 \div 50$$

$$= 8 \text{ hours}$$

Example 19

How far can you go if you travel for 3 hours at 10 miles per hour?

distance = speed × time

        = 10 × 3

        = 30 miles

Exercise 19S

1   Elizabeth walked for 3 hours at 4 miles per hour.
How far did she walk?

2   Andrew drove 100 miles in 4 hours.
At what average speed did he travel?

3   Karen drove 300 miles at an average speed of 60 miles per hour.
How long did her journey take her?

4   David was travelling by canal boat and went 30 miles in 8 hours.
At what average speed was he travelling?

5   Amanda rode her bike for 3 hours and travelled 21 miles.
At what average speed was she travelling?

6   Gerry ran for 2 hours and covered 16 miles.
At what average speed was he running?

7   Brigit swam for 3 hours and travelled 4 miles.
At what average speed was she swimming?

8   Alfred set off from home at 8 am. He travelled 200 miles by car and arrived at 11 am.
At what average speed was he travelling?

9   Jason set off for work at 07:55. He arrived at work at 08:10.
If he lives 5 miles from work, at what average speed did he travel?

10   Frances was using a keep fit treadmill. She ran for 40 minutes and travelled 10 kilometres.
At what average speed was she running?

Exercise 19T Mixed questions

1   Work out the perimeter of this shape.

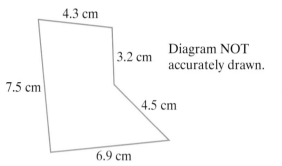

4.3 cm

3.2 cm

Diagram NOT accurately drawn.

7.5 cm

4.5 cm

6.9 cm

[E]

2   The diagram represents an L-shaped room whose corners are all right angles.

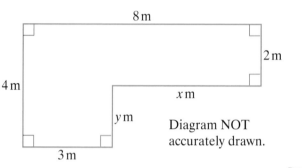

8 m

2 m

4 m

x m

y m

Diagram NOT accurately drawn.

3 m

  (a) **(i)** Work out the value of x.

    (ii) Work out the value of y.

  (b) Work out the perimeter of the shape.

  (c) Work out the area of the shape.

[E]

3   The diagram represents the babies' pool, with paving around, at a leisure centre. The pool is rectangular, 8 m long by 5 m wide and has a depth of 0.6 m throughout.

8 m

←2 m→

5 m

2 m

  (a) Work out the volume of the pool in m³.

The paving around the pool is 2 m wide.

  (b) Work out the area of the paving.

[E]

4   The diagram represents a tea packet in the shape of a cuboid.

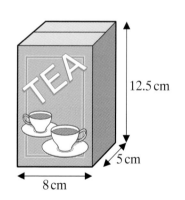

12.5 cm

5 cm

8 cm

  (a) Calculate the volume of the packet.

There are 125 grams of tea in a full packet.
Jason has to design a new packet that will contain 100 grams of tea when it is full.

  (b) **(i)** Work out the volume of the new packet.

    (ii) Express the weight of the tea in the new packet as a percentage of the weight of the tea in the packet shown.

[E]

5 The diagram represents a circular training track.
The diameter of the track, *AB*, is 70 metres.

Alisa and Bryony have a race.
Alisa runs along the diameter from *A* to *B* and back again.
Bryony starts at *A* and runs all the way round the track to
A again.

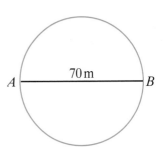

Work out how much further Bryony runs than Alisa.

[E]

6

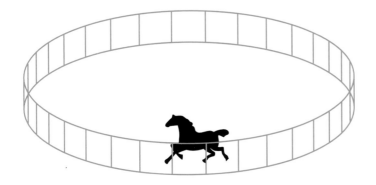

A new wire mesh fence is to be put around a circular
training ring.

The radius of the ring is 45 m.
(a) Calculate, to the nearest metre, the length of fence
needed.
(b) Calculate, to the nearest 10 square metres, the area of
the ring.

[E]

7

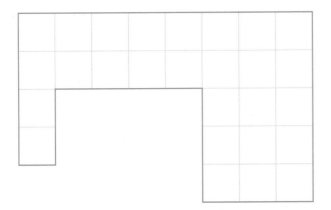

(a) Work out the perimeter of the shape.
(b) Work out the area of the shape.

8

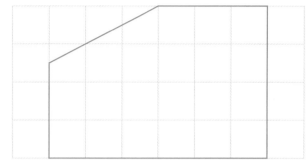

Measure and write down the perimeter of this shape.

Summary of key points

1 The perimeter of a shape is the distance around the edge of the shape.

2 The perimeter of a circle is called the circumference.

$C = \pi \times d$ where C is the circumference, d is the diameter
or
$C = 2 \times \pi \times r$ and r is the radius.

Part of the circumference is called an arc.

3 The area of a shape is a measure of the amount of space it covers. Typical units of area are square centimetres (cm^2), square metres (m^2) and square kilometres (km^2).

4 Area formulae for 2-D shapes:

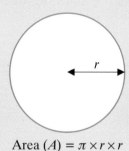

Area $(A) = \pi \times r \times r$

area of a rectangle	$= \text{length} \times \text{width}$ $= l \times w$
area of a square	$= \text{length} \times \text{length}$ $= l \times l = l^2$
area of a triangle	$= \frac{1}{2} \times \text{base} \times \text{vertical height}$ $= \frac{1}{2} \times b \times h$
area of a parallelogram	$= \text{base} \times \text{vertical height}$ $= b \times h$
area of a circle A	$= \pi \times r \times r$ $= \pi r^2$

5 The volume of a 3-D shape is a measure of the amount of space it occupies. Typical units of volume are cubic centimetres (cm^3) and cubic metres (m^3).

6 Volume formula for cuboids (3-D):

volume of a cuboid $= $ length $\times$ width $\times$ height
$= l \times w \times h$

volume of a cube $\quad = l \times l \times l = l^3$

7 $1 \text{ m}^2 = 10\ 000 \text{ cm}^2$
$1 \text{ m}^3 = 1\ 000\ 000 \text{ cm}^3$

8 For an object moving at a constant speed:

$$\text{speed} = \frac{\text{distance}}{\text{time}} \qquad \text{time} = \frac{\text{distance}}{\text{speed}}$$

distance $=$ speed $\times$ time

Typical units are miles per hour, and metres per second.

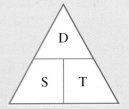

9 When an object's speed varies as it travels:

$$\text{average speed} = \frac{\text{total distance travelled}}{\text{time taken}}$$

20 Averages

The word average is often used. Think about these statements:

> 'Jenny is of average height.'
> 'A centre forward averages a goal a game.'
> 'Teenagers in Britain spend an average of £3 a week on hair care products.'

Here the word average means that something is typical, or describes something that typically happens.

In mathematics an **average** is usually a single value which is used to represent a set of data. It is in some way typical of the data and gives an idea of what the data is like.

Three different averages are commonly used:

- **the mode**   • **the median**   • **the mean**.

This unit shows you how each one is found and why it is useful.

On average a family size box contains 500 g of cereal. The weight of cereal in the box is only allowed to vary within a very small range.

20.1 The mode

■ **The mode of a set of data is the value which occurs most often.**

Example 1

The numbers of goals scored by a team in ten matches were:

    1, 5, 2, 0, 4, 2, 1, 2, 2, 3

Find the mode of this data.
The score that occurred most often was 2 goals, so the mode is 2.

It is possible to have more than one mode.

Example 2

Find the mode of another team's scores:

    3, 1, 0, 1, 4, 3, 6, 2, 1, 3

There are two modes: 1 and 3.

Example 3

Find the mode of a third team's scores:

    2, 5, 3, 4, 0, 1, 6

There is no mode since all scores occurred once only.

Exercise 20A

1 Write down the mode(s) in the following test results.

(**a**) English   32, 28, 41, 17, 28, 39, 19, 28, 40

(**b**) Maths   41, 29, 34, 28, 38, 41, 34, 37, 44

(**c**) History   16, 27, 35, 26, 32, 27, 33, 35, 29

(**d**) Science   26, 28, 36, 27, 31 24, 38, 28, 31

(**e**) Art   32, 28, 30, 35, 28, 39, 32, 36, 30

(**f**) What was the overall mode mark for all the data in parts (**a**) to (**e**) put together?

2 Here are some sets of test marks. Extend each set with a mark or marks so that the new set has the mode shown on the right.

(**a**) 8, 5 ,9, 6, 3, 7, 4   : mode 6

(**b**) 3, 7, 5, 8, 4, 3, 6, 9   : mode 5

(**c**) 2, 6, 7, 8, 5, 3, 2   : mode 4

(**d**) 11, 7, 15, 12, 15, 8, 9 : mode 11 and 15

(**e**) 6, 3, 8, 9, 4, 6, 5   : mode 7

(**f**) 8, 3, 6, 2, 7, 9, 7, 4   : mode 5 and 6

3 The bar charts below show a student's marks in different maths and English tests given on each day of the same week.

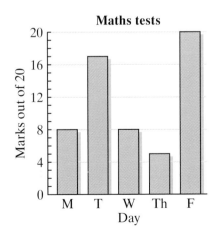

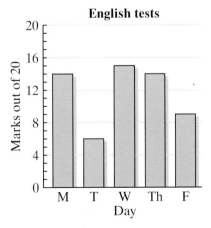

(**a**) Write down a list of each of the marks.

(**b**) Write down the mode for each list.

4 The table shows the number of times students were late handing in homework over a three-week period.

Number of times late	0	1	2	3	4	5
Number of students	7	12	4	2	3	2

(a) Write down the mode of the 'number of times late'. (Hint: the mode will be the 'number of times late' which occurs most often.)

(b) Use your result to write a sentence about the information the mode gives.

20.2 The median

■ **The median is the middle value when the data is arranged in order of size.**

Example 4

Find the median of Brian's homework marks for:

(a) English 5, 7, 9, 4, 1, 3, 7, 4, 6
(b) history 6, 8, 3, 7, 5, 3, 7, 2

(a) Arrange the English marks in order of size:

$$1, 3, 4, 4, 5, 6, 7, 7, 9$$

The middle mark is 5. Brian has four marks higher and four marks lower.

The median is 5.

(b) Arrange the history marks in order of size:

$$2, 3, 3, 5, 6, 7, 7, 8$$

The middle mark lies between the 5 and 6

so the median is $\dfrac{5 + 6}{2} = 5\frac{1}{2}.$

Notice that the median History mark is not a mark anyone has actually scored.

Because there is an even number of marks there is no 'middle mark' in the data.

Instead the value of a middle mark has been calculated.

Example 5

The table shows the shoe sizes of a group of students. Find the median shoe size.

Shoe size	7	$7\frac{1}{2}$	8	$8\frac{1}{2}$	9	$9\frac{1}{2}$
Number of students	2	2	5	4	3	3

To find the median you could list all the shoe sizes:

$$7, 7, 7\tfrac{1}{2}, 7\tfrac{1}{2}, 8, 8, 8, 8, 8, 8\tfrac{1}{2}, 8\tfrac{1}{2}, 8\tfrac{1}{2}, 8\tfrac{1}{2}, 9, 9, 9, 9\tfrac{1}{2}, 9\tfrac{1}{2}, 9\tfrac{1}{2}$$

and see the middle value or median is $8\tfrac{1}{2}$.

An easier way of finding the middle value is to add up the number of students. There are 19 students so the middle student is the 10th.

Starting at the left of the data table, the 10th student is in the $8\tfrac{1}{2}$ column, so the median shoe size is $8\tfrac{1}{2}$.

Exercise 20B

1   Rearrange the following marks in order and write down the median marks.

   (a) 5, 8, 3, 2, 7, 9, 6   (b) 15, 18, 9, 11, 17, 8, 12, 10, 9

   (c) 9, 4, 7, 3, 1, 6, 3, 8   (d) 8, 12, 18, 9, 14, 7, 10, 6

   (e) 9, 3, 7, 7, 2, 5, 8, 5   (f) 6, 14, 19, 8, 5, 16, 10, 15

2   Give examples of:

   (a) seven different marks with a median of 12

   (b) nine different marks with a median of 8

   (c) eight different marks with a median of 10

   (d) six different marks with a median of 4.5

3   What mark should Leo get in his next test to obtain the median stated?

   (a) 1, 8, 5, 2   : median 5

   (b) 8, 2, 4, 5, 3, 8, 4, 2, 9   : median 4

   (c) 7, 5, 3, 6, 8, 2, 7   : median $6\tfrac{1}{2}$

   (d) 7, 4, 3, 6, 2, 6, 8, 4   : median 5

4   The number of students getting marks in a quiz were:

Marks	0	1	2	3	4	5	6	7	8	9	10
Number of students	0	0	2	3	1	1	2	5	4	6	4

   (a) Write down the median mark.

   (b) Write two statements making use of your result.

5 The chart below shows the number of parking tickets issued by a traffic officer on seven days of a week.

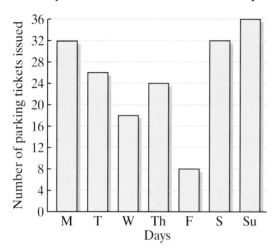

(a) Work out the median number of tickets issued per day.

(b) Write down the mode.

20.3 The mean

■ **The mean of a set of data is the sum of the values divided by the number of values.**

$$\text{mean} = \frac{\text{sum of the values}}{\text{number of values}}$$

Example 6

Find the mean of 3, 8, 4, 7, 7, 6, 4, 1

The sum of the values is:

$$3 + 8 + 4 + 7 + 7 + 6 + 4 + 1 = 40$$

There are 8 values, so divide 40 by 8:

$$40 \div 8 = 5$$

The mean is 5.

Example 7

The mean of three numbers is 5, the mean of four other numbers is 8. What is the mean of all seven numbers?

The sum of the three numbers is   15 $(\frac{15}{3} = 5)$

The sum of the four numbers is   32 $(\frac{32}{4} = 8)$

Mean of all seven numbers $= \dfrac{\text{Total sum}}{7} = \dfrac{15 + 32}{7} = \dfrac{47}{7} = 6.7$

Example 8

The mean of four numbers is 6. Three of the numbers are 4, 8 and 3. Find the value of the other number.

The sum of the four numbers is   24 $(\frac{24}{4} = 6)$

If ? is the missing number, $4 + 8 + 3 + ? = 24$

$$15 + ? = 24$$
$$? = 9$$

The other number is 9.

Exercise 20C

1 Calculate the mean for each set of data:
   (a) 12, 18, 9, 14, 8, 17   **(b)** 23, 15, 37, 26, 16, 21, 33, 23
   (c) 15, 25, 22, 34, 19, 20   **(d)** 25, 12, 31, 26, 31, 19, 30, 16
   (e) 16, 30, 15, 23, 12, 21   **(f)** 31, 18, 25, 33, 35, 19, 17, 24

2 Calculate the mean of:
   (a) £24, £34, £26, £18, £23
   (b) £42, £27, £36, £32, £28, £51

3 The heights of a group of students, in centimetres, are:

   158, 162, 172, 157, 161

   (a) Calculate the mean height.
   (b) Another student joins the group. His height is 159 cm. Calculate the new mean height.

4 Four parcels weigh in grams:

   515, 620, 542, 563

   (a) Calculate the mean weight.
   (b) A fifth parcel is added. If the new mean is 710 g, calculate the weight of the parcel.

5 What mark must Muriel get in her next test to obtain the mean given on the right:
   (a) 8, 3, 6, 7, 8, 4   : mean 7
   (b) 17, 14, 8, 11, 15, 17   : mean 14
   (c) 6, 3, 9, 7, 2, 6, 5   : mean 5.5
   (d) 23, 31, 20, 27, 32, 24   : mean 26

20.4 The range

The mean of a set of data is an average value. It does not tell you how spread out the data is.

One way of measuring the spread of a set of data is to find its **range**:

■ **The range of a set of data is the difference between the highest value and lowest value:**

Range = highest value − lowest value

Two sets of data may have the same mean but different ranges. Compare these two batsmen's scores after four innings:

Peter   0, 96, 100, 0   Mean is 49   Range is 100 − 0 = 100
Hanif   49, 51, 46, 50   Mean is 49   Range is 51 − 46 = 5

Both have the same mean or average score but Hanif has a much lower range. His scores are less spread out which shows he is a more consistent player.

Exercise 20D.

1   Paul's subject marks over a period of eight weeks were:
   History: 34, 56, 45, 46, 53, 62, 39, 48
   Maths: 73, 65, 68, 56, 65, 73, 77, 58
   English: 45, 51, 57, 44, 50, 48, 43, 49
   Art: 47, 51, 36, 78, 43, 20, 39, 27
   Science: 35, 38, 42, 55, 28, 43, 61, 54

   (a) Write down the range in each subject.
   (b) In which subject do you think Paul was:
       (i) best   (ii) most consistent   (iii) least consistent?
       Give reasons for your answers.

2   Copy and complete the table below, the first one is done for you.

Make	Standard	De Luxe	Price range
Ford 1.1	£7145	£7545	£400
Peugeot 1.5	£7460	£9190	
Rover 200	£9995	£13 895	
Citroen 1.9	£10 220		£3985
Volvo 1.6		£15 480	£3910
Audi 1.9	£17 693		£1821
Porsche 3.6	£58 995	£93 950	
Lada 1.3		£6495	£1500
Toyota 1.3	£8040		£4727

3 The table shows the maximum and minimum temperatures one day last year in degrees Fahrenheit.

City	Maximum	Minimum
London	64	42
Paris	69	47
Moscow	38	9
New York	72	59
Luxor	101	88

(a) Calculate the range in each city.

(b) List the cities in order, highest range first.

(c) Give a reason why someone might want to know this range.

20.5 Using appropriate averages

The different types of averages are useful in different situations:

The **mode** is useful when you need to know:

● which shoe size is most common
● which brand of cat food is most popular.

The **mean** is useful for finding a 'typical' value when most of the data is closely grouped, for example the height of a typical student in your class.

The mean height of the students in your class will be 'typical' unless some people are very much taller or shorter than the rest.

The mean may not give a very typical value if the data is very spread out, or if it includes a large value very different from the rest. For example:

> The Chairman of a company earns £200 000 a year and her 9 employees each earn £15 000.
> Their mean pay is £33 500.

In situations like this the **median** or middle value may be more typical. Median earnings in this company are £15 000.

Example 9

Find the mode, median and mean of these marks in a test:

> 1, 3, 17, 18, 19, 20, 21, 21, 24

The mode is 21.

The mean is $\frac{144}{9} = 16.$ Only two students scored less.

The median is 19.   Four students were above and four below this mark.

You can see that the two very low marks lowered the result for the mean. If the students scoring 1 and 3 had scored 15 and 16 the mean would have been 19, but the mode and median would have remained the same.

Example 10

In a maths coursework assessment the scores out of 40 were:

    38, 35, 35, 35, 30, 29, 28, 28, 11, 5

Find the mode, median and mean marks and the range. Comment on your results.

The mode mark is 35, the median mark is 29.5, the mean mark is 27.4 and the range is 33.

Here the median gives the best idea of a 'typical' score: half the students scored more and half less.

The mean has been distorted by the two very low scores and only 2 students have scored less.

The mode is not representative as only 1 student has scored a higher mark.

The range just tells you that there is a wide spread of marks.

Advantages and disadvantages of the three averages

	Advantages	Disadvantages
Mode	Easy to see or pick out. Not influenced by extreme values	Can be more than one mode. Cannot be used for further calculations
Median	Not influenced by extreme values	Actual value may not exist (Example 4, page 310)
Mean	Can easily be calculated. Uses all the data. Can be used for further calculation	Cannot be found graphically. Extreme values can distort the result

Exercise 20E

1 Find the mode, median and mean in each of the following and make a comment about each.

(a) The numbers of cars left in the station car-park overnight during one week were:

12, 6, 14, 9, 13, 6, 10

(b) The numbers of passengers boarding the train between 9 am and 12 am were:

24, 17, 32, 24, 35, 32, 28

(c) The costs of the first five tickets sold were:

£1.15, £12.40, £3.60, £3.60, £2.95

2 The weekly wages for the station staff were:

1 Station master   £385
2 Ticket office clerks   £200 each
2 Porters   £170 each
3 Cleaners   £140 each
2 Trainees   £85 each

(a) Calculate the total weekly wages bill.

(b) What is
  (i) the mode payment   **(ii)** the median payment
 (iii) the mean payment   **(iv)** the range?

(c) Comment on your answers to part **(b)**.

3 Find six numbers which have a mean of 5, a mode of 2 and a median of $4\frac{1}{2}$.

20.6 Averages from frequency tables

Sometimes it is convenient to collect data in a frequency table. (There is more about this on page 159.) This frequency table shows how many children there are in each family in a small village:

Number of children	Tally	Frequency
0					3
1	卌 |||	8
2	卌		6
3						4
4					3
5			1

You can find the mean, median and mode from a frequency table.

To find the mean, first find the total number of children:

$(0 \times 3) + (1 \times 8) + (2 \times 6) + (3 \times 4) + (4 \times 3) + (5 \times 1)$
$\quad 0 \quad + \quad 8 \quad + \quad 12 \quad + \quad 12 \quad + \quad 12 \quad + \quad 5 \qquad = 49$

The number of families is:

only one family
has 5 children

$3 + 8 + 6 + 4 + 3 + 1 = 25$

The mean number of children per family is $\dfrac{49}{25} = 1.96$

The mode, the most frequent number of children in a family, is 1. (Eight families had one child.)

To find the median number of children per family, find the 'middle family'. This is the 13th family when all the data is put in order:

$\downarrow$

0 0 0 1 1 1 1 1 1 1 1 2 2 2 2 2 2 3 3 3 3 4 4 4 5

So the median is 2.

A quicker way to find the median is to look for the 13th family in the frequency table:

Number of children	Tally	Frequency
0					3
1									8
2							6

the first 11 families

families 12, 13, 14, 15, 16 and 17

The 13th family is one of the six that have 2 children.

Example 11

The table shows the number of grade A–C passes obtained by Class 11B.

Grades A–C	1	2	3	4	5	6	7	8
Number of students	2	4	4	7	6	4	2	1

Find (a) the mode (b) the median (c) the mean.

(a) 7 students obtained 4 A–C grades so the mode is 4 passes at grades A–C.

(b) To find the median you could write out all the data in order:

Number of A–C grades

1 1 2 2 2 2 3 3 3 3 4 4 4 4 4 4 5 5 5 5 5 6 6 6 6 7 7 8

↑

There are 30 students so the middle of the data is between the 15th and 16th students. Both these got 4 passes so the median is 4.

(c) To find the mean you need: $\dfrac{\text{total number of passes}}{\text{number of students}}$

Total number of grades A–C is:

$$= \begin{array}{c} (2 \times 1) + (4 \times 2) + (4 \times 3) + (7 \times 4) + (6 \times 5) + (4 \times 6) + (2 \times 7) + (1 \times 8) \\ 2 \quad + \quad 8 \quad + \quad 12 \quad + \quad 28 \quad + \quad 30 \quad + \quad 24 \quad + \quad 14 \quad + \quad 8 \quad = 126 \end{array}$$

The number of students is 30 so the mean is $\dfrac{126}{30} = 4.2$

How to find the position of the median

An easy way to find the position of the median value in a set of data is to add 1 to the number of values, then divide the result by 2.

In the families example on page 317 there are 25 families:

$\dfrac{25 + 1}{2} = 13$ so the median family is the 13th

In Example 11 there are 30 students:

$\dfrac{30 + 1}{2} = 15.5$ so the median of the data is between the 15th and 16th students

This method works whether there is an odd or even number of items of data.

Exercise 20F

1 Find the mean of:
   (a) 5, 11, 23, 16 and 20
   (b) 114, 107, 134, 96 and 49
   (c) 3.2, 1.6, 4.5, 6.3 and 2.7

2 The mean of six numbers is 12. Five of the numbers are 11, 7, 21, 14 and 9. Calculate the sixth.

3   In training Fritz ran the 100 m in 10.8, 11.1, 10.7, 10.9 and 10.8 seconds.

   (a) What was his average (mean) time?

   (b) How fast must he run in the next race to bring his average down to 10.8 seconds?

4   Find: **(i)** the mode   **(ii)** the median of the following data:

   (a) 4, 2, 8, 8, 1, 4, 5, 8, 9

   (b) 6, 11, 5, 8, 4, 7, 3

   (c) 106, 111, 105, 108, 104, 107, 105

   (d) 3.1, 4.2, 2.6, 5.3, 3.1, 2.2, 4.6

   (e) 4, 8, 5, 2, 8, 4, 7

5   Find the mean of:

   (a) £4, £3.50, £1.36, £5.14 and £2.45

   (b) 1 hr 5 min, 3 hr 42 min, 2 hr 17 min, 5 hr 28 min

   (c) 2.1 m, 6.3 m, 5.4 m, 4.2 m, 3.6 m

6   The mean of five numbers is 7.
   If four numbers are 3, 12, 7 and 8, what is the fifth?

7   Use the information given to find the value of n in each of the following (where n is a missing number in the list):

   (a) 5, 7, 4, 1, n, 5   : the mean is 6

   (b) 3, 1, 4, 5, 4, n : the mode is 4

   (c) 1, 7, 2, 1, n, 4, 3   : the mode is 1 and 2

   (d) 2, n, 5, 7, 1, 3   : the median is $3\frac{1}{2}$

   (e) 2.6, 3.5, n, 6.2   : the mean is 4

   (f) 4, 7, 2, n, 2, 9, 6   : the median is 5

8   A stallholder bought 5 boxes of fruit at £12.40 a box and 3 boxes at £16.20.
   What was the average cost of the boxes?

9   Attendance of Class 11B for the first 10 weeks was:
      20, 23, 25, 21, 20, 24, 20, 23, 24, 20.
   Find: **(a)** the mean **(b)** the mode **(c)** the median.

10 Rashid brought packs of cans of coke from a cash and carry to sell to friends. His sales over the first ten days were:
      4, 11, 25, 31, 31, 25, 23, 32, 31, 32

   (a) Find: **(i)** the mean **(ii)** the mode **(iii)** the median.

   (b) Comment on your results.

11 The bar chart shows the number of cans of cola bought by Class 10C during a two-week period.

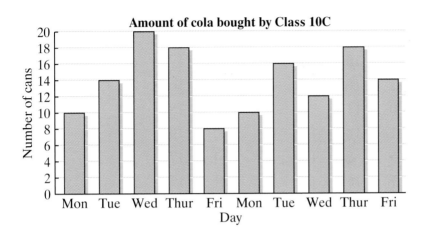

Find:

(a) the mode

(b) the median

(c) the mean number of cans bought.

12 Jez chews bubble gum until the taste has gone. He cannot decide between two different brands. He measures his chewing times on six occasions for each brand.
This table shows his chewing times in minutes.

Brand A	11	10	13	9	17	12
Brand B	11	13	12	12	14	10

The mean and range of the chewing time for Brand A are given in the next table. Calculate the mean and range of the chewing time for Brand B.

	Mean	Range
Brand A	12	8
Brand B		

Which brand should Jez decide to use, assuming they cost the same?
Give a clear explanation.

[E]

13 The numbers of students attending sports and other clubs
after school in a two-week period are given in this table.

Class	Mon	Tues	Wed	Thur	Fri	Mon	Tues	Wed	Thur	Fri
11A	22	18	20	21	12	20	18	20	21	11
11B	20	15	17	18	10	21	15	17	18	8
11C	16	19	17	16	8	17	16	16	15	7
10A	26	22	20	11	24	24	20	18	9	22
10B	19	18	21	14	17	18	20	19	11	16
10C	17	13	15	8	13	17	14	17	6	12
9A	22	20	17	16	19	21	21	13	16	20
9B	18	12	16	18	17	18	12	15	18	17
9C	23	14	18	20	18	22	14	17	20	16
9D	16	11	15	12	10	16	11	14	8	10

Find: **(i)** the mode   **(ii)** the median   **(iii)** the mean
(a) for each class
(b) for each Monday
(c) for both Thursdays together.

14 Using the table in question **13**:
(a) find the range for each class
(b) find the overall range for the classes put together.

15 This table shows the number of students gaining certificates
for outstanding sporting achievement during the year.

Number of certificates	Tally	Frequency
1	IIII	4
2	IHT IIII	9
3	IHT III	8
4	IHT IHT	10
5	IHT I	6
6	III	3

(a) Calculate: **(i)** the mean
                 (ii) the mode
                 (iii) the median.
(b) Which one of these results could be used to encourage
students to take part?
Explain the reasons for your choice.

20.7 Stem and leaf diagrams

When an exact item of data is required for further calculations it is often useful to draw a **stem and leaf diagram** instead of a frequency chart.

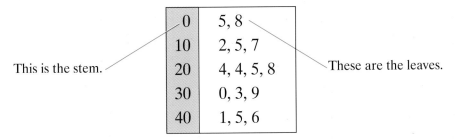

This is the stem.   These are the leaves.

0	5, 8
10	2, 5, 7
20	4, 4, 5, 8
30	0, 3, 9
40	1, 5, 6

The data represented is 5, 8, 12, 15, 17, 24, 24, 25, 28, 30, 33, 39, 41, 45, 46.

Example 12

A year 11 class took a general knowledge test and scored these marks.

36	11	28	42	17	39	24	12	33	9
22	43	31	6	26	12	32	46	25	43
16	38	25	34	45	23	40	7	14	28
48	9	32	27	10	29	19	36	27	39

(a) Show this data on a frequency chart.

(b) Represent the data as a stem and leaf diagram.

(a)

Class interval	Tally	Frequency
0–9	||||	4
10–19	⦀⦀ |||	8
20–29	⦀⦀ ⦀⦀ |	11
30–39	⦀⦀ ⦀⦀	10
40–49	⦀⦀ |||	7

(b)

Stem	Leaves	Frequency
0	9, 6, 7, 9	4
10	1, 7, 2, 2, 6, 4, 0, 9	8
20	8, 4, 2, 6, 5, 5, 3, 8, 7, 9, 7	11
30	6, 9, 3, 1, 2, 8, 4, 2, 6, 9	10
40	2, 3, 6, 3, 5, 0, 8	7

This leaf represents 48

You should write the leaves in order:

Stem	Leaves	Frequency
0	6, 7, 9, 9	4
10	0, 1, 2, 2, 4, 6, 7, 9	8
20	2, 3, 4, 5, 5, 6, 7, 7, 8, 8, 9	11
30	1, 2, 2, 3, 4, 6, 6, 8, 9, 9	10
40	0, 2, 3, 3, 5, 6, 8	7

There are 40 marks in the class. $\dfrac{40 + 1}{2} = 20\frac{1}{2}$, so the median value is halfway between the 20th and the 21st marks.

If you had only draw a frequency chart you would not be able to find the median mark. You can count across your stem and leaf diagram to find the 20th mark, 27, and the 21st mark, 28. The median mark is 27.5.

Exercise 20G

1 Forty people were asked 'What are the last two digits of your telephone number?' The results were:

```
45   15   55   26   43   27   22   36   98   81
17   36   24   36   55   30   43   08   24   26
25   08   23   45   72   29   57   17   67   69
44   53   68   14   90   26   36   49   52   37
```

(a) Using class intervals 0–9, 10–19, etc draw a frequency chart to represent this data.

(b) Copy and complete this stem and leaf diagram. The 10–19 class interval is done for you.

Stem	Leaves	Frequency
00		
10	4, 5, 7, 7	4
20		
30		
40		
50		
60		
70		
80		
90		

(c) In which class interval is: (i) the mode
　　　　　　　　　　　　　　　　　　 (ii) the median?

(d) What is the median?

2 Forty-one pupils recorded the time taken in seconds to solve a puzzle at the school fête. The times were:

25	38	50	9	35	48	9	12	47	34
52	11	32	41	36	29	7	44	18	23
39	22	17	4	49	38	57	15	33	58
14	8	35	27	17	43	20	37	6	26	24

(a) Represent this data using a stem and leaf diagram.
(b) What is the median time?

3 Mrs Bridge asked the pupils in her class how long, in minutes, the journey to school took. The times were:

9	15	38	27	35	18	25	14	37	8	34	6	20	
10	29	12	36	8	23	14	17	25	14	32	5		

(a) Draw a stem and leaf diagram to represent this data.
(b) What was the median time?
(c) Calculate the mean time taken, giving your answer to 1 decimal place.

4 Marks in a mock examination out of 60 were:

25	42	54	37	18	35	29	53	47	53
44	56	35	26	34	43	37	15	55	34
52	35	9	43	58	27	52	45	50	20
24	43	55	46	35	14	38	27	44	32
35	19	36	28	46	34	45	34	40	59

(a) Using class intervals 0–19, 10–19, etc draw a frequency chart to represent this data.
(b) Draw a stem and leaf diagram to represent this data.
(c) In which class interval is: (i) the mode; (ii) the median?
(d) Calculate the mean of the class interval 30–39. Give your answer to 1 decimal place.
(e) What is the median?
(f) If the pass mark was 2 above the median, what percentage passed?

5 The weights, to the nearest kilogram, of 48 men are given below.

64	84	70	92	82	54	68	81	65	70	92	86
86	52	63	80	68	72	58	75	81	57	60	77
93	76	60	74	80	58	68	85	66	82	65	83
69	81	55	77	66	71	65	74	95	74	82	70

(a) Represent this data using a stem and leaf diagram. Include the frequency for each class interval on your diagram.
(b) What is the median weight?
(c) A person is overweight if he is more than 5 kg over the median weight. How many people are overweight?

Summary of key points

1   The mode of a set of data is the value which occurs most often.

2   The median is the middle value when the data has been arranged in order of size.

3   The mean of a set of data is the sum of the values divided by the number of values:

$$\text{mean} = \frac{\text{sum of the values}}{\text{number of values}}$$

4   The range of a set of data is the difference between the highest value and lowest value:

$$\text{range} = \text{highest value} - \text{lowest value}$$

21 Algebra 4

This unit shows you how to use word formulae and equations to help solve problems.

21.1 Using word formulae

Example 1

David works in a factory. He is paid per hour and is given an extra bonus at the end of the week. His pay can be calculated using this **word formula**:

Pay = Rate of pay × hours worked + bonus

Work out his pay when he works for 40 hours at a rate of pay of £4 and earns a bonus of £5.

Pay = £4 × 40 + £5

= £160 + £5 = £165

He earns £165.

Exercise 21A

1 To work out his pay Keith uses the word formula:

Pay = Rate of pay × hours worked + bonus

Work out his pay when he works for 30 hours at a rate of pay of £4 and earns a bonus of £10.

2 To work out her pay Freda uses the word formula:

Pay = Rate of pay × hours worked + bonus

Work out her pay when she works for 35 hours at a rate of pay of £3.50 and earns a bonus of £15.

3 To work out the distance around his bicycle wheel Devon uses the formula:

Distance = 3 × diameter

Work out the distance around the wheel if the diameter is 60 cm.

4   To work out the distance around her bicycle wheel Helen uses the formula:

Distance = 3 × diameter

Work out the distance around the wheel if the diameter is 50 cm.

5   Use the formula:

Cost of pens = Cost of one pen × number of pens

to work out the cost of 17 pens if one pen cost 25p.

Writing word formulae

Sometimes you will need to write a word formula from information you are given.

Example 2

Jill worked for 40 hours at a rate of pay of £5 an hour. How much should she get paid?

You can write a word formula to find her pay for any number of hours she works:

Pay = Hours worked × rate of pay
    = 40 × £5
    = £200

Example 3

Rashmi buys 24 pens at 45 pence each. Work out the total cost of the pens.

Here is a suitable word formula:

Cost = Number of pens × cost of one pen
     = 24 × 45p
     = 1080p = £10.80

■   **A word formula represents a relationship between quantities using words.**

Exercise 21B

Write down a word formula to help solve each of these problems. Use your formula to work out the answers.

1   Daniel works for 40 hours at a rate of pay of £4 an hour. How much should he get paid?

2 Helen works for 20 hours at a rate of pay of £3.25 an hour. How much should she get paid?

3 Abdul works for 30 hours at a rate of pay of £5.75 an hour. How much should he get paid?

4 Susan buys 12 pens at £1.20 each. Work out the total cost of the pens.

5 Roger buys 24 pens at 50p each. Work out the total cost of the pens.

6 Rachel buys 12 books at 28p each. Work out the total cost of the books.

7 Keith buys 15 stamps at 25p each. Work out the total cost of the stamps.

8 Andy buys 20 stamps at 19p each. Work out the total cost of the stamps.

9 Karen sold 50 cakes at 40p each. How much money does she collect?

10 Louise sold 45 loaves of bread at 92p each. How much money does she collect?

11 Mark adds together his age and the age of his sister Pauline. He gets a total of 28. If Mark is 16 how old is Pauline?

12 James loses some £1 coins from his money bag. He had £12 to start with and now only has £7. How many coins has he lost?

13 A chocolate bar machine holds 48 bars of chocolate. After 23 are sold how many are left?

14 At Anne's birthday party there were 48 cans of drink. Everybody at the party had 4 cans. How many people were at the party?

15 Naomi shared a bag of sweets equally between herself and her 6 friends. There were 56 sweets in the bag. How many sweets did the 7 people have each?

16 Norman and his 9 friends were playing football and smashed a window. The cost of repairing the window was £126. The 10 friends decide to split the cost equally between them. How much do they each have to pay?

21.2 Using algebraic formulae

Section 21.1 showed you how to use word formulae. You also need to know how to use **algebraic formulae** in which quantities are represented by letters.

The next two exercises provide practice in finding the value of an algebraic expression.

Example 4

Let $a = 3; b = 2; c = 5$

Work out the value of:

(a) $a + b$ $a + b = 3 + 2 = 5$
(b) ab $ab = a \times b = 3 \times 2 = 6$
(c) $3c - 2a$ $3c - 2a = 3 \times c - 2 \times a$
                     $= 3 \times 5 - 2 \times 3$
                     $= 15 - 6 = 9$

Exercise 21C

In this exercise $a = 3, b = 2, c = 5$ and $d = 0$
Work out the value of:

1 $a + c$	**2** $b + c$	**3** $a + b + c$
4 $2a$	**5** $3c$	**6** $5d$
7 ac	**8** ad	**9** $5b + 2a$
10 $4c - 2b$	**11** $5a + 2b$	**12** $2a + 3b + 4c$
13 $c - a$	**14** $c - 2d$	**15** $ab - c$
16 $4b + 2d$	**17** $3c - 2b$	**18** $5a - 3c$
19 $ac - ab$	**20** abc	

Exercise 21D

In this exercise $p = 2, q = 4, r = 3$ and $s = 0$
Work out the value of:

1 $3p$	**2** $4q$	**3** $5r$	**4** $10s$
5 $p + q$	**6** $q + r$	**7** $r + s$	**8** $p + q + r$
9 $2p + 4q$	**10** $3s + 3p$	**11** $5r - 2q$	**12** $4r - 3q$
13 pq	**14** qr	**15** rs	**16** pqr
17 $5q - 8p$	**18** $3p - 2r$	**19** $4p - 2s$	**20** $3q - pr$

Example 5

The distance around the outside of a rectangle is called the perimeter.

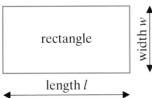

A word formula for the perimeter is:

Perimeter = 2 × length + 2 × width

If the length is represented by the letter l and the width by w an algebraic formula for the perimeter is:

$P = 2l + 2w$

Suppose $l = 6$ and $w = 4$

$P = 2 \times 6 + 2 \times 4 = 12 + 8 = 20$

■ **An algebraic formula represents a relationship between quantities using letters.**

Exercise 21E

1 The formula for the perimeter of an equilateral triangle is $P = 3l$. Work out the value of P when:
(a) $l = 5$ (b) $l = 7$ (c) $l = 4$ (d) $l = 12.4$

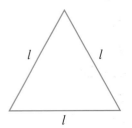

2 The formula for the area of a rectangle is $A = lw$. Find the value of A when:
(a) $l = 6$ and $w = 4$ (b) $l = 5$ and $w = 3$
(c) $l = 10$ and $w = 5.6$ (d) $l = 7.5$ and $w = 3.4$

3 James uses the formula $d = st$ to work out the distance travelled, where the time taken is t and the speed is s. Find the value of d when:
(a) $s = 40$ and $t = 4$ (b) $s = 50$ and $t = 2.5$
(c) $s = 70$ and $t = 3$ (d) $s = 15.8$ and $t = 5$

4 Ayesha uses the formula $v = u + at$ to work out velocity. Find the value of v when:
(a) $u = 3, a = 2$ and $t = 5$
(b) $u = 5, a = 10$ and $t = 3$
(c) $u = 0, a = 10$ and $t = 6$
(d) $u = 2.5, a = 3$ and $t = 1.5$

Remember BIDMAS. You must do at first and then u.

5 Alex uses the formula $P = rh + b$ to work out his pay, where r is his rate of pay per hour, h is the number of hours worked, and b is his bonus. Work out the value of P when:

   (a) $r = 4, h = 12$ and $b = 5$

   (b) $r = 5, h = 40$ and $b = 10$

   (c) $r = 1.5, h = 4$ and $b = 1$

   (d) $r = 2.5, h = 6$ and $b = 2.5$

21.3 Using negative numbers

Sometimes you will need to use negative numbers in algebraic formulae.

The next four exercises provide practice in using negative numbers.

Exercise 21F

Work out:

1 $2 - 5$	**2** $-2 + 4$	**3** $-2 - 4$	**4** $3 + (-7)$
5 $-3 + 5$	**6** $-5 - 3$	**7** $2 - (-3)$	**8** $6 - (-5)$
9 $10 + (-10)$	**10** $-1 - 8$	**11** $1 - (-7)$	**12** $-4 - (-2)$

Remember:

2 like signs next to each other are replaced by a $+$

2 unlike signs next to each other are replaced by a $-$

So

$+ \quad + \quad = \quad +$

$+ \quad - \quad = \quad -$

$- \quad + \quad = \quad -$

$- \quad - \quad = \quad +$

Example 6

$a = 4 \quad b = 6 \quad c = -3$

Find: **(a)** $a - b$ $a - b = 4 - 6 = -2$

         (b) $c + b$ $c + b = -3 + 6 = 3$

         (c) $a + c$ $a + c = 4 + (-3) = 4 - 3 = 1$

         (d) $a - c$ $a - c = 4 - (-3) = 4 + 3 = 7$

Exercise 21G

In this exercise let $a = 3, b = -2, c = -5$ and $d = 0$
Work out the value of:

1 $a + b$	**2** $a + c$	**3** $b + c$	**4** $c + d$
5 $a - b$	**6** $c - a$	**7** $b - c$	**8** $c - d$
9 $b - a$	**10** $a - c$	**11** $c - b$	**12** $d - c$
13 $a + b + c$	**14** $a + b - c$	**15** $a + b - d$	**16** $d - a$
17 $a - b + c$	**18** $a - b - c$	**19** $d - a + b$	**20** $d - b + a - c$

Exercise 21H

Work out:

1	4×-3	**2**	-2×6	**3**	10×-1	**4**	-2×-2
5	3×-7	**6**	-8×4	**7**	-6×-5	**8**	-3×-9

Example 7

$a = -5 \quad b = 3 \quad c = -3$

Find: (a) $2a$ $2a = 2 \times a = 2 \times -5 = -10$
       (b) ab $ab = a \times b = -5 \times 3 = -15$
       (c) ac $ac = a \times c = -5 \times -3 = 15$

Exercise 21I

In this exercise let $a = -3, b = 2, c = -5$ and $d = 0$
Work out the value of:

1	$a + c$	**2**	$b + c$	**3**	$a + b + c$	**4**	$2a$
5	$3c$	**6**	$5d$	**7**	ac	**8**	ad
9	$5b + 2a$	**10**	$4c - 2b$	**11**	$5a + 2b$	**12**	$2a + 3b + 4c$
13	$c - a$	**14**	$c - 2d$	**15**	$ab - c$	**16**	$4b + 2d$
17	$3c - 2b$	**18**	$5a - 3c$	**19**	$ac - ab$	**20**	abc

Negative numbers in algebraic formulae

Example 8

The formula $v = u + at$ is used in maths and science to work
out velocities.

Work out the value of v when $u = 5, a = -2$ and $t = 10$.

$$v = u + at$$
$$v = 5 + -2 \times 10$$
$$v = 5 + -20$$
$$v = -15$$

Exercise 21J

1   Using the formula $v = u + at$ to work out velocity, find the
    value of v when:
    (a) $u = 3, a = -2, t = 5$ **(b)** $u = -5, a = 10, t = 3$
    (c) $u = 0, a = -10, t = 6$ **(d)** $u = -2.5, a = -3, t = 1.5$

2   Using the formula $P = a(b - c)$, find the value of P when:
   (a) $a = 2, b = 3, c = 3$ (b) $a = 2, b = 3, c = -3$
   (c) $a = 3, b = -2, c = 3$ (d) $a = 3, b = -2, c = -3$

3   Using the formula $r = st + (t - s)$, find the value of r when:
   (a) $s = 2, t = -3$ (b) $s = 3, t = -2$
   (c) $s = -2, t = 3$ (d) $s = -2, t = -3$
   (e) $s = 4, t = 2$ (f) $s = 5, t = -4$

21.4 Substituting into more complicated formulae

All the formulae you have used so far have been chosen because the letter only appears once.

You are now going to learn how to deal with formulae containing powers.

Example 9

Use the formula $s = \frac{1}{2}at^2$ to find s when $t = 4$ and $a = 10$.

$s = \frac{1}{2} \times 10 \times 4^2$

$s = \frac{1}{2} \times 10 \times 16$

$s = 5 \times 16 = 80$

Remember: BIDMAS
Indices are worked out
before $\times$.

Exercise 21K

1   The formula for working out how far a ball has fallen when dropped off a cliff is given by:

   $s = 5t^2$

   Find the value of s when t is:
   (a) 1   (b) 2   (c) 5   (d) 10   (e) 3.5

2   Draw this table of values in your book.

x	−3	−2	−1	0	1	2	3
$y = 3x^2 + 4$							

   Complete the table by substituting the values of x into the formula to find the values of y.

3 The speed of a car is given by the formula:

$$v = \sqrt{(u^2 + 2as)}$$

By substituting in values, find v when:

(a) $u = 10, a = 2, s = 5$ (b) $u = 5, a = 5, s = 10$

(c) $u = 5, a = 0, s = 10$ (d) $u = -5, a = 2, s = 5$

(e) $u = -10, a = -5, s = 10$ (f) $u = 0, a = 10, s = 10$

4 If $a = 2, b = 8$ and $c = -4$ work out the value of:

(a) $a^2 + b$ (b) $c^2 - b$

(c) $b^3 - a^2$ (d) $b^2 - c^2$

(e) $b - (a^2 + c)$ (f) $c^2 + (a - c^2)$

(g) $2(a + b)^2$ (h) $c(a + b)^3$

(i) $(a + b)^2 + (a + c)^2$ (j) $2(b + c)^2 - 3(b - c)^2$

21.5 Using algebraic equations

You can solve some problems using **algebraic equations**. Though they can look similar an equation is different from a formula:

formulae can be used to calculate a result:

Pay = rate of pay × hours worked

You could put **any** values into these parts of the formula and get a result for the amount of pay.

equations may be true for one value or several values but are not generally true for any value:

$3x = 6$

This equation is **only** true when $x = 2$.
$x = 2$ is called a solution of the equation.

■ **An algebraic equation can be used to help find an unknown quantity.**

Example 10

The perimeter of this rectangle is 30 cm.
The length is x cm and the width is 4 cm.
Work out the value of the length.

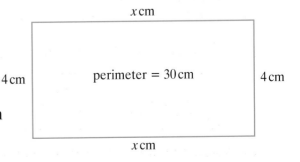

x cm

perimeter = 30 cm

4 cm   4 cm

x cm

$$x + x + 4 + 4 = 30$$
$$2x + 8 = 30$$
$$2x + 8 - 8 = 30 - 8 \quad \text{(Subtract 8 from}$$
$$2x = 22 \quad\quad \text{each side.)}$$
$$x = 11$$

The length x is 11 cm.

Example 11

Suzanne thought of a number, multiplied it by 4 then took away 3. The answer was 13. What was the number she thought of?

Call the number n. Suzanne can make the following equation:

$$4 \times n - 3 = 13$$
$$4n - 3 = 13$$
$$4n - 3 + 3 = 13 + 3 \quad \text{(Add 3 to each side.)}$$
$$4n = 16$$
$$n = 4$$

The number was 4.

Exercise 21L

Use the information in these diagrams to find the values of the letters:

1 **(a)**

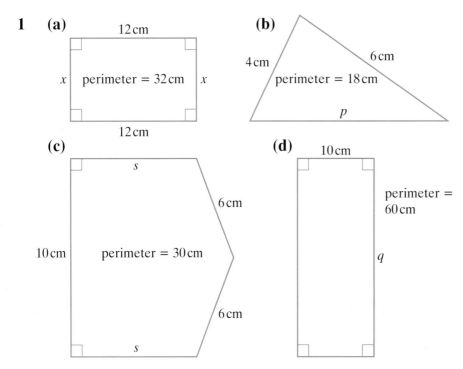

2 The perimeter of this isosceles triangle is 30 cm. Work out the lengths of the sides marked y.

3 Gail thought of a number. She multiplied it by 5 then added 4. The answer was 14. What number did she first think of?

4 Andrew thought of a number. He multiplied it by 6 then took away 15. The answer was 27. What number did Andrew first think of?

5 Darren started a new book and read x pages of the book for each of the first four days. On the fifth day he read 12 pages and finished the book. The book had 100 pages. Work out the value of x.

6 Julia thought of a number, took 5 away from it, then multiplied the answer by 6 to get a final answer of 30. What number did Julia first think of?

7 There were 25 chocolates in a box. Four friends had an equal number of chocolates and this left 9 chocolates in the box. How many chocolates did each of the friends have?

8 Trevor thought of a number, added 7 to it and then multiplied the answer by 5. This gave an answer of 55. What number did Trevor first think of?

9 Sigourney thought of a number, took 12 away from it and then divided her answer by 3 to get a final answer of 4. What number did Sigourney first think of?

21.6 Inequalities

There are lots of apples, but there are only a few bananas. If a is the number of apples and b is the number of bananas, then a is greater than b. You can write this as $a > b$.

■ > means *greater than*
< means *less than*

The thinner end points towards the smaller number.

The thicker end points towards the larger number.

Example 12

Put the correct sign between these pairs of numbers to make a true statement:

(a) 6, 7

(b) 8, 5

(a) 6 is less than 7

6 < 7

(b) 8 is greater than 5

8 > 5

Example 13

Write down the values of x that are whole numbers and satisfy these inequalities:

$x > 2$ and $x < 7$

$x > 2$ so the numbers must be bigger than 2:

3, 4, 5, 6, 7, 8 ...

$x < 7$ so the numbers must stop before 7.

The answer is 3, 4, 5, 6.

> A number **satisfies** an inequality if it makes that inequality true.
>
> $x = 3$ **satisfies** the inequality $x < 7$ because $3 < 7$.

Exercise 21M

1   Put the correct sign between these pairs of numbers to make a true statement:

    (a) 4, 6   **(b)** 5, 2   **(c)** 12, 8   **(d)** 6, 6

    (e) 15, 8   **(f)** 3, 24   **(g)** 10, 3   **(h)** 0, 0.1

    (i) 6, 0.7   **(j)** 4.5, 4.5   **(k)** 0.2, 0.5   **(l)** 4.8, 4.79

2   Write down whether these statements are true or false. If they are false write down the correct sign between the numbers.

    (a) $6 > 4$ **(b)** $2 > 6$ **(c)** $6 > 6$ **(d)** $6 > 8$

    (e) $6 < 4$ **(f)** $8 = 14$ **(g)** $7 < 6.99$ **(h)** $6 > 6.01$

    (i) $7 < 0$ **(j)** $4 < 4$ **(k)** $6 = 4$ **(l)** $6 > 0.84$

3   Write down the values of x that are whole numbers and satisfy these inequalities:

    (a) $x > 4$ and $x < 6$ **(b)** $x > 3$ and $x < 8$

    (c) $x > 0$ and $x < 4$ **(d)** $x > 3$ and $x < 6$

    (e) $x > 1$ and $x < 4$ **(f)** $x > 2$ and $x < 6$

    (g) $x < 4$ and $x > 1$ **(h)** $x < 7$ and $x > 3$

    (i) $x < 5$ and $x > 0$ **(j)** $x < 8$ and $x > 2$

    (k) $x < 10$ and $x > 5$ **(l)** $x < 8$ and $x > 6$

Inequalities on a number line

You can show inequalities by shading a number line.

Example 14

Draw a number line from 0 to 10. Shade in the inequality

$x > 4$

x is greater than 4. You shade all the numbers to the right of 4:

Example 15

Draw a number line from 0 to 10. Shade in the inequalities

$x > 3$ and $x < 8$

x is greater than 3 and less than 8. You shade in the numbers between 3 and 8

> You can write these inequalities as
>
> **$3 < x < 8$**
>
> This means x lies between 3 and 8.

Exercise 21N

1   Draw 6 number lines from 0 to 10. Shade in these inequalities:

(a) $x > 6$ (b) $x > 5$
(c) $x < 4$ (d) $x > 8$
(e) $x < 6$ (f) $x > 9$

2   Draw 6 number lines from 0 to 10. Shade in these inequalities:

(a) $x > 3$ and $x > 7$ (b) $x > 5$ and $x < 8$
(c) $x > 4$ and $x < 6$ (d) $x > 7$ and $x < 9$
(e) $x > 3$ and $x < 5$ (f) $x > 2$ and $x < 8$

3   Draw 6 number lines from 0 to 10. Shade in these inequalities:

(a) $4 < x < 7$ (b) $2 < x < 5$
(c) $5 < x < 6$ (d) $7 < x < 9$
(e) $1 < x < 6$ (f) $5 < x < 7$

4 Write down the inequalities represented by the shading on these number lines:

(a)

(b)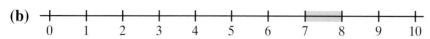

Summary of key points

1 A word formula represents a relationship between quantities using words:

Pay = Rate of pay × hours worked

2 An algebraic formula represents a relationship between quantities using letters. For example, the perimeter of a rectangle is related to its length l and width w by:

$P = 2l + 2w$

3 An algebraic equation can be used to help find an unknown quantity.

4 > means *greater than*
< means *less than*.

22 Transformations

Karen decided to rearrange the furniture in her bedroom.

First she moved her stereo further along the wall by sliding it.

Then she turned her bed a quarter turn.

This slide is a **translation**.

This turn is a **rotation**.

The mirror was now in a position where she could see her own **reflection** close up without the bed being in the way.

The photograph on the stereo was rather small, so she had an **enlargement** made of this.

Reflections, rotations, translations and enlargements are all **transformations**. To help distinguish between the 'before' and 'after' positions of a shape they have special names: The starting shape is called the **object** and the finishing position of the shape is called the **image**.

22.1 Translation

The rook in these pictures has moved 4 squares forward.

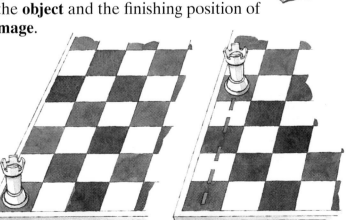

The knight has moved 2 squares forward and 1 square to the right.

■ **Sliding movements like this are called translations.**

Here are some examples:

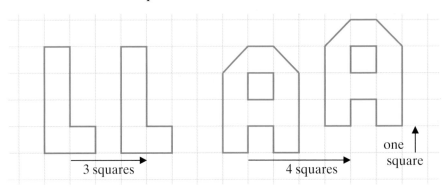

3 squares   4 squares   one square

Example 1

Draw the image of *ABCD* after a translation of 3 squares to the right and 1 square up.

In a translation all the points of the shape have to move exactly the same amount. The vertices are all going to be displaced 3 squares to the right and 1 square up.

Translate each vertex.

Join up the vertices.

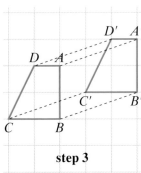

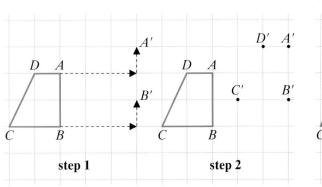

step 1   step 2   step 3

Vertices are corner points.

You could check that this is right by making the shape out of card or paper and sliding it from object to image.

Exercise 22A

1 Draw each shape on squared paper and translate it by the amount shown.

(a)   **(b)**   **(c)**   **(d)**

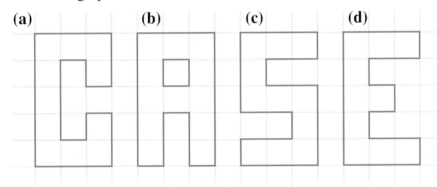

   2 squares right   3 squares up   4 squares down   1 square left

2 Draw these shapes on squared paper and their images after translating by the amount shown.

(a)   **(b)**   **(c)**   **(d)**

2 squares right and 3 squares up	2 squares left and 4 squares down	4 squares right and 1 square down	3 squares left and 4 squares up

3 Perform the four translations of question **2** on this shape.

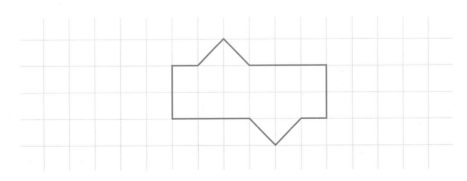

4   Describe the translation for each object-image pair.

(a)   **(b)**   **(c)**

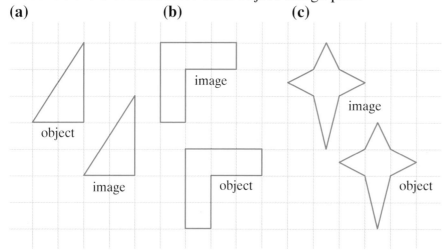

22.2 Rotation

■   **Images of a shape which are formed by turning are called
rotations of the shape. The rotation can be described as a
fraction of a full turn or as an angle. If no direction is
given the turn is in an anticlockwise direction.**

Object and image are
congruent.

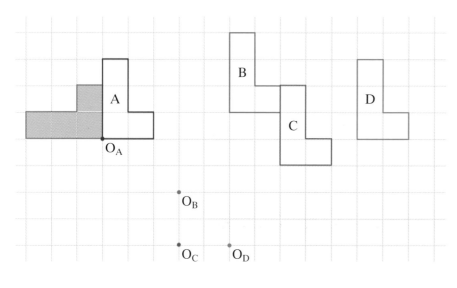

The shaded shape
rotated around point
O_A becomes A.

The shaded shape
rotated around point
O_B becomes B.

The shaded shape is the object. All the others are images and
they are all $\frac{3}{4}$ turns anticlockwise but they are not the same
rotation.

The point about which the turning occurs is important.

■   **The point about which the turning occurs is called the
centre of rotation.**

Example 2

Draw the image of each shape after it has been rotated $\frac{1}{4}$ turn using the point O as centre.

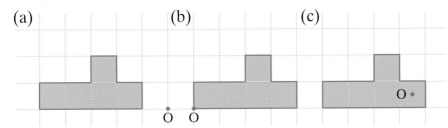

(a)   (b)   (c)

Tracing paper makes this task much easier. Once the tracing has been done fix the centre of rotation with a pencil so that point of the tracing paper does not move. The tracing paper can then be rotated (turned) $\frac{1}{4}$ turn.

With a little practice this is quite easy to do.

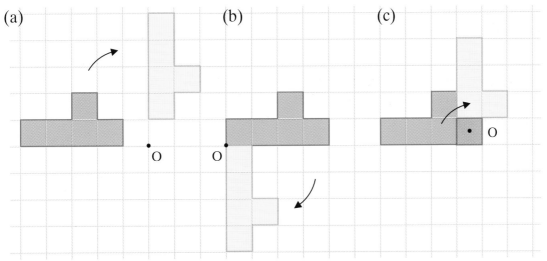

(a)   (b)   (c)

Notice that each line of the object has turned $\frac{1}{4}$ turn. This is obvious on this shape which is drawn on grid lines but would be true for diagonal lines as well.

Exercise 22B

1   Draw separate images for each shape after a rotation of 90° clockwise about each of the centres marked.

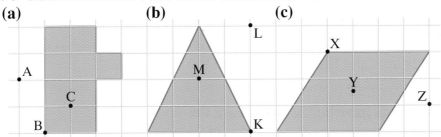

(a)   (b)   (c)

2 Copy these shapes and perform $\frac{1}{2}$ turns using the centres marked.

(a)   **(b)**   **(c)**

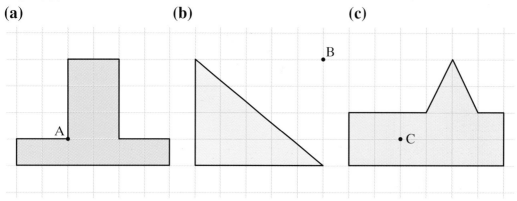

3 Each picture shows an object and its image. Copy each picture and write down how much the object has been rotated and the direction of the rotation. Try to identify the centre of the rotation.

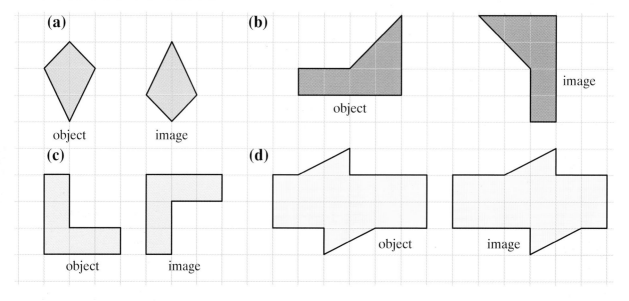

22.3 Reflection

When you next look in a mirror check to see that your image appears to be as far behind the mirror as you are in front. You can test this by moving nearer and further away from the mirror.

■ **Mathematical reflections have images which are the same distance behind the mirror line as the object is in front.**

Example 3

Reflect shape A in the mirror line drawn.

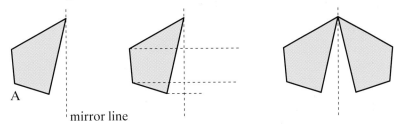

Step 1: Identify the image points. These are the same distance on the other side measured at right angles.

Step 2: Join up the points.

■ **Mirror lines are two-way. The mirror line may go through the object requiring reflections to go both ways.**

Here is a reflection which is two-way.

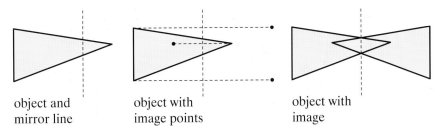

object and mirror line

object with image points

object with image

Example 4

Reflect the shape in the sloping mirror line.

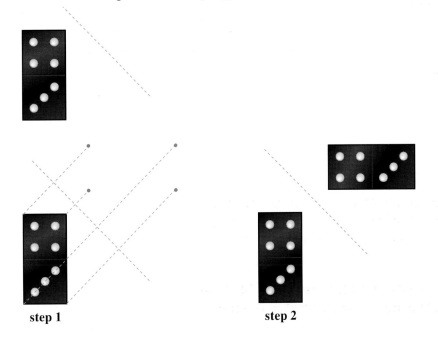

step 1

step 2

When the mirror line is sloping it is a good idea to turn the paper until the mirror line is vertical.

This way it is easier to find the images of the vertices.

Exercise 22C

1 Copy each shape onto squared paper and draw the image after reflection using the dotted line as the mirror line.

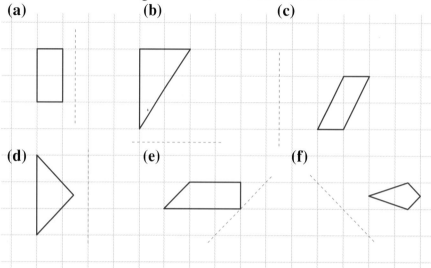

(a)   (b)   (c)

(d)   (e)   (f)

2 Each diagram shows an object with its image. Copy the diagrams onto squared paper and draw in the mirror line.

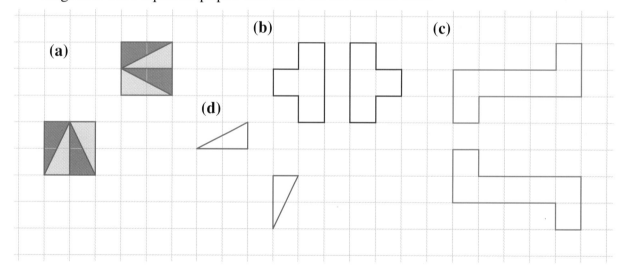

(a)   (b)   (c)

(d)

22.4 Enlargement

■ **In an enlargement all angles stay the same and lengths are changed in the same proportion.**

■ **The scale factor is the multiplier for lengths and perimeter.**

■ **The centre of enlargement determines the final position of the enlarged figure.**

When one shape is an enlargement of another the shapes are called **similar shapes**.

Example 5

Enlarge the shape *ABCD* by a scale factor of 2 using *A* as the centre of enlargement.

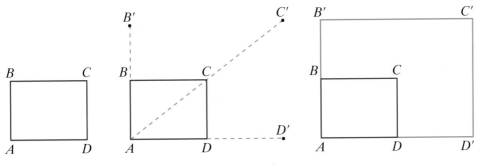

A' is at *A*.

Point *A* is the centre and is fixed.

All other points move 2 times as far away.

A'D' is the image of *AD*.

$A'D' = 2AD$ $A'C' = 2AC$ $A'B' = 2AB$

Example 6

Enlarge shape *EFGHJ* by a scale factor of 2 with the centre *C* marked inside the shape.

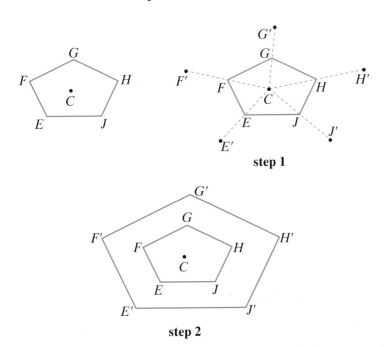

Again, all points move to image positions which are 2 times as far away from the fixed centre.

Example 7

In this example the centre O is outside the shape.

The scale factor is still 2.

Again, all points have their images twice as far away from the fixed centre as the object point.

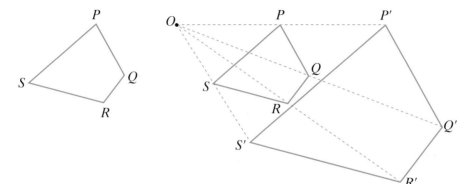

■ **In an enlargement image lines are parallel to their corresponding object lines.**

If the shape is drawn on a grid you can see if the lines which go along the grid lines are the right length by counting. Be sure to use the lengths.

You can check sloping lines by regarding them as a combination of sideways movement and up/down movement.

Example 8

Enlarge shape $KLMN$ by a scale factor of 2 and with the centre of enlargement at the origin.

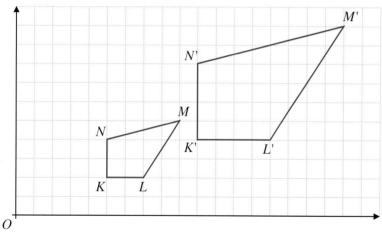

LM is 2 across and 3 up.

The image $L'M'$ will be 4 across and 6 up.

MN is 4 left and 1 down.

The image $M'N'$ will be 8 left and 2 down.

Exercise 22D

In this exercise each question has a scale factor and three possible centres of enlargement. Copy each diagram and draw the three enlargements either on the same diagram or on separate diagrams.

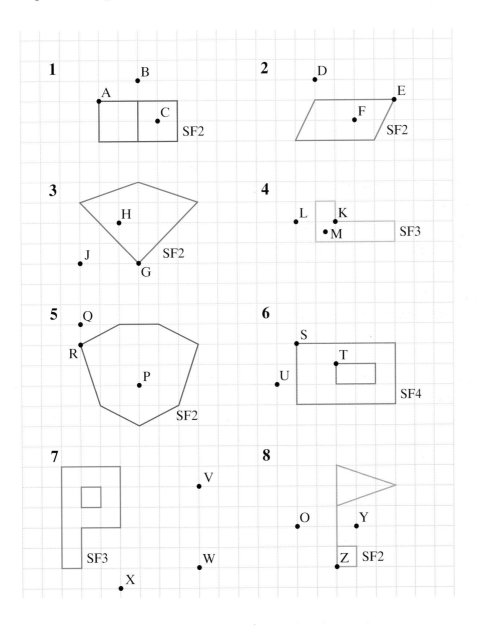

Exercise 22E

1 On a copy of the grid, enlarge the shaded shape by a scale factor of 3.
   Start your enlargement at point B.

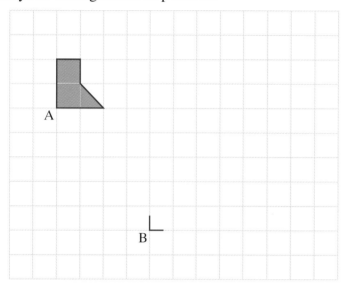

[E]

2 Copy and reflect each of the shapes in the mirror lines given.

   (a)

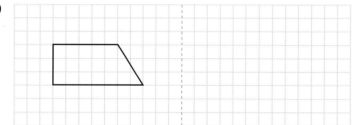

   (b)

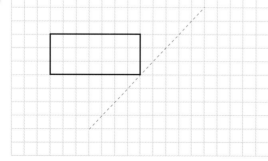

[E]

3 One drawing is an enlargement of the other.

(**a**) Find the centre of enlargement and label this point C.

(**b**) State the scale factor of the enlargement.

[E]

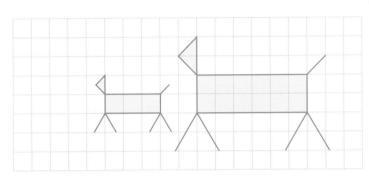

4 Make copies of the figures to answer the questions.

(**a**) Reflect the shape A in the mirror line.

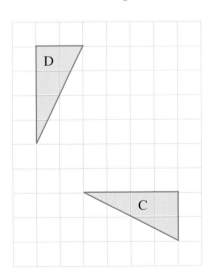

mirror line

Label the reflection B.

(**b**) Describe the transformation which maps the triangle C onto the triangle D.

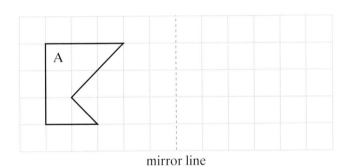

[E]

Summary of key points

1 Sliding movements are called translations.

2 Images of a shape which are formed by turning are called rotations of the shape.

3 The point about which the turning occurs is called the centre of rotation.

4 Mathematical reflections have images which are the same distance behind the mirror line as the object is in front.

5 Mirror lines are two-way. The mirror line may go through the object requiring reflections to go both ways.

6 In an enlargement all angles stay the same and lengths are changed in the same proportion.

7 The scale factor is the multiplier for the lengths and perimeter in an enlargement.

8 The centre of enlargement determines the final position of the enlarged figure.

9 In an enlargement, image lines are parallel to their corresponding object lines.

23 Probability

This morning there is a **chance** of heavy rain with the **possibility** of thunder. In the afternoon the rain will die away and it is likely that the sun will break through the clouds, **probably** towards evening.

Weather forecasts are made by studying weather data and using a branch of mathematics called probability.

Probability uses numbers to represent how likely or unlikely it is that an event such as 'a thunderstorm' will happen.

Probability is used by governments, scientists, economists, medical researchers and many other people to **predict** what is likely to happen in the future by studying what has already happened.

23.1 The probability scale

■ **An event which is certain to happen has a probability of 1.**
For example, the probability that night will follow day is 1.

■ **An event which cannot happen has a probability of 0.**
For example, the probability that you will grow to be 5 metres tall is 0.

All probabilities *must* have a value greater than or equal to 0 and less than or equal to 1.

This can be shown on a **probability scale**:

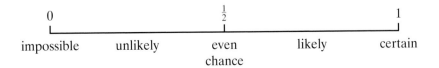

Exercise 23A

1 Draw a 0 to 1 probability scale and mark on it the probability that:
   (a) it will rain tomorrow
   (b) the sea will disappear
   (c) Boxing Day follows Christmas Day
   (d) you will buy a new pair of shoes soon
   (e) you will have homework tonight
   (f) the sun will not rise next week
   (g) a member of the class will be late tomorrow
   (h) if you toss a coin it will be tail uppermost
   (i) you will see a TV star on your way home.

2 Give two examples where you think the event is:
   (a) impossible   (b) unlikely
   (c) about an even chance   (d) likely
   (e) certain

23.2 Using numbers to represent probabilities

In Exercise 23A you may have found it difficult to know where to put some of the statements on the probability scale. It becomes easier if the probability is given a value.

When you toss a coin there are two possible outcomes: either a head or tail. One of these outcomes is tossing a head.

The probability P of tossing a head can be written:

Probability of a head $P(H) = \dfrac{\text{number of successful outcomes}}{\text{total number of possible outcomes}}$

$$= \frac{1 \text{ outcome (head)}}{2 \text{ possible outcomes (head or tail)}} = \frac{1}{2}$$

Another way of writing this is:

P(head) = $\frac{1}{2}$

Probabilities can be written as fractions, decimals or percentages.

P(head) = $\frac{1}{2}$
= 0.5
= 50%

■ **The probability that something will happen is:**

Probability = $\dfrac{\textbf{number of successful outcomes}}{\textbf{total number of possible outcomes}}$

assuming that the outcomes are all equally likely.

Probability is a measure of *how likely* it is that something will happen on a scale 0 to 1. If you toss a coin 10 times that does not mean you will get *exactly* 5 heads, but if you toss a coin 500 times it is *likely* that you will get about 250 heads.

If getting a head is a *success* then getting a tail is a *failure*.

Example 1

Find the probability of getting a 6 when a fair dice is rolled.

Probability P(6) = $\dfrac{\text{number of successful outcomes}}{\text{total number of possible outcomes}}$ = $\dfrac{1}{6}$

This is the number of successful outcomes.

There are six possible outcomes: 1, 2, 3, 4, 5, 6.

The word *fair* is used because each number has an equal chance of turning up: the outcomes are *equally likely*.

Exercise 23B

1 A fair six-sided dice is rolled.
   What is the probability of getting:

   (a) a 4   (b) a 1 or 2
   (c) an odd number   (d) an even number
   (e) a 5 or 6   (f) 3 or more
   (g) less than 5   (h) a prime number
   (i) more than 6   (j) $2\frac{1}{2}$?

2   A card is selected from a pack of 52 cards.
What is the probability it will be:

(a) an ace   (b) the ace of spades

(c) a black card   (d) a 5 or 6

(e) smaller than a 4   (f) a picture card

(g) a 3, 4 or 5   (h) a diamond

(i) a club or spade   (j) any card other than a club?

3   A token is taken from a bag containing 6 red and 5 blue tokens.
What is the probability that the token will be:

(a) a blue   (b) a red

(c) a yellow   (d) a red or blue?

23.3 Certain and impossible events

In question 3 of the previous exercise it is *impossible* to pick a yellow token. It is *certain* that you will pick either a red or blue token because they are the only colours available.

If you write these probabilities as fractions you find:

P(yellow) which is *impossible* is $\frac{0}{11}$ which is 0.

P(red or blue) which is *certain* is $\frac{11}{11}$ which is 1.

■ **The probability of an event happening is always greater than or equal to 0 (impossible) and less than or equal to 1 (certain). This can be written:**

$$0 \leq \textbf{Probability} \leq 1$$

Remember:
$a < 5$ a is less than 5
$a \leq 5$ a is less than or equal to 5

$5 > a$ 5 is greater than a
$5 \geq a$ 5 is greater than or equal to a

Exercise 23C

1   Write down the probability of the following:

(a) you will grow to be 5 metres tall

(b) Christmas Day is the 25th December

(c) it will rain tomorrow

(d) you will live to be 150 years old

(e) you will die

(f) if you toss a coin it will be a tail

(g) if you roll a dice it will be an odd number

2   Write three statements for each of the following.
The first one is done for you.

(a) a probability of 0
'A baby will be born with false teeth' has a
probability of 0.

(b) a probability of 1

(c) a probability of about $\frac{1}{2}$

(d) a probability of about $\frac{3}{4}$

3   Here are the nets of two differently numbered dice:

Dice A   Dice B

If both dice are rolled what is the probability that using
(i) Dice A   (ii) Dice B   you will get:

(a) a 6   (b) a 5

(c) a score of 4 or more   (d) a 2

(e) an even number   (f) an odd number

(g) a prime number   (h) a square number?

4   A bowl of fruit contains 3 apples, 4 bananas, 2 pears and 1
orange. Norma takes one piece of fruit without looking.
Write (i) as a fraction (ii) as a decimal and (iii) as a
percentage, the probability that she takes:

(a) an apple   (b) a banana

(c) a pear   (d) an orange.

The first one is done for you.

(a) P(apple) is   (i) $\frac{3}{10}$ (ii) 0.3   (iii) 30%

5   A hundred raffle tickets are sold. Gary buys 8 tickets,
Susan 5 tickets and Raman 12 tickets. What is the
probability that the first prize will be won by:
(a) one of these three   (b) Gary
(c) Susan   (d) somebody other than Raman?
Write each answer in three ways: as a fraction, a decimal,
and a percentage.

6   Meryl, James and Gita are playing Monopoly. The
probability that Meryl will win is $\frac{1}{3}$. The probability that
James will win is $\frac{1}{4}$. What is the probability that Gita will win?

23.4 The probability that something will *not* happen

The probability of rolling a 6 on a fair dice is $P(6) = \frac{1}{6}$,

The probability of *not* getting a 6 is:

$$P(\text{not a 6}) = \frac{5}{6}$$

There are 5 ways of not getting 6: 1, 2, 3, 4, 5

The six possible outcomes are still: 1, 2, 3, 4, 5, 6

Notice that: $P(6) = \frac{1}{6}$

$P(\text{not a 6}) = \frac{5}{6} = 1 - \frac{1}{6} = 1 - P(6)$

■ **If the probability of an event happening is p then the probability of it *not* happening is $1 - p$.**

Exercise 23D

1 (a) The probability it will rain tomorrow is $\frac{1}{3}$.
   What is the probability it will not rain?
   (b) The probability that it will snow on Christmas day is 0.2.
   What is the probability it will not snow on Christmas day?
   (c) The probability that Rovers will get to the next round of a competition is 30%.
   What is the probability that they will not get to the next round?

2 Susan rolls a fair six-sided dice. What is the probability that she will:
   (a) get a 2   (b) *not* get a 2
   (c) get 3 or more   (d) *not* get 3 or more
   (e) *not* get an even number?

3 The probability of Ahmed winning his game of chess is 0.62. The probability of him drawing is 0.24.
   (a) What is the probability of him losing?
   (b) If Ahmed plays two games what is the probability he will:
      (i) win both games
      (ii) draw both games?

23.5 Finding an estimated probability by experimenting

Sometimes you will need to estimate the probability that an event will happen. For example, you can estimate the probability of getting a head when you toss a coin by carrying out a trial:

Toss it several times and keep a record of:

- the number of successful trials (heads)
- the total number of trials (how many tosses altogether).

- **The estimated probability that an event will happen in a game or experiment is:**

  $$\text{Estimated probability} = \frac{\text{number of successful trials}}{\text{total number of trials}}$$

The estimated probability may be different from the theoretical probability.

- **The estimated probability is called the relative frequency that the event will happen.**

Example 2

If you toss a coin 20 times and get 12 heads and 8 tails the **estimated probability** of getting a head is: $\frac{12}{20} = \frac{3}{5}$

The **theoretical probability** is: $\frac{10}{20} = \frac{1}{2}$

From the experiment it would seem that you are more likely to get a head than a tail. Next time you do the same experiment you may well get a different result but the theoretical probability is *always* $\frac{1}{2}$.

If the estimated probability of Adrian winning a tennis match is $\frac{3}{4}$ then if he plays 24 matches he would expect to win 18 times.

So:
$\frac{3}{4} \times 24 = 18$

Exercise 23E

1 (a) Roll a dice 36 times and record your results in a frequency table like this. Work out the probability of rolling each number from your results.

Number	Tally	Frequency	Probability
1			$\overline{36}$
2			$\overline{36}$
3			$\overline{36}$
4			$\overline{36}$
5			$\overline{36}$
6			$\overline{36}$
		Total	$\overline{36}$

Use the table to answer these questions.
   (b) What is the probability of rolling:
       (i) a three
       (ii) a five?
   (c) What is the probability of rolling an even number?
   (d) What was the total of all the probabilities?
   (e) Explain your result to part (d).

2 Roll a dice another 36 times and complete a table as in question **1**.
   Compare your results with the previous question and make comments.

3 Combine your results for questions **1** and **2** in a table and comment on the probabilities you obtain.

4   Roll two dice 36 times, add the total spots on the uppermost faces and record your results in the table like this:

Number	Tally	Frequency	Probability
2			
3			
4			
5			
12			

Total 8

(a)   What is the probability of a score being:
    (i)   even   (ii)   greater than 8
    (iii)   a square number   (iv)   12
    (v)   7   (vi)   3
    (vii) 11   (viii) 1?

(b)   What is the total of the probabilities?

(c)   What is the probability that you will not score 12?

5   Roll two dice a further 36 times and complete a table as in question **4**.
Compare the results with those in question **4**.

6   Combine your results for questions **4** and **5** in a table and make comments on your results.

7   Using your frequency column data from question **6** draw a bar chart to show your result.

(a)   Which number occurred the most times?

(b)   Which number occurred the least?

(c)   Write two sentences about the shape of your chart.

(d)   Which of these numbers is most likely to occur:
    (i) 12   (ii) 6   (iii) 2?

(e)   Why did the number 1 not occur?

23.6 Using a sample space diagram to find theoretical probabilities

If you roll a red dice and a blue dice what are all the outcomes? If the red dice shows 1 the blue dice could show 1, 2, 3, 4, 5 or 6. You can record these outcomes as ordered pairs putting the red result first, then the blue result like this:

(1, 1) (1, 2) (1, 3) (1, 4) (1, 5) (1, 6) ⟍ red dice 1

blue dice 6

If the red dice showed 2 first the outcomes could be:

(2, 1) (2, 2) (2, 3) (2, 4) (2, 5) (2, 6)

You can represent all the outcomes on a **sample space diagram**:

```
      6 ┤ (1,6) (2,6) (3,6) (4,6) (5,6) (6,6)

      5 ┤ (1,5) (2,5) (3,5) (4,5) (5,5) (6,5)

      4 ┤ (1,4) (2,4) (3,4) (4,4) (5,4) (6,4)
Red dice
      3 ┤ (1,3) (2,3) (3,3) (4,3) (5,3) (6,3)

      2 ┤ (1,2) (2,2) (3,2) (4,2) (5,2) (6,2)

      1 ┤ (1,1) (2,1) (3,1) (4,1) (5,1) (6,1)
        └──┬─────┬─────┬─────┬─────┬─────┬──
           1     2     3     4     5     6
                    Blue dice
```

The sample space diagram shows that there are 36 possible outcomes.

■ **You can represent all possible outcomes on a sample space diagram.**

Example 3

Use a sample space diagram to find the probability of getting a total score of 7 when two dice are rolled.

```
      6 ┤ (1,6) (2,6) (3,6) (4,6) (5,6) (6,6)

      5 ┤ (1,5) (2,5) (3,5) (4,5) (5,5) (6,5)

      4 ┤ (1,4) (2,4) (3,4) (4,4) (5,4) (6,4)
Red dice
      3 ┤ (1,3) (2,3) (3,3) (4,3) (5,3) (6,3)

      2 ┤ (1,2) (2,2) (3,2) (4,2) (5,2) (6,2)

      1 ┤ (1,1) (2,1) (3,1) (4,1) (5,1) (6,1)
        └──┬─────┬─────┬─────┬─────┬─────┬──
           1     2     3     4     5     6
                    Blue dice
```

There are 6 ways of scoring 7:

(1, 6) (2, 5) (3, 4) (4, 3) (5, 2) (6, 1)

There are 36 possible outcomes so:

$$P(\text{scoring } 7) = \tfrac{6}{36}$$

Exercise 23F

1   Use the sample space diagram in Example 3 to find the
    probability of:
    (a) a total score of **(i)** 5 **(ii)** 9 **(iii)** 12
    (b) **not** rolling a score of **(i)** more than 4 **(ii)** less than 10
    (c) rolling a 2 on either one or both of the dice
    (d) rolling a double
    (e) rolling a double 6.

2   This sample space diagram shows the outcomes when the
    Ace, King, Queen, Jack, and ten from both the spades and
    hearts suits are placed in two separate piles and one card
    is taken from each pile:

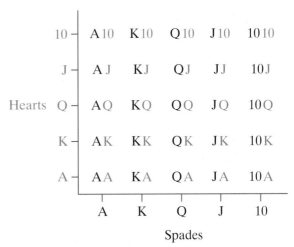

    Find the probability that:
    (a) both cards will be Kings
    (b) both of the cards could be either an Ace or a King
    (c) both cards will be a pair
    (d) at least one card will be an Ace
    (e) neither card will be a 10
    (f) neither card will be a King or Jack
    (g) one card will be a spade
    (h) both cards will be hearts.

3   Write out all the possible outcomes for each of the
    following:
    (a) tossing a coin twice
    (b) obtaining an odd or even number when a dice is rolled
        twice
    (c) when a coin is tossed and a dice is rolled

4 Work out the number of possible outcomes:
   (a) when a dice is thrown twice
   (b) when two discs are taken from a bag containing two red discs, two blue discs and one yellow disc, one at a time.

5 Four discs: red, green, yellow and blue are placed in a bag. One disc is taken at random, its colour recorded and then replaced. A second disc is then taken.
   Draw a sample space diagram to show all the possible outcomes.

6 (a) Copy and complete this table to show the possible outcomes when a red dice and a blue dice are rolled:

Total score	Ordered pairs	Theoretical probability
2	(1,1)	$\frac{1}{36}$
3	(1,2) (2,1)	$\frac{1}{36}$
4	(1,3) (2,2) (3,1)	$\frac{1}{36}$
5		
6		
7		
11		
12		

   (b) Which was the most likely score?
   (c) What is the sum of all the probabilities? Explain your answer.

7 Nicola is twice as likely to turn right at a T-junction as turn left. List all the possible outcomes after she has passed two junctions.

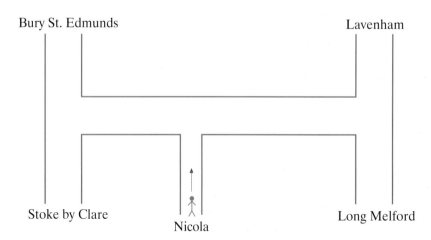

23.7 Using two-way tables to find probabilities

Another way to show possible outcomes is to use a two-way table. You can do this even if you do not know all the outcomes.

Example 4

This table shows some information of how staff got to school. None of them came by car.

	Walk	Bus	Cycle	Total
Monday	4	p	3	12
Tuesday	7	q	u	t
Total	r	9	s	25

(a) Fill in the missing numbers. The boxes are labelled to help you.

(b) What is the probability that one of these members of staff:
  (i)   walked to school on Monday
  (ii) cycled on Tuesday
  (iii) came by bus?

(a) Monday: $4 + p + 3 = 12$
        so the number who travelled by bus p is 5.

  Tuesday: The total who travelled by bus is $5 + q = 9$
        so the number who travelled by bus q is 4.

  The total r who walked is $4 + 7 = 11$.

  The total who walk + bus + cycle is 25
            $11 + 9 + s = 25$

  so the total s who cycle is 5.

  The total t for Tuesday is $25 - 12 = 13$

  The number u who cycle on Tuesday is $13 - (7 + 4) = 2$

The completed table is:

	Walk	Bus	Cycle	Total
Monday	4	5	3	12
Tuesday	7	4	2	13
Total	11	9	5	25

(b) (i) 4 walked out of a total of 12 arriving at school so the probability is $\frac{4}{12} = \frac{1}{3}$.

(ii) 2 cycled out of 13 so probability is $\frac{2}{13}$.

(iii) A total of 9 bus journeys over the two days so probability is $\frac{9}{25}$.

Exercise 23G

1 The number of clients interviewed by Jackie and Sam about their holiday is shown in this table:

	Monday	Tuesday	Total
Jackie	25	35	
Sam			
Total	55		100

(a) How many people did Jackie question?

(b) How many people did Sam question on Monday?

(c) How many people were questioned on Tuesday?

(d) Copy and complete the table.

(e) If a person is chosen at random what is the probability that they were questioned by:
   (i) Jackie on Monday
   (ii) Sam on Tuesday   (iii) Jackie
   (iv) Sam on Monday or Jackie on Tuesday?

> A **random** selection is one in which each person has the same chance of being chosen.

2 Jackie and Sam asked the 100 clients how they travelled to France and in which month they travelled. Some of the information is recorded in this table:

	Air	Le Shuttle	Boat	Total
July	8		10	20
August	15	5		50
September		8		
Total	30			100

(a) Copy and complete the table.

(b) What is the probability that a client selected at random travelled:

   (i) in July   (ii) in August or September   (iii) by air
   (iv) by boat   (v) by Le Shuttle in August   (vi) by boat in August
   (vii) not by air   (viii) not by boat in July?

3 The same 100 people were asked if they had their holiday in France, went on to Italy or went elsewhere. Some details are given in this table:

	France	Italy	Elsewhere	Total
July		10	4	20
August	18		8	
September		16		30
Total	28	50		

(a) Copy and complete the table.

(b) What is the probability that a randomly picked holidaymaker:
  (i) went to Italy
  (ii) went to France
  (iii) went in July
  (iv) did not go to Italy
  (v) went elsewhere in August
  (vi) went to France in September
  (vii) went to Italy in September?

4 The same 100 people were also asked in what type of accommodation they stayed. Part of the information is given in this table:

	Hotel	Caravan	Camping	Other	Total
July	11	4	3		
August		14		6	
September		7	4	3	30
Total	49		15	11	100

(a) Copy and complete the table.

(b) What is the probability a person selected at random:
  (i) stayed in a hotel in August
  (ii) stayed in a caravan in July
  (iii) stayed in a hotel
  (iv) did not stay in a hotel
  (v) stayed in a caravan
  (vi) used other accommodation
  (vii) went camping in August?

■ **Two-way tables can be used to help solve probability problems.**

Summary of key points

1 An event which is certain to happen has a probability of 1.

2 An event which cannot happen has a probability of 0.

3 The probability that an event will happen is:

$$\text{Probability} = \frac{\text{number of successful outcomes}}{\text{total number of possible outcomes}}$$

assuming all outcomes are equally likely.

4 The probability of an event happening is always greater than or equal to 0 (impossible) and less than or equal to 1 (certain). This can be written:

$$0 \leq \text{Probability} \leq 1$$

5 If the probability of an event happening is p then the probability of it not happening is $1 - p$.

6 The estimated probability that an event will happen in a game or experiment is:

$$\text{Estimated probability} = \frac{\text{number of successful trials}}{\text{total number of trials}}$$

The estimated probability is called the relative frequency that the event will happen.

7 Sample space diagrams and two-way tables can be used to help solve probability problems.

24 Calculators and computers

This unit shows you some ways of using scientific calculators, graphical calculators and computers to help you solve mathematical problems.

The examples will work on most Casio calculators and computers. Your teacher will tell you if you need to change any of the instructions.

24.1 How well do you know your calculator?

You can use your calculator to help you with fractions, decimals, percentages and money problems. Try these examples:

Examples 1 to 7 apply only to scientific calculators.

Example 1   Fractions and decimals

Work out $\frac{2}{5} + 1\frac{1}{4}$

Press the keys [2] [a%c] [5] [+] [1] [a%c] [1] [a%c] [4] [=]

Answer: $1\frac{13}{20}$

Example 2   Fractions and money

Work out $\frac{3}{8}$ of £6

Press the keys [3] [a%c] [8] [×] [6] [=]

Press [a%c] again to get £2.25

Example 3   Percentages

Work out 20% of 6 metres.

Press [6] [×] [2] [0] [SHIFT] [=]%

Answer: 1.2 metres

Example 4   Percentage increase or decrease

Reduce £18 by 10%

Press [1] [8] [×] [1] [0] [SHIFT] [=]% [−]

The calculator display is 16.2
The answer is £16.20

You can make your calculator show pounds and pence correctly by using [MODE] [FIX] [2]

This will always display two decimal places. Don't forget to return your calculator to *normal* mode after doing this.

Converting to fractions or decimals

Press [SHIFT] [a%c]% to convert your answer to an improper (top-heavy) fraction.

Press [a%c] again to convert the answer to a decimal.

Example 5   Constant calculations

Work out   £2.80 × 1.2
£8.90 × 1.2   Find out what happens when you press
£7.65 × 1.2   any of the keys +, −, × or ÷ *twice*

Press   `1` `.` `2` `×` `×` `2` `.` `8` `0` `=`
`8` `.` `9` `0` `=`
`7` `.` `6` `5` `=`

Answers: £3.36, £10.68, £9.18

Example 6   Memory calculations

Find the total cost of the following items, and calculate the change from £10

2 at £1.15,   3 at 98p,   and 1 at £1.79

Press   `2` `×` `1` `.` `1` `5` `=` `Min`
`3` `×` `0` `.` `9` `8` `=` `M+`
`1` `.` `7` `9` `=` `M+`
`MR`

The total is £7.03

Now press   `1` `0` `−` `MR` `=`

The change is £2.97

Example 7   Squares and powers

(a) Work out 47^2.

(b) Work out 7^6.

(a) Press   `4` `7` `x²`
The answer is 2209.

Try experimenting with different calculator keys. Can you work out what the `1/x`, `x¹ᐟʸ`, and `x!` keys do?

(b) Press   `7` `xʸ` `6` `=`
The answer is 117 649.

Brackets

You can use the brackets keys to make sure the calculator performs the operations in the right order.

On most scientific calculators the brackets keys look like this:

Example 8

Work out $\dfrac{7 + 11 + 6}{3}$

(a) by hand

(b) on a calculator.

(a) $\dfrac{7 + 11 + 6}{3} = \dfrac{24}{3} = 8$

(b) Press [(--- 7 + 1 1 + 6 ---)] ÷ 3 =

The answer is 8.

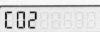

Try pressing

7 + 1 1 +
6 ÷ 3 =

The calculator is working out

$7 + 11 + \dfrac{6}{3}$.

Example 9

Work out $46(19 + 7)$

Press 4 6 × [(--- 1 9 + 7 ---)] =

The answer is 1196.

You can use brackets inside brackets.

Example 10

Work out $12 \left(7 + \dfrac{9 + 13}{2} \right)$

Press 1 2 × [(--- 7 + [(--- 9
+ 1 3 ---)] ÷ 2 ---)] =

The answer is 216.

Your calculator tells you how many brackets you've used:

[C02 88688]

This means you've used two sets of brackets.

Exercise 24A

1 Write down the keys you need to press to work out $1\frac{2}{3} - \frac{4}{5}$.

2 Use your calculator to work out:

(a) $\frac{9}{23} + \frac{6}{7}$ (b) $4\frac{1}{2} - \frac{7}{8}$ (c) $1\frac{1}{30} - \frac{5}{16}$ (d) $42\frac{4}{7} + 1\frac{2}{11}$

3 Work out the following fractions and percentages using a calculator.

(a) 10% of 309   (b) $\frac{3}{7}$ of 1000   (c) 19% of 46

(d) 72% of 4609   (e) $\frac{9}{10}$ of 1701   (f) $\frac{6}{17}$ of 320

4 Use your calculator to work the total cost of:

27 apples at 19p each
3lb carrots at 58p per lb
25 packets of batteries at £4.25 per packet
7 pints of milk at 42p per pint

Remember: You can use the memory keys so you don't need to write down any working.

5   Use your calculator to work out the following, giving your answers to 2 decimal places.

(a) $\dfrac{19 + 3}{7}$ (b) $\dfrac{42 + 16 + 9}{12}$ (c) $\dfrac{108 + 18}{81}$

(d) $6(13 + 3)$ (e) $14(12 + 6)$ (f) $29(18 + 21)$

(g) $\dfrac{4(6 + 19)}{3}$ (h) $71\left(6 + \dfrac{4 + 21 + 9}{18}\right)$ (i) $\dfrac{12(7 + 19)}{16}$

24.2 Using memories to represent formulae

Look at this problem:

Examples and exercises in Sections 24.2 to 24.6 apply only to graphical calculators.

Find the area and perimeter of a rectangle 9.83 cm by 5.21 cm using the calculator memories.

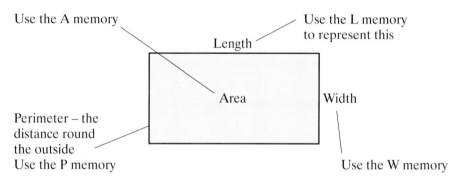

Use the A memory

Length

Use the L memory to represent this

Area

Width

Perimeter – the distance round the outside
Use the P memory

Use the W memory

Example 11

Use memories L, W, A and P to calculate the area and perimeter of a rectangle 9.83 cm by 5.21 cm.

Remember to press the
ALPHA key before each memory.

It is useful, but not essential, to store each answer in a memory for future use like this:

(a) Store the number 9.83 in memory L.
Press the keys   9 . 8 3 → L EXE

(b) Store the number 5.21 in memory W.
Press the keys   5 . 2 1 → W EXE

(c) Multiply memories L and W and store the answer in memory A.
Press the keys   L W → A EXE

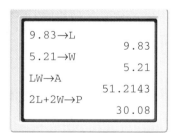

```
9.83→L
              9.83
5.21→W
              5.21
LW→A
           51.2143
2L+2W→P
             30.08
```

(d) Calculate the perimeter and store it in memory P.
Press the keys   2 L + 2 W → P EXE

Now try pressing the keys   L W EXE

This produces the same result for the area but the answer is not stored in a memory.

Example 12

Use the memories set up in Example 8.

Entering formulae:
Formulae are typed exactly as you would write them. Multiplication signs are not required between letters or before brackets.

(a) To find the area of a rectangle again, press:

(i) `L` `W` `EXE`   (ii) `W` `L` `EXE`

(iii) `L` `×` `W` `EXE`   (iv) `W` `×` `L` `EXE`

(b) To find the perimeter of a rectangle, press:

(i) `2` `L` `+` `2` `W` `EXE`

(ii) `2` `×` `L` `+` `2` `×` `W` `EXE`

(iii) `2` `(` `L` `+` `W` `)` `EXE`

(iv) `2` `×` `(` `L` `+` `W` `)` `EXE`

(a) All four answers are the same.

(b) All four answers are the same.

Exercise 24B

1 Calculate the area A and perimeter P of a rectangle 73 cm by 49 cm using any of the methods described above. Write down the sequence of keys as you press them.

2 Calculate the area A and circumference C of a circle whose radius R is 12 cm. You should store the value 12 in memory R, then use the formulae

$$\pi R^2 \rightarrow A$$

and $2 \pi R \rightarrow C$

To get π you might have to press first.
To get R^2 you might have to press

3 Calculate the area A of a triangle whose base B is 15 cm and whose height H is 9 cm, using the formula

$$(BH) \div 2 \rightarrow A$$

4 Calculate the volume V of a rectangular box whose length L is 120 cm, width W is 60 cm and height H is 20 cm using the formula

$$LWH \rightarrow V$$

24.3 Saving formulae so you can use them again

Find out how to *write*, *run* and *clear* programs on your graphical calculator.

If you want to use the same formula *several times*, you should *save* it as a **program**.

Then you can use it again with different numbers.

Select the WRT mode on the calculator, and choose a program number, say Prog 3.

Now type $\quad \pi R^2 \rightarrow A$

Select the RUN mode on the calculator and then press

You can improve this program by adding more lines which tell you what to do:

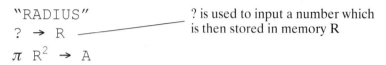

"RADIUS"

? → R ——————— ? is used to input a number which is then stored in memory R

π R^2 → A

To run this new program, press `Prog` `3` `EXE`

When the ? appears press

`1` `0` `EXE` `EXE` ——————— Notice that `EXE` is pressed twice.

`1` `1` `EXE` `EXE` — The first `EXE` enters the radius and Prog 3 produces the area.

`1` `2` `EXE` — The second `EXE` runs the program again.

You can do the same thing by writing a computer program in BASIC:

```
10 PRINT "RADIUS"
20 INPUT R
30 LET A = PI * R ^ 2
40 PRINT A
50 END
```

```
RUN
RADIUS
?
10
314.1592654
RUN
```

Exercise 24C

1 Write a program for your graphical calculator or computer which will calculate the volume V of a cube whose sides are of length L. Use your program to find the volumes of cubes whose sides are 5 cm, 6 cm and 7 cm.

2 Write a program for your graphical calculator or computer which will convert a temperature in degrees Fahrenheit (°F) to degrees Celsius (°C) using the formula
$C = (F - 32) \times \frac{5}{9}$
Use your program to convert 86° F, 50° F, 68° F to degrees Celsius.

24.4 Solving problems by trial and improvement

Look at this problem.

Find the radius of a circle whose area is 100 cm^2.

radius r = ?

The area A is 100cm^2

Sometimes you can solve problems like this by guessing a solution and trying it out. You can find the radius by guessing it and trying it in the formula $A = \pi r^2$

You can improve your guess and try again until you get a good enough answer.

This method of solving problems is called **trial and improvement**.

You can use a graphic calculator or computer to help you do this.

> You can solve problems by trial and improvement using a normal calculator. You have to record your results by hand in a table.

Example 13

Use a trial and improvement method to find the radius of a circle whose area is $100\,\text{cm}^2$.

Your answer should be accurate to 2d.p.

Method:
Guess a value for R
Calculate πR^2
Compare your answer with 100
Decide whether your chosen value for R was too big or too small.
Make a better choice for R and repeat the process until you achieve the accuracy you need.

Solution:

Type   ? → R : πR^2 EXE

You *must* include the colon `:` or replace it by SHIFT EXE

> Notice that EXE is pressed twice.
>
> The first EXE enters the radius and the area is calculated.
>
> The second EXE repeats the process and another ? appears

When the ? appears, enter your first guess. Suppose this is $R = 10\,\text{cm}$.

Type 1 0 EXE EXE

The answer, 314.159 265 4 is too big, so now try a smaller value for R

Type 5				EXE EXE	giving	78.539 816 34	too small
Type 7				EXE EXE	giving	153.938 04	too big
Type 6				EXE EXE	giving	113.097 335 5	too big
Type 5 · 5				EXE EXE	giving	95.033 177 77	too small
Type 5 · 7 5				EXE EXE	giving	103.868 907 1	too big
Type 5 · 6 5				EXE EXE	giving	100.287 491 5	too big
Type 5 · 6 4				EXE EXE	giving	99.932 805 67	too small
Type 5 · 6 4 2				EXE EXE	giving	100.003 692 6	too big

You can now see that the radius R lies between 5.64 cm and 5.642 cm, so:

$\quad R = 5.64\,\text{cm}$ to 2d.p.

You can do the same thing by writing a computer program in BASIC:

```
10  INPUT R
20  PRINT PI * R ^ 2
30  GOTO 10
40  END
```

```
RUN
?
5
78.53981634
?
7
153.93804
?
```

Try it using the numbers 5, 7, 6, 5.5, 5.75 and so on until you decide when to stop.

Exercise 24D

1   Use a trial and improvement method to find the length L of the side of a cube whose volume V is 100 cm³.
The formula is $V = L^3$.
Hint: Find out how to enter L^3 (L raised to the power 3) on your graphical calculator.
Your answer should be accurate to 2 d.p.

2   Use ? → F : (F − 32) × 5 ÷ 9   to convert a temperature of 22° C to Fahrenheit (°F) by trial and improvement. Stop at 1 d.p. accuracy.

24.5 Using the [Ans] and [EXE] keys to produce number sequences

You can use these keys together to generate number sequences. For example:

```
1
                        1.
Ans + 1
                        2.
                        3.
                        4.
                        5.
```

Press [1] [EXE]

Press [Ans] [+] [1] [EXE] [EXE] [EXE] ... keep pressing [EXE]

The calculator appears to be 'counting'.

Each time [EXE] is pressed, 'Ans + 1' is calculated, where Ans is the *last displayed answer*.

Example 14

[Ans] recalls the most recent answer

(a) Use the [Ans] and [EXE] keys to produce the Even Numbers, starting with 2.

(b) Show how the [Ans] and [EXE] keys can be used to produce the sequence 2, 6, 18, 54, ...

(a) [2] [EXE] [Ans] [+] [2] [EXE] [EXE] [EXE] . . .

(b) [2] [EXE] [Ans] [×] [3] [EXE] [EXE] [EXE] . . .

[EXE] performs (or repeats) the most recent calculation(s).

Exercise 24E

Write down the key presses, including  and , which will generate the following sequences:

1 1, 3, 5, 7, 9, . . .
2 5, 10, 15, 20, 25, . . .
3 2, 4, 8, 16, 32, . . .
4 3, 9, 27, 81, 243, . . .
5 10, 9, 8, 7, 6, . . .
6 16, 8, 4, 2, 1, . . .
7 200, 20, 2, 0.2, 0.02, . . .
8 −5, −7, −9, −11, −13, . . .

24.6 Drawing graphs on your calculator

Choose the default ranges for x and y by pressing

`Range` `INIT` `Range`

On some calculators you may need to press

`Range` `SHIFT` `Mcl` `Range`

Clear the graphics screen by pressing `SHIFT` `Cls`

Have a look at the graphics screen by pressing `G↔T`.

Both axes are marked off in intervals of 1.

Find out how to set the ranges of x and y to your own chosen values.

Example 15

Draw the graphs of $y = x + 1$ and $y = x - 1$

Press the `Graph` key, then press `x` `+` `1` `EXE`

You will see graph $y = x + 1$

The first graph is drawn. To see the second graph, press the

`Graph` key again and press `x` `−` `1` `EXE`

If your calculator has a key marked `x, θ, T` press it to get `x`.

Notice that the second graph is drawn on the same axes as the first without erasing the first. It is superimposed. This will continue to happen with further graphs until you *clear the graphics screen*.

Clear the screen by pressing `SHIFT` `Cls`.

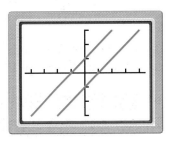

Example 16

Draw the graphs of $y = 2x + c$ for several different values of c in the range -4 to 4.

Type "C" : ? → C : Graph Y = 2X + C `EXE`

When the ? appears, enter your values for c

Press: `-` `4` `EXE` (wait until the graph is drawn)

`EXE` `-` `2` `EXE` (wait until the graph is drawn)

`EXE` `0` `EXE` (wait until the graph is drawn)

`EXE` `2` `EXE` (wait until the graph is drawn)

`EXE` `4` `EXE` (wait until the graph is drawn)

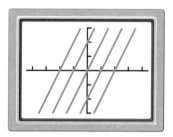

Clear the screen to finish.

Exercise 24F

In each question draw all the graphs on the same axes and write down what you see.

Clear the screen before starting the next question.

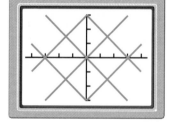

1   $y = x,$ $y = x + 3,$ $y = x - 3,$
    $y = -x,$ $y = -x + 3,$ $y = -x - 3$

2   $y = x,$ $y = 2x,$ $y = 3x,$
    $y = -x,$ $y = -2x,$ $y = -3x$

3   $y = x^2,$ $y = x^2 + 1,$ $y = x^2 - 1,$
    $y = -x^2,$ $y = -x^2 + 1,$ $y = -x^2 - 1$

24.7 Investigating number sequences with a spreadsheet

Examples and exercises in Sections 24.7 to 24.9 apply only to spreadsheets.

You can generate many number sequences on the same spreadsheet and compare them.

Find out how to enter numbers and formulae in the cells of your spreadsheet.

Find out how to copy a formula from one cell to other cells.

Example 17

Generate the whole numbers from 1 to 10 in column A.

Put the number 1 in cell A1.
Put the formula A1 + 1 in cell A2.
Copy the formula in A2 down column A as far as A10.

Example 18

Generate the even numbers in column B.

Put the number 2 in cell B1.
Put the formula B1 + 2 in cell B2.
Copy the formula in B2 down column B as far as B10.

Example 19

Generate the odd numbers in column C.

Put the number 1 in cell C1.
Put the formula C1 + 2 in cell C2.
Copy the formula in C2 down column C as far as C10.

Formula
A2 = A1 + 1

Formula
B2 = B1 + 2

Formula
C2 = C1 + 2

	A	B	C	D
1	1	2	1	
2	2	4	3	
3	3	6	5	
4	4	8	7	
5	5	10	9	
6	6	12	11	
7	7	14	13	
8	8	16	15	
9	9	18	17	
10	10	20	19	

Example 20

Generate the triangular numbers in column D.

Put the number 1 in cell D1.
Put the formula D1 + A2 in cell D2.
Copy the formula in D2 down column D as far as D10.

Example 21

Add consecutive odd numbers and put the answers in column E.

Put the formula C1 + C2 in cell E2.
Copy the formula in E2 down column E as far as E10.

You should now be able to see that the numbers in column E are multiples of 4.

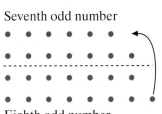

Seventh odd number

Eighth odd number

Example 22

Add consecutive triangular numbers and put the answers in column F.

Put the formula D1 + D2 in cell F2.
Copy the formula in F2 down column F as far as F10.

What is the name of the sequence of numbers in column F?

Fourth triangular number

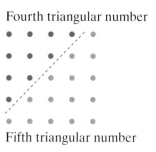

Fifth triangular number

| Formula | | | | Formula | | Formula | |
| D2 = D1 + A2 | | | | E2 = C1 + C2 | | F2 = D1 + D2 | |

	A	B	C	D	E	F	G
1	1	2	1	1			
2	2	4	3	3	4	4	
3	3	6	5	6	8	9	
4	4	8	7	10	12	16	
5	5	10	9	15	16	25	
6	6	12	11	21	20	36	
7	7	14	13	28	24	49	
8	8	16	15	36	28	64	
9	9	18	17	45	32	81	
10	10	20	19	55	36	100	

24.8 Problem solving with a spreadsheet

Look at this problem:

A farmer has 200 metres of fencing. He wants to use all the fencing to enclose a rectangular area of his field for his animals to graze. Find the length and width of the rectangle which give his animals the maximum grazing area.

You can solve problems like this on a computer by using a spreadsheet.

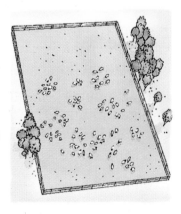

Example 23

Think of all the rectangles you can draw whose perimeters are 200 metres. For example, some could be long and thin; others short and wide.

Use a spreadsheet to find the length L and width W of the rectangle which has the maximum area. (There is more about this in Section 24.2.)

The perimeter of each rectangle is 200 metres, so

$$2L + 2W = 200$$

You can divide by 2 to make this equation simpler

$$L + W = 100$$

Use your spreadsheet to try lots of values for L from $L = 0$ to $L = 100$ metres.

Increase L by 10 metres each time. For each value of L calculate a value for W using

$$W = 100 - L$$

Then multiply the values of L and W to get the area of each rectangle.

		A	B	C	D
	1	L	W	LW	
	2	0	100	0	
	3	10	90	900	
	4	20	80	1600	
	5	30	70	2100	
	6	40	60	2400	
	7	50	50	2500	
	8	60	40	2400	
	9	70	30	2100	
	10	80	20	1600	
	11	90	10	900	
	12	100	0	0	

Formula
A3 = A2 + 10

(W = 100 − L)
Formula
B2 = 100 − A2

(A = L × W)
Formula
C2 = A2 × B2

Make sure your spreadsheet has at least 12 rows and 3 columns. Use column A for the length, column B for the width and column C for the area of each rectangle.

You can now find the maximum value for the area in column C quite easily.

Find out how your spreadsheet produces graphs.

Drawing graphs of the data in the spreadsheet can give you a better understanding of how the area changes as the lengths and widths of the rectangles change.

Rectangles can be long and thin or short and wide, but the rectangle with the biggest area is SQUARE!

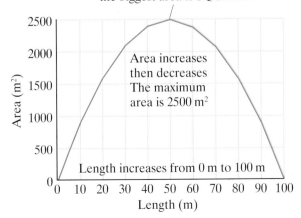

Area increases then decreases
The maximum area is 2500 m²

Length increases from 0 m to 100 m

24.9 Trial and improvement on a spreadsheet

Example 24

The width of a rectangle is 2 cm less than the length. Use a trial and improvement method to find the length when the area of the rectangle is 30 cm². Your answer should be accurate to 2 d.p.

Use L for the length and W for the width of the rectangle.

Put the length in column A, the width in column B and the area of the rectangle in column C.

Make sure your spreadsheet has at least 12 rows and 3 columns.

Use your spreadsheet to try lots of values for L.

For each value of L calculate a value for W using: $W = L - 2$

Then multiply each value of L by the corresponding value of W to find the area of each rectangle.

Now you can look for the two areas which are nearest to 30 cm².

Use these answers to try new values for L and repeat the process until you achieve the accuracy you need.

This is much easier than it looks.
Remember that the computer is doing all the really hard work.

Values of L from 0 to 10 in steps of 1

	A	B	C
1	L	W	LW
2	0	–2	0
3	1	–1	–1
4	2	0	0
5	3	1	3
6	4	2	8
7	5	3	15
8	6	4	24 too small
9	7	5	35 too big
10	8	6	48
11	9	7	63
12	10	8	80

Values: A2 = 0
Formulae: A3 = A2 + 1
B2 = A2 – 2
C2 = A2 × B2

Values of L from 6 to 7 in steps of 0.1

	A	B	C
1	L	W	LW
2	6	4	24
3	6.1	4.1	25.01
4	6.2	4.2	26.04
5	6.3	4.3	27.09
6	6.4	4.4	28.16
7	6.5	4.5	29.25 too small
8	6.6	4.6	30.36 too big
9	6.7	4.7	31.49
10	6.8	4.8	32.64
11	6.9	4.9	33.81
12	7	5	35

Values: A2 = 6
Formulae: A3 = A2 + 0.1
B2 = A2 – 2
C2 = A2 × B2

Values of L from 6.5 to 6.6 in steps of 0.01

	A	B	C
1	L	W	LW
2	6.5	4.5	29.25
3	6.51	4.51	29.360
4	6.52	4.52	29.470
5	6.53	4.53	29.580
6	6.54	4.54	29.691
7	6.55	4.55	29.802
8	6.56	4.56	29.913 too small
9	6.57	4.57	30.024 too big
10	6.58	4.58	30.136
11	6.59	4.59	30.248
12	6.6	4.6	30.36

Values: A2 = 6.5
Formulae: A3 = A2 + 0.01
B2 = A2 – 2
C2 = A2 × B2

Values of L from 6.56 to 6.57 in steps of 0.001

	A	B	C
1	L	W	LW
2	6.56	4.56	29.913
3	6.561	4.561	29.924
4	6.562	4.562	29.935
5	6.563	4.563	29.946
6	6.564	4.564	29.958
7	6.565	4.565	29.969
8	6.566	4.566	29.980
9	6.567	4.567	29.991 too small
10	6.568	4.568	30.002 too big
11	6.569	4.569	30.013
12	6.57	4.57	30.024

Values: A2 = 6.56
Formulae: A3 = A2 + 0.00
B2 = A2 – 2
C2 = A2 × B2

You can now see that the length L lies between 6.567 cm and 6.568 cm.

So $L = 6.57$ cm correct to 2 d.p.

25 Scatter diagrams

Statements such as 'Smoking can cause lung cancer' and 'Drink driving causes accidents' are often made in the media. Sometimes they are supported by data showing, for example, whether there is a relationship between a smoking habit and the chance of getting lung cancer.

This section shows you how to compare two sets of data to see whether there is a relationship between them. For example, is there a relationship between people's heights and their hand spans? Daniel did an experiment to find out. Here is his data:

Height (cm)	177	175	165	164	172	160	170	180	164	175	171	169
Hand span (cm)	26	25	20	19	25	18	22	27	21	23	24	23

He plotted each pair of values (177, 26), (175, 25) and so on, on a graph. This is called a **scatter diagram** or **scatter graph**.

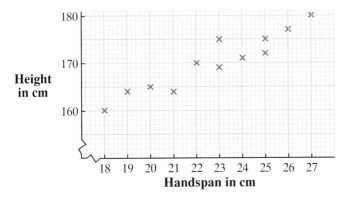

There does appear to be a relationship between height and hand span: the taller you are, the larger your hand span.

Here is another example. On a journey Chandra noted down how many miles there were still to go. She did this every ten minutes:

Time (mins)	10	20	30	40	50	60	70	80	90	100
Miles to go	72	61	54	46	40	32	24	18	7	0

Here is a scatter diagram showing this data:

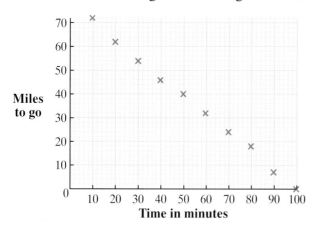

There is a relationship between the time she has been travelling and the number of miles still to go: the greater the time travelled the fewer miles there are to go.

Sometimes there is no relationship between two sets of data: Every Monday Trish recorded the temperature in °F and the rainfall in mm. Her results were:

Temperature (°F)	74	70	63	68	65	64	60	51	54	56	50
Rainfall (mm)	1	0	2	7	5	1	8	2	11	4	6

A scatter graph of this data looks like this:

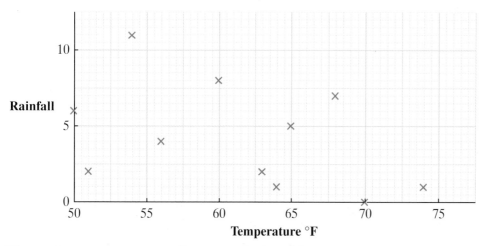

There does not appear to be any relationship.

Correlation

■ **The relationship between two sets of data is called a correlation.**

You need to be able to recognise different types of relationships between two sets of data. Look at these three scatter graphs:

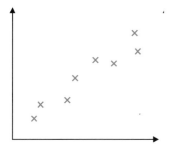

As one value increases the other one also increases. There is a **positive correlation**.

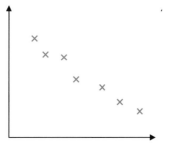

As one value increases the other decreases. There is **negative correlation**.

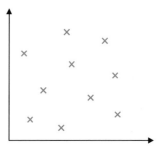

The parts are randomly and widely spaced out. There is **no correlation**.

Line of best fit

When there is positive or negative correlation on a scatter diagram you can draw a *line of best fit* to describe the relationship between the two sets of data.

■ **The line of best fit is a straight line that passes through or is as close to as many of the plotted points as possible.**

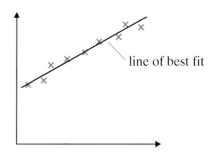

line of best fit

For your GCSE exam you should use a ruler and draw your line of best fit by eye.

Example 1

Jade counted the number of ice-creams sold at a kiosk each day for two weeks and compared it with the average temperature.

No. of ice-creams sold	2	7	1	55	48	20	30	15	32	42	37	23	25	11
Average temperature (°C)	12	13	10	20	21	16	17	13	19	20	19	15	16	14

(a) Draw a scatter graph to illustrate this data.

(b) Draw and label the line of best fit on your scatter graph.

(c) Use your line of best fit to predict the average temperature on a day when 17 ice creams are sold.

(a), (b)

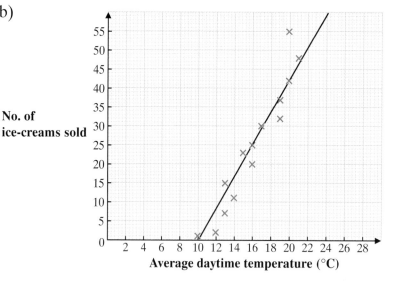

(c) Read across from 17 to the line of best fit, and then down to the temperature axis. Using the line of best fit you can predict that the temperature will be 14 °C.

Exercise 25A

1 Pupils' marks in two maths papers were:

Paper I	37	57	69	74	50	85	45	70	52	65	30	80
Paper II	25	50	70	68	45	76	36	72	48	60	24	74

   (a) Draw a scatter graph to represent this data.
   (b) What type of correlation do you find?

2 What type of correlation would you expect if you compared the following data:
   (a) heights of people and shoe size
   (b) ages of cars and their selling prices
   (c) time taken to get to school and marks in French
   (d) sizes of gardens and the numbers of birds in them
   (e) marks in science and marks in maths?

3 Write down an example which you might expect would give each of the following correlations:
   (a) negative
   (b) none
   (c) positive.

4 A group of students went to a fitness centre. Their heights and weights were recorded and they were rated 0–40 on a number of activities. The results are recorded in this table:

Weight	65	60	72	66	61	56	68	61	62	58	57	64
Height	175	172	182	179	174	165	176	168	172	170	162	175
Agility	24	19	32	26	23	18	34	12	35	28	16	21
Strength	29	22	32	35	26	18	25	14	30	26	20	33
Reaction	35	17	28	33	15	27	19	22	29	37	31	25
Skipping	26	34	20	18	36	28	22	36	32	28	38	16

(a) Draw scatter graphs for each of the following:
  (i) weight and height   **(ii)** weight and agility
  (iii) height and skipping   **(iv)** strength and reaction
  (v) agility and reaction   **(vi)** strength and skipping.

(b) For each of your scatter graphs state what type of correlation if any is shown.

(c) If possible, draw and label a line of best fit on each of your scatter graphs.

> **Remember:** You can't draw a line of best fit if there is no correlation.

5 Here are sketches of six scatter diagrams:

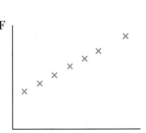

Which diagram(s) show:

(a) positive correlation

(b) negative correlation

(c) no correlation?

6   This is a scatter diagram showing students' percentage
scores in Paper 1 and Paper 2 of a mathematics
examination:

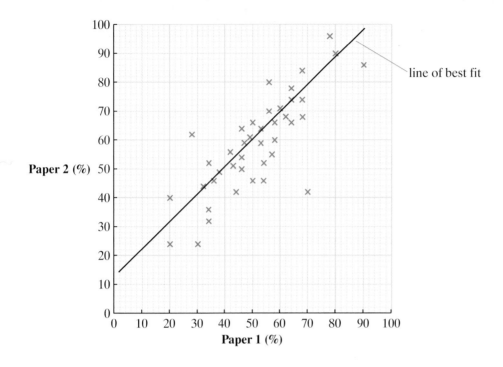

(a) What type of correlation does this diagram show?

Student A scored 43% on Paper 1, but did not take
Paper 2.

(b) Use the line of best fit to estimate the percentage the
student might have scored on Paper 2.

[E]

7   For each pair of variables below, state whether you think
there would be:

   positive correlation

or **negative** correlation

or **no** correlation.

Give a brief reason for your choice.

(a) *The amount of rain falling*   and   *the number of people
outdoors*.

(b) *The amount of apples a person ate*   and   *the person's
results in mathematics tests*.

[E]

8 This table gives you the marks scored by pupils in a French test and in a German test:

French	15	35	34	23	35	27	36	34	23	24	30	40	25	35	20
German	20	37	35	25	33	30	39	36	27	20	33	35	27	32	28

(a) Draw a scatter graph of the marks scored in the French and German tests on a grid like this:

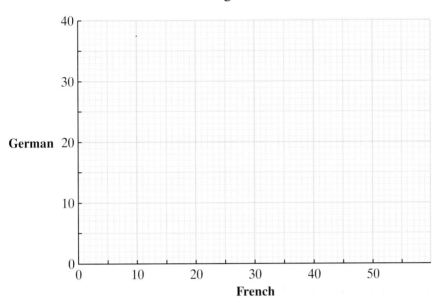

(b) Describe the correlation between the marks scored in the two tests.

(c) Draw and label the line of best fit on your scatter graph.

(d) Use your line of best fit to predict the German mark of a pupil who scored 28 in their French test.

[E]

Summary of key points

1 The relationship between two sets of data is called a correlation.

2 The line of best fit is a straight line that passes through or is as close to as many of the plotted points as possible.

Examination practice paper

Non-calculator

1 There were 3604 people watching Accrington Stanley playing football last Saturday.
(a) Write this number in words.
(b) Write this number to the nearest hundred.

(2)

2 Work out the value of:
(a) $3a + b$ when $a = 4$ and $b = 5$
(b) cd when $c = 2$ and $d = 7$

(3)

3 (a) Write down the name of these shapes:

(i)   (ii)

(4)

(b) Draw in any lines of symmetry that the shapes may have.

(2)

(iv)

(c) Write down any order of rotational symmetry that the shapes may have.

(iii)

(2)

4 Jenny and Simon threw a dice 30 times and obtained the following scores:

5, 6, 1, 4, 3, 6, 3, 5, 3, 2, 4, 6, 5, 4, 3,
3, 2, 1, 3, 4, 6, 4, 6, 2, 1, 4, 5, 6, 3, 4

(a) Design a suitable table and use it to work out the frequency of each score.
(b) Draw a bar graph to show your results.
(c) Work out the:
  (i) modal score   (ii) mean score   (iii) range.

(10)

5 In the local elections the following votes were cast.

George Evans 3641, Shenna Read 5643,
Precious Davies 4904, Sybil Jones 3667

Arrange the votes cast in order, largest first.

(2)

6 Work out
    (a) $3^2 + 2^3$
    (b) $2 \times 7 + 3$

                                                   (3)

7 **(a)** Write down the two missing numbers in this number pattern.

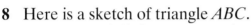

          $3, 8, 13, \ldots, \ldots, 28$
    (b) Write down in words the rule that you used to work out your answers.
    (c) Work out the 20th member of the pattern.
    (d) Write down the algebraic rule to give the nth term of the pattern.

                                                   (8)

8 Here is a sketch of triangle ABC.
    (a) Draw a full size accurate drawing of the triangle.
    (b) Measure the size of angle B.
    (c) By taking suitable measurements from your drawing work out its:
       (i) perimeter
      (ii) area.

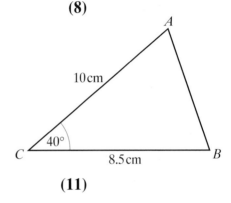

                                                  (11)

9 Jasmine has a bag containing 12 beads. 5 of them are white, 4 are black and the rest are red. If one bead is taken from the bag write down the probability that it will be:
    (a) white
    (b) not black
    (c) blue.

                                                   (4)

10 Work out the area of this shape by counting squares.

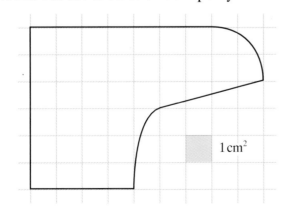

1 cm²

                                                   (2)

11 A coin and a dice are thrown at the same time. Make a list of all the possible outcomes of the event.

(2)

12 In this question do *not* use a calculator but show *all* of your working.
   (a) Simon buys 24 packets of stamps, each of which contains 144 stamps. How many stamps does he have in total?
   (b) The shopkeeper has a box with 600 stamps in it. He makes packets of stamps that each contain 36 stamps from the stamps in the box. How many packets of stamps can he make?

(6)

13 Solve these equations to find the value of the letter.
   (a) $2a + 3 = 9$
   (b) $4(2s - 5) = 7$
   (c) $2(3c + 4) = 5(2c - 7)$

(6)

14 (a) Work out:
     (i) $\frac{3}{5}$ of 270
     (ii) $\frac{3}{5} + \frac{7}{10}$
   (b) Simplify fully $\frac{6}{15} \times \frac{5}{12}$

(6)

15 (a) Complete these tables for the lines
    $y = 2x - 1$ and $y = x^2 - 3$

x	−3	−2	−1	0	1	2	3
$y = 2x - 1$	−7		−3		1		5

x	−3	−2	−1	0	1	2	3
$y = x^2 - 3$	6		−2	−3			6

   (b) Use the tables to draw the graphs of the two lines. Use an x-axis labelled from −3 to +3 using a scale of 2 cm for one unit and a y-axis labelled from −8 to +6 using a scale of 1 cm for one unit.

(7)

16 Boris travelled 200 miles at an average speed of 50 miles per hour.
   (a) How long did the journey take him?

He completed the return journey in 4 hours 30 minutes.

   (b) At what average speed did he travel on the return journey?
   (c) How many kilometres was the whole journey?

(6)

17 On a copy of the probability line below mark the probability of the following events happening:
   (a) it snows in London in August, with an S
   (b) a coin when thrown comes down heads, with a C
   (c) you have been born, with a B.

(3)

18

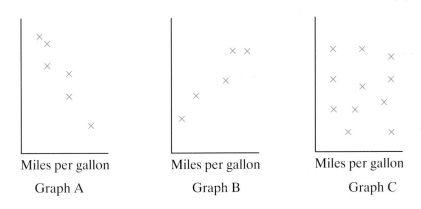

Graph A   Graph B   Graph C

   (a) For each of the three graphs write down the types of correlation they show.

   The 3 scatter graphs connect the number of miles travelled on one gallon of petrol with 3 different measurements.

   (b) Write down possible measures for the vertical axes in Graphs A, B and C.

(6)

19 Share £250 in the ratio 1 : 4.

(2)

20 The scale of a map is 1 : 50 000. Two villages are 12 cm apart on the map. How far apart are they in real life?

(3)

Examination practice paper

Calculator

1 Read the scales:

(a)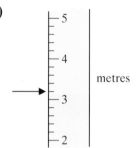

(b)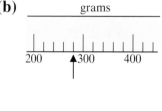

(c)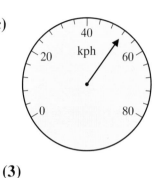

(3)

2 **(a)** Copy and complete this bill:

£   p

2 tins of paint @ £7.98 per tin
3 paint brushes @ £1.20 per brush
1 bottle of brush cleaner @ £1.50 per bottle   _____
                          Total

(b) How much change is left from £25.00?

(4)

3 The table gives the highest and lowest temperatures in several cities during one year.

	London	Manchester	Exeter	Newcastle
Highest	31 °C	29 °C	33 °C	30 °C
Lowest	−12 °C	−16 °C	−9 °C	−14 °C

(a) Which city recorded the lowest temperature?

(b) Which city recorded the smallest difference between the highest and lowest temperatures?

(2)

4 **(a)** Give the mathematical name of shape *BCDE*.

(b) What type of angle is *BXA*?

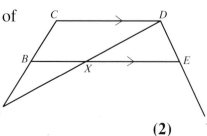

(2)

5 Find the numbers in the cloud, write down:

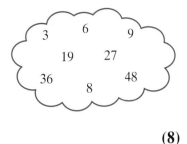

    (a) the square numbers

    (b) the factors of 12

    (c) the multiples of 4

    (d) the prime numbers.

    (8)

6 Use your calculator to work these out, giving your answers to 1d.p.

    (a) $(15.7)^2 + \sqrt{3.76} - 1.9$

    (b) $\dfrac{13.6 \times 2.17}{0.805}$

    (5)

7 Simplify:

    (a) $12a - 3b - 7a + b$

    (b) $3(2a + 3b)$

    (c) $x(2x + y)$

    (6)

8 Work out:

    (a) the bearing of B from A

    (b) the bearing of A from B.

    (3)

9 Storage boxes are available in 4 colours: red, green, blue and grey. The table shows the probability that a customer choosing a box at random will select a particular colour:

Colour	Red	Green	Blue	Grey
Probability		0.1	0.2	0.1

What is the probability of red being chosen?

    (2)

10 In a survey, the method of travel of 720 students travelling
to school was recorded. The table shows the information.

Method of travel	Number of students
Walk	
Car	280
Bus	176
Bicycle	
Total	720

Use the information to copy and complete this table and
pie chart.

(4)

11 Work out the value of *a*.
Give reasons for each stage of your working.

(3)

12 6 marker pens weigh 220 g.
Find the weight of 15 marker pens.

(3)

13 From the stem and leaf diagram, work out:
(a) the median
(b) the mode
(c) the range.

Stem	Leaf
10	8, 8, 9
20	0, 5, 5, 5, 7, 9
30	1, 3, 6, 8, 9
40	0, 2, 5

(5)

14 Henri has the choice of buying the same training shoes in three different shops. Work out the cost of the shoes in each shop and give the name of the shop with the best buy.

Ben's shoes

£30 with

25% off

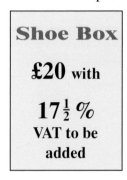

Shoe Box

£20 with

$17\frac{1}{2}\%$

VAT to be added

Saleo'Shoes

£34 with

$\frac{1}{3}$ off

(7)

15 (a) Draw the 4th diagram.

1st   2nd   3rd

(b) Copy and complete the table:

Diagram No.	1	2	3	4	5	6
Number of dots	5					

(c) How many dots are there in the 101st diagram?

(3)

16 (a) The distance between London and Exeter is 170 miles. Calculate an estimate for the distance in kilometres.
(b) William weighs 83 kg. Calculate an estimate for William's weight in lbs.

(4)

17 (a) Describe fully the transformation that maps A onto B.
(b) Translate B by 2 units to the right and 4 units up. Call the new position C. Draw a diagram to show C.
(c) Write down the coordinates of the centre of the rotation that maps C onto A?

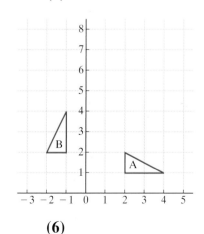

(6)

18 Write down, in its simplest form, an expression for the perimeter of the rectangle shown.

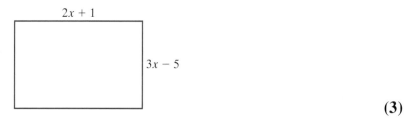

2x + 1

3x − 5

(3)

19 Factorise:

(a) $4a + 10b$

(b) $p^2 + 2pq$

(3)

20 Draw plan and elevation views of the shape in the diagram.

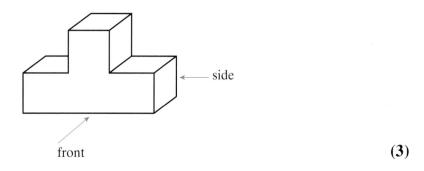

side

front

(3)

21 **(a)** Work out the coordinates of the mid-point of AB.

(b) Write down the coordinates of two points, C and D, which would complete the rectangle $ABCD$.

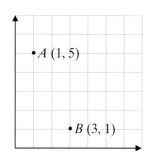

•A (1, 5)

•B (3, 1)

(4)

22 **(a)** Anita goes on holiday to Malta.
The exchange rate is £1 = 0.61 Lire.
She changes £227 into Lire.
How many Lire should she get?

(b) When she comes back she has 15 Lire to change back
into £.
The exchange rate is the same.
How much should she get?

(4)

23 The table shows the engine size and the cost of some cars
when new.

Engine size (cc)	1600	1800	2000	3000	1400	1700	1000
Cost (£)	14 500	18 000	21 000	32 000	13 000	15 000	9500

(a) Draw a scatter graph for the information.

(b) Draw a line of best fit on your scatter diagram.

(c) Use your line of best fit to estimate the cost, when
new, for a car with engine size 2400 cc.

(5)

24

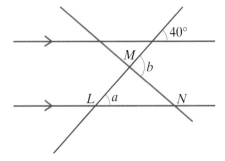

(a) **(i)** Find the value of angle *a*.

(ii) Give a reason for your answer.

LMN is an isoceles triangle.

(b) **(i)** Work out angle *b*.

(ii) Give reasons for your answer.

(5)

25 The diameter of each of the wheels on this rickshaw is
    45 cm.

    Work out the circumference of each wheel.

(3)

Formulae sheet: Foundation tier

Area of trapezium $= \dfrac{1}{2}(a + b)h$

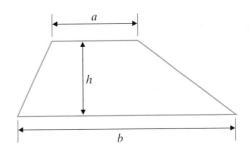

Answers

Unit 1 Number

Exercise 1A

1

	Ten thousands	Thousands	Hundreds	Tens	Units
(a)		4	*x*	*x*	*x*
(b)				3	*x*
(c)	*x*	*x*	1	*x*	*x*
(d)			*x*	*x*	9
(e)		*x*	*x*	0	*x*
(f)	*x*	*x*	4	*x*	*x*
(g)			7	7	7
(h)		6	*x*	*x*	6

Exercise 1B

1 **(a)** Thirty six **(b)** Ninety five **(c)** Five hundred and ninety eight
(d) Two hundred and forty six **(e)** Five thousand six hundred
and twenty three **(f)** One thousand two hundred and fifty one
2 **(a)** Seven hundred and nine **(b)** Eight hundred and ninety
(c) Six thousand and fifty four **(d)** Nine thousand two hundred
and one **(e)** Twenty six thousand and seven **(f)** Forty thousand
two hundred **(g)** Thirty two thousand **(h)** Seventy thousand
and ninety **(i)** Thirteen thousand four hundred and six
3 **(a)** 63 **(b)** 708 **(c)** 7000 **(d)** 18 600
(e) 75 000 **(f)** 809 000 **(g)** 4 000 000 **(h)** 1 001 000
(i) 9020 **(j)** 40 600
4 **(a)** Nine million nine hundred and ten thousand
(b) Four hundred and three thousand
(c) Thirty seven million nine hundred thousand
(d) Nine million nine hundred and thirteen thousand
(e) Fifty four million two hundred thousand
5 **(a)** 69   1010   2306 **(b)** 45 270   83 663   80 774
(c) 294 000   682 000   2 600 000   3 990 000
6 **(a)** 200   114   104   88   86   79
(b) 330 000   30 300   3033   3003   3000
(c) 6 102 000   6 000 006   990 000   660 000   600 006
7 7   23   56   93   234   469   614
8 Ford Fiesta   £9 010   Rover 623   £19 435
Volkswagen Polo   £9 110   Ford Granada   £20 270
Vauxhall Astra   £10 530   Jaguar Sport   £30 479
Citroën Xantia   £17 065   Marcos Mantera   £31 007
Volvo 940   £17 815   Mercedes Benz   £31 394
9 **(a)** 1984 **(b)** 1961 **(c)** 1982 **(d)** 1984
10 **(a)** Belgium   Thirty thousand five hundred and thirteen
Luxembourg   Two thousand five hundred and seventy six
Spain   Five hundred and four thousand seven
                   hundred and eighty two
Portugal   Ninety two thousand and eighty two
France   Five hundred and forty seven thousand and
                   twenty six
(b) France   Spain   Portugal   Belgium   Luxembourg

Exercise 1C

1 **(a)** 7 **(b)** 9 **(c)** 19 **(d)** 13 **(e)** 15
2 **(a)** 9 **(b)** 1 **(c)** 11 **(d)** 3 **(e)** 7
3 **(a)** 4 **(b)** 8 **(c)** 2 **(d)** 2 **(e)** 3
4 **(a)** 3 **(b)** 7 **(c)** 9 **(d)** 15 **(e)** 11
5 **(a)** increase of 5
(b) increase of 6
(c) decrease of 5
(d) decrease of 9
(e) decrease of 20

Exercise 1D

1 43 **2** 80 **3** 331 **4** 270 **5** 45 **6** 96
7 84 **8** 216 **9** 401 **10** 416

Exercise 1E

1 1003 **2** 3166 **3** 305 **4** 12
5 **(a)** 19 **(b)** 17 **(c)** 36
6 **(a)** 88 **(b)** 59 **(c)** 133
7 **(a)** £21 898 **(b)** £22 960 **(c)** £17 874

Exercise 1F

1 1080 **2** 4494 **3** 456 **4** 4558 **5** 2268
6 **(a)** 1500 m **(b)** 2800 m **(c)** 18 000 m   56 000 m
7 36   108   216
8 **(a)** 204 miles **(b)** 2040 miles **(c)** 10 608 miles

Exercise 1G

1 **(a)** 21 **(b)** 15 **(c)** 40 **(d)** 20 **(e)** 14
2 £480 **3 (a)** 10 **(b)** 30 **(c)** 36
4 **(a)** 7 **(b)** 13 **(c)** 36
5 **(a)** 12 **(b)** 41 **(c)** 80

Exercise 1H

1 33 **2** 368 **3** 2016 **4** 49
5 **(a)** 3 **(b)** 2 **(c)** 8 **(d)** 12
6 **(a)** 90 **(b)** 19
7 **(a)** 561 **(b)** 23 **(c)** 6 **(d)** 1683
8 **(a)** 576 **(b)** 458

Exercise 1I

1 **(a)** 60 **(b)** 60 **(c)** 190 **(d)** 190 **(e)** 990
(f) 2410
2 **(a)** 300 **(b)** 700 **(c)** 2400 **(d)** 3100 **(e)** 8800
(f) 29 500
3 **(a)** 2000 **(b)** 36 000 **(c)** 29 000 **(d)** 322 000
(e) 717 000 **(f)** 2 247 000
4 **(a)** 1030   1020   850   960   990   820
(b) 83 700   80 800   67 000   65 900   53 300   45 300
(c) 30   30   20   30   40   30

5

Area	Population
132 000	9 800 000
301 000	57 400 000
34 000	14 300 000
357 000	80 200 000
70 000	3 500 000

6 **(a)** 150 to nearest fifty **(b)** 300 to nearest hundred
(c) 6300 to nearest hundred **(d)** 35 000 to nearest thousand
(e) £45 000 000 to nearest million pounds
(f) 13 300 to nearest hundred

Exercise 1J

1 **(a)** 10 **(b)** 50 **(c)** 4 **(d)** 200
(e) 5000 **(f)** 500 **(g)** 3000 **(h)** 70
(i) 6000 **(j)** 9000 **(k)** 80 000 **(l)** 80 000
2 200 **3** 2000

4 (a) $\dfrac{60 \times 60}{30} = 120$ **(b)** $\dfrac{200 \times 300}{150} = 400$

(c) $\dfrac{10 \times 30 \times 100}{300} = 100$ **(d)** $\dfrac{2000}{10 \times 100} = 2$

(e) $\dfrac{500}{10 \times 50} = 1$ **(f)** $\dfrac{100 \times 90}{20 \times 30} = 15$

5 (a) 378.35294 $\dfrac{200 \times 100}{50} = 400$

(b) 22.155242 $\dfrac{10 \times 1000}{500} = 20$

(c) 107.91304 $\dfrac{140 \times 50}{70} = 100$

(d) 9.2592593 $\dfrac{1000}{10 \times 10} = 10$

(e) 5.6772083 $\dfrac{5200}{130 \times 10} = 4$

(f) 6.3631905 $\dfrac{900 \times 80}{40 \times 300} = 6$

6 (a) $50 \times 100 = 5000$ **(b)** Bigger – numbers both rounded up
7 $1350 \div 50 = 27$ **8** $900 \times 400 \div 1000 = 360$ Answer wrong
9 $16 \times 2 \times 195 = 30 \times 200 = 6000$ **10** $120 \times 40 \times 50 = 240\,000$

Exercise 1K

1 2 18 1110 73 536 500 000 **2** 537 811 36 225
3 17 19 23 29 31 37 41
4 (a) 450 **(b)** 71 **(c)** 71

Exercise 1L

1 (a) 3 or 9 **(b)** 2, 3, 4, 6, 8, 12 or 24 **(c)** 2, 4, 8, 16 or 32
(d) 5, 11 or 55 **(e)** 2, 3, 4, 6, 9, 12, 18, 27, 36, 54 or 108
(f) 5, 25, 125 or 625
2 1, 2, 3, 4, 6, 12
3 (a) 1, 2, 4, 8, 16, 32 **(b)** 1, 2, 4, 5, 8, 10, 20, 25, 40, 50, 100, 200
(c) 1, 2, 4, 5, 10, 17, 20, 34, 68, 85, 170, 340
(d) 1, 2, 4, 5, 8, 10, 20, 25, 40, 50, 100, 125, 200, 250, 500, 1000
4 (a) 1 and 2 **(b)** 1 and 3 **(c)** 1 and 5
(d) 1 and 2 **(e)** 1 and 7
5 (a) 1, 2, 5 and 10 **(b)** 1, 2, 5 and 10 **(c)** 1, 2 and 4
(d) 1, 2, 3 and 6 **(e)** 1, 3 and 9
6 (a) 1, 2, 5 or 10 **(b)** 1, 2 or 4 **(c)** 1, 5, 7 and 35
(d) 1, 2, 3 and 6 **(e)** 1, 2, 3 and 6
7 9, 18, 27 . . . **8** 99
9 (a) 54, 60, 66, etc **(b)** 1020, 1040, 1060, etc
(c) 105, 120, 135, etc
10 (a) 16, 24, 32, etc **(b)** 150, 300, 450, etc.
(c) 3, 15, 27, etc
11 10 20 ~~30~~ 40 50
  ~~60~~ 70 80 ~~90~~ 100
  110 ~~120~~ 130 140 ~~150~~
  ~~160~~ 170 ~~180~~ 190 200
  ~~210~~ ~~220~~ 230 ~~240~~ 250
(f) 180

Exercise 1M

1 (a) 16, 25, 36 **(b)** 100, 400, 900 **(c)** 81, 144, 196
(d) 64, 125, 216 **(e)** 512, 1000, 8000 **(f)** 64, 125, 216
2 ① 2 3 ④ 5 6 7 ~~8~~ ⑨ 10
  11 12 13 14 15 ⑯ 17 18 19 20
  21 22 23 24 ㉕ 26 ~~27~~ 28 29 30
  31 32 33 34 35 ㊱ 37 38 39 40
  41 42 43 44 45 46 47 48 ㊾ 50
  51 52 53 54 55 56 57 58 59 60
  61 62 63 ㉔ 65 66 67 68 69 70
  71 72 73 74 75 76 77 78 79 80
  ㊁ 82 83 84 85 86 87 88 89 90
  91 92 93 94 95 96 97 98 99 ⑩⓪
(d) 1 and 64
3 (a) 2 **(b)** 10 **(c)** 6 **(d)** 5 **(e)** 20
4 (a) 6 **(b)** 9 **(c)** 15 **(d)** 3 **(e)** 8

Exercise 1N

1 (a) 25 **(b)** 8 **(c)** 10 000 **(d)** 1 000 000 **(e)** 27
2 (a) 16 **(b)** 100 **(c)** 125 **(d)** 10 000
(e) 10 000 000 **(f)** 1 **(g)** 1 **(h)** 9
3

Power of 10	Index	Value	Value in words
10^1	**1**	10	**Ten**
10^2	**2**	**100**	A hundred
10^3	3	**1000**	**A thousand**
10^4	4	**10 000**	**Ten thousand**
10^5	5	**100 000**	**A hundred thousand**
10^6	6	1 000 000	**A million**
10^7	7	**10 000 000**	**Ten million**
10^8	**8**	**100 000 000**	A hundred million

Exercise 1O

1 (a) 5, –10 **(b)** 0, –13 **(c)** 13, –15 **(d)** –2, –21
2 (a) 0, –1 **(b)** –2, –5 **(c)** 3, 7 **(d)** –7, –12
(e) –15, –24 **(f)** –1, 2
3 (a) 7 **(b)** –3 **(c)** –1 **(d)** –5 **(e)** –6 **(f)** 3
(g) 0 **(h)** –7 **(i)** –5 **(j)** 0
4 (a) –40 **(b)** –70 **(c)** 30 **(d)** –30 **(e)** –160
(f) 270 **(g)** –30 **(h)** 0 **(i)** 120 **(j)** –400
5 (a) Minsk **(b)** Minsk **(c)** Tripoli
6 18°C **7** 6°C

Exercise 1P

1 (a) 2 **(b)** 2 **(c)** –1 **(d)** 0 **(e)** –3 **(f)** 3
(g) –6 **(h)** –6 **(i)** –7 **(j)** –9 **(k)** –5 **(l)** –9
2 (a) 2 **(b)** –2 **(c)** 2 **(d)** –3 **(e)** –3 **(f)** –2
(g) 1 **(h)** 7 **(i)** –1 **(j)** 3 **(k)** 2 **(l)** 3
3 (a) 6 **(b)** –20 **(c)** 10 **(d)** –12 **(e)** 4 **(f)** 18
(g) 30 **(h)** 21 **(i)** –30 **(j)** –14 **(k)** –42 **(l)** 15

Unit 2 Algebra 1

Exercise 2A

1 $a + 3$ **2** $x + 4$ **3** $x + 7$ **4** $b - 2$ **5** $c - 3$
6 $q - p$ **7** $x + y$ **8** $a + 4$ **9** $3b - 6$

Exercise 2B

1 $6a$ **2** $4p$ **3** $5b$ **4** $6q$ **5** $2c$ **6** $3n$ **7** $5w$
8 $5y$ **9** $7z$ **10** $7a$ **11** $5b$ **12** $8p$
13 $a + a + a + a$ **14** $p + p$ **15** $p + p + p + p + p$
16 $a + a + a + a + a + a + a$ **17** $y + y$ **18** $q + q + q + q$
19 $c + c + c$ **20** $d + d + d + d + d$
21 $a + a + a + a + a + a + a + a + a + a$
22 $h + h + h + h + h$ **23** $g + g + g + g + g + g$
24 $z + z + z + z$

Exercise 2C

1 $6a$ **2** $7b$ **3** $7c$ **4** $2d$ **5** $4e$ **6** $4f$ **7** $7a$
8 $8a$ **9** $12c$ **10** $14g$ **11** $4g$ **12** $3s$ **13** $14q$
14 $18p$ **15** $6p$

Exercise 2D

1 $7a + 6b$ **2** $9m + 7n$ **3** $p + 5q$ **4** $6c + 16e$
5 $12p + 2y$ **6** $4a + 13g$ **7** $9k + q$ **8** $d + 10f$

9 $3h + 10$ **10** $2e + 5f$ **11** $4g + 7n$ **12** $4 + 2g$
13 $14a$ **14** $5b$ **15** $4c$ **16** $5d$ **17** $7a + b$
18 $8b + 6c$ **19** $p + 1$ **20** $14y + 4z$ **21** $17a$
22 0 **23** $7s$ **24** a

Exercise 2E

1 $p \times q$ **2** $r \times s \times t$ **3** $2 \times e \times f$ **4** $5 \times a \times b \times c$
5 $7 \times k \times l \times m$ **6** $9 \times a \times b$ **7** $15 \times a \times b \times c$
8 $3 \times p \times q \times r \times s$ **9** $16 \times s \times t$ **10** $6 \times y \times z$
11 $8 \times d \times e \times f \times g$ **12** $20 \times a \times b \times c \times d$ **13** pq
14 efg **15** rst **16** $2ef$ **17** $2cd$ **18** hds **19** $2sf$
20 $3def$ **21** $4pq$ **22** $3hj$ **23** $5kv$ **24** $12rst$

Exercise 2F

1 $8ab$ **2** $15cd$ **3** $12pq$ **4** $20st$ **5** $30fg$
6 $28pq$ **7** $36mn$ **8** $24abc$ **9** $24rst$ **10** $20pq$
11 $12ab$ **12** $40pqr$ **13** $12st$ **14** $28pt$ **15** $45ce$
16 $64gqr$ **17** $48dr$ **18** $60st$ **19** $30ty$ **20** $40rst$

Exercise 2G

1 (a) b^3 (b) p^2 (c) r^7 (d) s^5 (e) q^4 (f) c^5
2 (a) a^4 (b) s^3 (c) t^6 (d) v^3 (e) f^5 (f) y^5
3 (a) $a \times a \times a$ (b) $a \times a \times a \times a$ (c) $d \times d$
 (d) $e \times e \times e \times e \times e$ (e) $f \times f \times f \times f$
 (f) $p \times p \times p \times p \times p$ (g) $a \times a \times a \times a \times a \times a \times a$
 (h) $s \times s$ (i) $k \times k \times k \times k \times k \times k$ (j) $n \times n \times n$
 (k) $n \times n \times n \times n \times n \times n \times n \times n$
 (l) $a \times a \times a \times a \times a \times a \times a \times a \times a \times a$
4 (a) a^5 (b) c^6 (c) d^2 (d) e^3 (e) f^7
5 (a) 9 (b) 8 (c) 16 (d) 625 (e) 32
 (f) 64 (g) 16 (h) 128 (i) 216 (j) 125
 (k) 25 (l) $100\,000$

Exercise 2H

1 (a) 9 (b) 1 (c) 25 (d) 13 (e) 21 (f) 15
 (g) 6 (h) 4 (i) 8 (j) 5 (k) 6 (l) 2
 (m) 1 (n) 5 (o) 16
2 (a) $4 + 5 = 9$ (b) $4 \times 5 = 20$ (c) $(2 + 3) \times 4 = 20$
 (d) $(3 - 2) \times 5 = 5$ (e) $(5 - 2) \times 3 = 9$ (f) $5 \times 2 \times 3 = 30$
 (g) $5 \times 4 + 5 + 2 = 27$ (h) $5 \times 4 + 5 - 2 = 23$
3 (a) 49 (b) 25 (c) 243 (d) 123 (e) 72 (f) 17
 (g) 22 (h) 7 (i) 7 (j) 0 (k) 7 (l) 0

Exercise 2I

1 $2p + 2q$ **2** $3c + 3d$ **3** $5y - 5n$ **4** $3t + 3u$
5 $14p + 7q$ **6** $6a - 4b$ **7** $8a + 4b$ **8** $3a - 6b$
9 $12r - 15s$ **10** $10a - 70b$ **11** $24s + 16t$
12 $30p + 20q - 10r$ **13** $36a + 48b$ **14** $28s - 35t$
15 $15a - 12b + 6c$

Exercise 2J

1 $5a + 5b$ **2** $11a + 6b$ **3** $18a - b$ **4** $7p + 7q$
5 $26a + 2c$ **6** $18p$ **7** $s + 44t$ **8** $2d + 33e$
9 $9b + 7z$ **10** $8a + 11b + c$ **11** $17a - b$ **12** $12p + 4q$
13 $22a$ **14** $5g + 24h$ **15** $21a + 5b + 10c$

Exercise 2K

1 (a) 5 (b) 17 (c) 10 (d) 4 (e) 4 (f) 4
 (g) 0 (h) 14
2 (a) $-p - q$ (b) $-p + q$ (c) $-a - b - c$ (d) $-a - b + c$
 (e) $-r - s$ (f) $-r + s$ (g) $-p - q + r$ (h) $-p + q - r$

Exercise 2L

1 $3a + 2b$ **2** $4p + q$ **3** $3a + 7$ **4** $y - 4z$
5 $7r + 10s$ **6** 7 **7** $6s + 7t$ **8** $6a - 11b$ **9** $5n$
10 $12h - 23k$ **11** c **12** $a - 5b$

Exercise 2M

1 $3a + 4b$ **2** $4p + 3q$ **3** $3a + 17$ **4** $y - 2z$
5 $7r + 14s$ **6** 23 **7** $6s + 23t$ **8** $6a + 19b$ **9** $7n$
10 $12h - 7k$ **11** $c + 12d$ **12** $a + 5b$

Exercise 2N

1 $5a$ **2** a^6 **3** $6p$ **4** p^5 **5** $11p$ **6** $7k$
7 $11s$ **8** $13t$ **9** $-2d$ **10** $-3j$ **11** $2a + 2b$
12 $15p - 5q$ **13** $4s + 8t$ **14** $36p - 8p + 20r$
15 $23m + 13n$ **16** $18s + 2t$ **17** $12a - 23b$ **18** $4p + 11d$
19 $9a - 2b$ **20** $4b$ **21** $11t - 30$ **22** $5p - 12q$ **23** $5n$
24 $25a - 22b$ **25** $-a - 7b$

Exercise 2O

1 2×10 or 4×5 **2** 2×3 **3** 2×2 **4** $p \times q$
5 $p \times q \times r$ **6** $x \times y$ **7** $3 \times a \times b$ **8** $4 \times a$
9 $a \times a$ **10** $b \times b \times b$ **11** $6 \times r \times s$ **12** $8 \times e \times f$

Exercise 2P

1 (a) $x(x + 3)$ (b) $a(a - b)$ (c) $p(p + q)$ (d) $3(a + 4b)$
 (e) $5(a + 2)$ (f) $2(b - 2c)$ (g) $4(1 + 2a)$ (h) $2(a - 1)$
 (i) $3(a + 3)$ (j) $5(p + 5)$ (k) $4(a + 4)$ (l) $4(p - 2)$
 (m) $7(x - 2)$ (n) $7(y + 1)$ (o) $y(7y + 1)$ (p) $5(q - 3)$
 (q) $x(x + 2)$ (r) $y(y + 3)$ (s) $3(a - 1)$ (t) $a(2a^3 + 3)$
 (u) $x(3y - 4z)$ (v) $a(4a - 5)$ (w) $a(5a^4 - 4)$ (x) $x(5x + 4)$

Exercise 2Q

1 $6(x + 7)$ **2** $5(y + 4)$ **3** $6(p + 3)$ **4** $3(x + 9)$
5 $5(q + 6)$ **6** $5(4t + 1)$ **7** $4(2t - 9)$ **8** $x(2x + 3)$
9 $p(2p + 1)$ **10** $t(t - 3)$ **11** $y(5y - 7)$ **12** $c(4 - 3c)$

Exercise 2R

1 $12, 15; +3$ **2** $25, 30; +5$ **3** $5, 6; +1$ **4** $35, 42; +7$
5 $18, 24; +6$ **6** $40, 50; +10$ **7** $13, 15; +2$
8 $16, 19; +3$ **9** $23, 28; +5$ **10** $17, 21; +4$

Exercise 2S

1 $9, 6; -3$ **2** $12, 8; -4$ **3** $15, 10; -5$ **4** $21, 14; -7$
5 $13, 10; -3$ **6** $17, 12; -5$ **7** $13, 11; -2$ **8** $13, 9; -4$
9 $18, 13; -5$ **10** $17, 10; -7$

Exercise 2T

1 $27, 81; \times 3$ **2** $64; \times 4$ **3** $5, 625; \times 5$
4 $1000, 10\,000; \times 10$ **5** $24, 48; \times 2$ **6** $54, 162; \times 3$
7 $128, 512; \times 4$ **8** $2000, 20\,000; \times 10$ **9** $250; \times 5$
10 $375; \times 5$

Exercise 2U

1 $16, 2; \div 2$ **2** $16; \div 4$ **3** $1000, 100; \div 10$ **4** $25; \div 5$
5 $12, 6; \div 2$ **6** $18, 6; \div 3$ **7** $32, 8; \div 4$ **8** $200, 20; \div 10$
9 $10; \div 5$ **10** $15; \div 5$

Exercise 2V

1 add 2, 11   **2** add 4, 21   **3** cube numbers, 216
4 add 2, 12   **5** add 3, 17   **6** add 4, 23
7 add two previous numbers, 26
8 add two previous numbers, 39   **9** add 2, 15   **10** add 3, 19
11 add 5, 27   **12** add 5, 28

Exercise 2W

1 $3n$, 60   **2** $5n$, 100   **3** n, 20   **4** $7n$, 140
5 $6(n-1)$, 114   **6** $10n$, 200   **7** $2n+3$, 43   **8** $3n+1$, 61
9 $5n-2$, 98   **10** $4n-3$, 77

11 (a)

　　　13 matches　　　　　　　16 matches

(b)

Term number	1	2	3	4	5
Matches used	4	7	10	**13**	**16**

(c) (term number) × 3 + 1 = (6 × 3) + 1   **(d)** $3n+1$
12 $5n+1$

Unit 3 Angles and turning

Exercise 3A

1 **(a)**, **(b)**, **(c)** and **(f)**
2 **(a)** South   **(b)** North   **(c)** West   **(d)** West
3 **(a)** North-West   **(b)** North-East   **(c)** South-East
4 **(a)** $\frac{1}{4}$ turn   **(b)** $\frac{1}{2}$ turn   **(c)** $\frac{1}{4}$ turn

Exercise 3B

1 a is acute
2 b is a right angle
3 c is obtuse
4 d is right, e is acute
5 f is obtuse, g is acute, h is obtuse, i is acute
6 j is obtuse, k is acute
7 50°   **8** 145°   **9** 40°
10 135°   **11** 85°   **12** 90°

Exercise 3C

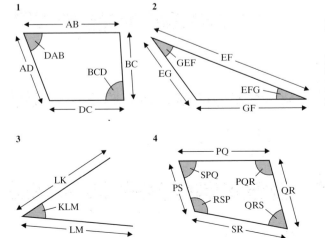

Exercise 3D

1 $ABC = 50°$ **2** $EFG = 135°$
3 $SPQ = 142°, PQR = 65°, QRS = 98°, RSP = 55°$
4 $ZXY = 62°, XYZ = 80°, YZX = 38°$
5 $ABC = 50°, BCA = 83°, CAB = 47°$
6 $RST = 124°, STU = 59°, TUR = 121°, URS = 56°$
7 $RPQ = 105°, PQR = 45°, QRP = 30°$
8 $ZXY = 47°, XYZ = 53°, YZX = 80°$

Exercise 3F

1 $a = 85°$ **2** $b = 102°$ **3** $c = 46°$ **4** $d = 99°$
5 $e = 60°$ **6** $f = 45°$

Exercise 3G

1 $a = 90°$ **2** $b = 120°, c = 60°, d = 60°$ **3** $e = 38°$
4 $f = 261°$ **5** $g = 33°, h = 147°, i = 147°$
6 $j = 132°, k = 48°, l = 48°$ **7** $m = 135°$
8 $n = 60°, p = 60°, q = 120°$ **9** $r = 36$

Exercise 3H

1 $a = 30°$ **2** $b = 70°$ **3** $c = 97°$ **4** $e = 25°$
5 $f = 74°$ **6** $g = 66°$ **7** $h = 40°$ **8** $i = 85°$ **9** $j = 60°$

Exercise 3I

1 A/B C/E D/F G/H I/J K/L
3 b and e
4 c and e
5 **(a)** $a = 50°$, alternate angles
　　(b) $b = 35°$, corresponding angles
　　(c) $c = 110°$, alternate angles
　　(d) $d = 140°$, corresponding angles
　　(e) $e = 115°$, angles on a straight line; $f = 115°$, alternate angles; $g = 65°$, angles on a straight line
　　(f) $h = 55°$, angles on a straight line; $i = 55°$, alternate angles; $j = 55°$, corresponding angles (or opposite angles)
6 **(a)** **(i)** 109°   **(ii)** Angles on a straight line add up to 180°
　　(b) **(i)** 24°   **(ii)** Alternate angles

Exercise 3J

1 063°   **2** 110°   **3** 320°   **4** 145°   **5** 230°   **6** 080°
7 **(a)** 230°   **(b)** 090°   **(c)** 270°
8 **(a)** 298°   **(b)** 080°   **(c)** 208°

Unit 4 Fractions

Exercise 4A

1 $\frac{1}{2}, \frac{1}{2}, \frac{1}{4}, \frac{3}{4}, \frac{2}{5}, \frac{3}{5}, \frac{3}{10}, \frac{7}{10}, \frac{5}{8}, \frac{3}{8}, \frac{4}{9}, \frac{5}{9}$

2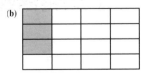
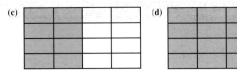

3 (a) (b) (c)

Exercise 4B

1 (a) $\frac{15}{28}$ (b) $\frac{13}{28}$ **2** (a) 75 (b) $\frac{28}{75}$ (c) $\frac{47}{75}$

3 (a) $\frac{37}{51}$ (b) $\frac{10}{51}$ (c) $\frac{4}{51}$ **4** (a) $\frac{198}{217}$ (b) $\frac{19}{217}$

5 (a) 16 (b) $\frac{7}{16}$ (c) $\frac{3}{16}$ (d) $\frac{5}{16}$ (e) $\frac{1}{16}$

6 (a) 17 (b) $\frac{5}{17}$ (c) $\frac{6}{17}$ (d) $\frac{4}{17}$ (e) $\frac{2}{17}$

7 (a) $\frac{6}{11}$ (b) $\frac{4}{11}$ (c) $\frac{1}{11}$

Exercise 4C

1 (a) $2\frac{1}{2}$ (b) $1\frac{3}{4}$ (c) $1\frac{2}{7}$ (d) $1\frac{1}{8}$ (e) $1\frac{1}{8}$
(f) $3\frac{1}{5}$ (g) $2\frac{3}{10}$ (h) $4\frac{4}{5}$ (i) $2\frac{2}{7}$ (j) $2\frac{2}{5}$
(k) $6\frac{2}{3}$ (l) $1\frac{7}{9}$ (m) $9\frac{3}{4}$ (n) $5\frac{2}{5}$ (o) $2\frac{8}{9}$
(p) $1\frac{7}{10}$

2 (a) $\frac{3}{2}$ (b) $\frac{11}{2}$ (c) $\frac{11}{4}$ (d) $\frac{5}{3}$ (e) $\frac{13}{4}$ (f) $\frac{22}{5}$
(g) $\frac{37}{10}$ (h) $\frac{26}{5}$ (i) $\frac{31}{4}$ (j) $\frac{9}{4}$ (k) $\frac{19}{10}$ (l) $\frac{28}{3}$
(m) $\frac{17}{6}$ (n) $\frac{43}{8}$ (o) $\frac{29}{8}$ (p) $\frac{109}{100}$

Exercise 4D

1 (a) $\frac{2}{3}$ (b) $\frac{1}{2}$ (c) $\frac{1}{2}$ (d) $\frac{1}{3}$ (e) $\frac{2}{5}$ (f) $\frac{2}{3}$
(g) $\frac{2}{3}$ (h) $\frac{3}{4}$ (i) $\frac{7}{11}$ (j) $\frac{6}{7}$ (k) $\frac{3}{4}$ (l) $\frac{5}{6}$

2 $\frac{2}{4}, \frac{1}{2}; \frac{4}{8}, \frac{1}{2}; \frac{4}{10}, \frac{2}{5}; \frac{3}{12}, \frac{1}{4}; \frac{6}{9}, \frac{3}{3}; \frac{6}{16}, \frac{3}{8}$

3 (a) $\frac{2}{7}$ (b) $\frac{5}{7}$ **4** (a) $\frac{2}{7}$ (b) $\frac{1}{2}$ (c) $\frac{3}{14}$

5 (a) $\frac{8}{15}$ (b) $\frac{1}{6}$ (c) $\frac{1}{5}$ (d) $\frac{1}{10}$

Exercise 4E

1 (a) 35 (b) £26 (c) 24 kg (d) 64 (e) £21
(f) 45p (g) £1.70 (h) 63p (i) £2.80

2 (a) 34 (b) 51

3 (a) 130 000 (b) 30 000 (c) 40 000 **4** 46

5 £74.40 **6** 375 kg

7 140 ham, 224 salad, 70 tuna, 126 cheese

8 (a) 305 (b) 427 (c) 2318

9 (a) 320 (b) 120 (c) 40

Exercise 4F

1 (a) $\frac{1}{2}, \frac{5}{10}, \frac{50}{100}, \frac{4}{8}, \frac{6}{12}, \frac{3}{6}$ (b) $\frac{2}{3}, \frac{4}{6}, \frac{6}{9}, \frac{40}{60}, \frac{60}{90}, \frac{20}{30}$
(c) $\frac{7}{10}, \frac{14}{20}, \frac{35}{50}, \frac{70}{100}, \frac{1400}{2000}, \frac{28}{40}$
(d) $\frac{3}{4}, \frac{12}{16}, \frac{15}{20}, \frac{21}{28}, \frac{300}{400}, \frac{90}{120}$
(e) $\frac{1}{8}, \frac{3}{24}, \frac{4}{32}, \frac{8}{64}, \frac{60}{480}, \frac{7}{56}$

2 (a) $\frac{2}{10}, \frac{3}{15}, \frac{4}{20}, \frac{5}{25}, \frac{6}{30}$ (b) $\frac{4}{10}, \frac{6}{15}, \frac{8}{20}, \frac{10}{25}, \frac{12}{30} \cdots$
(c) $\frac{10}{12}, \frac{15}{18}, \frac{20}{24}, \frac{25}{30}, \frac{30}{36} \cdots$ (d) $\frac{14}{16}, \frac{21}{24}, \frac{28}{32}, \frac{35}{40}, \frac{42}{48} \cdots$
(e) $\frac{6}{40}, \frac{9}{60}, \frac{12}{80}, \frac{15}{100}, \frac{18}{120} \cdots$

3 (a) $\frac{3}{6}, \frac{2}{6}$ (b) (i) $\frac{12}{30}, \frac{15}{30}$ (ii) $\frac{7}{70}, \frac{10}{70}$
(iii) $\frac{3}{12}, \frac{10}{12}$ (iv) $\frac{5}{10}, \frac{6}{10}$ (v) $\frac{16}{24}, \frac{3}{24}$ (vi) $\frac{15}{20}, \frac{12}{20}$

Exercise 4G

1 (a) $\frac{3}{6}$ (b) $\frac{1}{7}$ (c) $\frac{5}{6}$ (d) $\frac{3}{5}$ (e) $\frac{2}{3}$ (f) $\frac{3}{4}$

2 $\frac{2}{10}, \frac{2}{5}, \frac{1}{2}, \frac{3}{4}, \frac{7}{6}$

Exercise 4H

1 (a) $\frac{7}{8}$ (b) $\frac{7}{9}$ (c) $\frac{1}{2}$ (d) $\frac{8}{9}$ (e) $\frac{3}{4}$ (f) $\frac{5}{8}$
(g) $1\frac{3}{8}$ (h) $\frac{5}{6}$ (i) $1\frac{1}{6}$ (j) $\frac{7}{10}$ (k) $1\frac{1}{3}$ (l) $1\frac{1}{10}$

2 (a) $\frac{19}{24}$ (b) $\frac{17}{20}$ (c) $\frac{31}{36}$ (d) $1\frac{31}{40}$ (e) $\frac{17}{30}$
(f) $1\frac{1}{12}$ (g) $\frac{23}{24}$ (h) $1\frac{1}{18}$

3 (a) $\frac{5}{6}$ (b) $\frac{17}{30}$ (c) $\frac{33}{40}$ (d) $\frac{31}{36}$ (e) $1\frac{11}{42}$
(f) $1\frac{13}{70}$ (g) $1\frac{11}{30}$ (h) $1\frac{7}{20}$

4 (a) $3\frac{5}{6}$ (b) $5\frac{3}{4}$ (c) $4\frac{1}{8}$ (d) $9\frac{1}{12}$ (e) $5\frac{3}{16}$
(f) $3\frac{2}{3}$ (g) $7\frac{5}{6}$ (h) $7\frac{4}{15}$ **5** $\frac{1}{2}$ **6** $\frac{49}{80}$

Exercise 4I

1 (a) $\frac{2}{11}$ (b) $\frac{2}{9}$ (c) $\frac{6}{8} = \frac{3}{4}$ (d) $\frac{2}{12} = \frac{1}{6}$

2 (a) $\frac{1}{4}$ (b) $\frac{1}{8}$ (c) $\frac{1}{8}$ (d) $\frac{5}{8}$ (e) $\frac{1}{2}$ (f) $\frac{1}{4}$
(g) $\frac{1}{2}$ (h) $\frac{1}{5}$

3 (a) $\frac{11}{24}$ (b) $\frac{9}{20}$ (c) $\frac{13}{36}$ (d) $\frac{7}{18}$ (e) $\frac{13}{30}$ (f) $\frac{2}{5}$
(g) $\frac{9}{20}$ (h) $\frac{7}{40}$

4 (a) $\frac{1}{6}$ (b) $\frac{7}{24}$ (c) $\frac{1}{30}$ (d) $\frac{13}{30}$ (e) $\frac{2}{15}$ (f) $\frac{3}{20}$
(g) $\frac{1}{30}$ (h) $\frac{3}{20}$

5 (a) $5\frac{3}{20}$ (b) $7\frac{1}{6}$ (c) $1\frac{1}{4}$ (d) $2\frac{1}{5}$ (e) $2\frac{7}{10}$
(f) $\frac{9}{10}$ (g) $\frac{3}{4}$ (h) $2\frac{4}{5}$

6 $\frac{9}{16}$ **7** $\frac{3}{5}$ **8** $\frac{1}{12}$ acre

---

Exercise 4J

1 (a) $\frac{3}{8}$ (b) $\frac{3}{32}$ (c) $\frac{8}{25}$ (d) $\frac{9}{32}$ (e) $\frac{5}{36}$ (f) $\frac{21}{40}$
(g) $\frac{9}{50}$ (h) $\frac{4}{9}$

2 (a) $\frac{2}{5}$ (b) $\frac{3}{5}$ (c) $\frac{1}{2}$ (d) $\frac{6}{25}$ (e) $\frac{5}{8}$ (f) $\frac{1}{8}$
(g) $\frac{4}{15}$ (h) $\frac{4}{7}$

3 (a) $3\frac{1}{2}$ (b) $3\frac{1}{3}$ (c) $4\frac{4}{5}$ (d) 6 (e) 14
(f) 6 (g) 4 (h) 10

4 (a) $1\frac{5}{8}$ (b) 3 (c) $1\frac{1}{9}$ (d) $1\frac{3}{4}$ (e) $5\frac{1}{4}$
(f) $5\frac{13}{24}$ (g) $4\frac{1}{5}$ (h) $5\frac{1}{4}$

5 Jamie $11\frac{1}{4}$ hours, Claire $8\frac{3}{4}$ hours **6** $82\frac{1}{2}$ mins **7** $7\frac{5}{16}$ cm^2

8 $47\frac{1}{4}$ cubic inches

9 (a) $12\frac{4}{7}$ cm (b) $20\frac{3}{7}$ cm (c) $2\frac{5}{14}$ cm

---

Exercise 4K

1 (a) $1\frac{1}{3}$ (b) $\frac{3}{4}$ (c) $1\frac{1}{2}$ (d) $\frac{5}{7}$ (e) $3\frac{1}{3}$
(f) $1\frac{7}{8}$ (g) $1\frac{1}{9}$ (h) $\frac{7}{8}$

2 (a) 5 (b) $1\frac{3}{10}$ (c) $1\frac{2}{3}$ (d) $\frac{39}{76}$ (e) $\frac{1}{2}$ (f) 2
(g) $\frac{17}{27}$ (h) $\frac{21}{40}$

3 (a) $\frac{3}{32}$ (b) $\frac{5}{12}$ (c) $\frac{1}{10}$ (d) $\frac{4}{25}$ (e) $\frac{1}{3}$
(f) $\frac{13}{24}$ (g) $\frac{17}{60}$ (h) $\frac{1}{6}$

4 (a) 16 (b) 16 (c) 10 (d) $9\frac{1}{7}$ (e) 5
(f) $1\frac{5}{7}$ (g) 15 (h) 24

Unit 5 Two-dimensional shapes

Exercise 5A

1 (c) 2, 2 (d) 90° (e) 4 (f) 90°, equal
(g) equal, 90° (h) sides, equal (i) opposite, equal,
opposite (j) sides, equal, opposite (k) parallel

2 (a) trapezium (b) parallelogram (c) trapezium
(d) kite (e) square (f) trapezium

3 (b) L (c) B, S, Q, T (d) rectangle (e) E, M
(f) F, K (g) P (h) isosceles (i) L
(j) C, G, H, U, R

Exercise 5C

6 $EG = 5 \text{ cm} \pm 2 \text{ mm}$
7 $XZ = 5.8 \text{ cm} \pm 2 \text{ mm}, YZ = 4.5 \text{ cm} \pm 2 \text{ mm}$

Exercise 5D

1 A and D **2** A and C, B and D **3** B and D
4 A and D, B and C **5** B and C **6** A and B

Exercise 5E

1 **(a) (i)** congruent **(ii)** SSS **(b) (i)** congruent **(ii)** RHS
   (c) (i) not necessarily congruent
   (d) (i) congruent **(ii)** SAS
   (e) (i) not necessarily congruent
   (f) (i) congruent **(ii)** ASA
2 A and B; SAS **3** A and B; RHS
4 A and C; ASA **5** A and B; SAS

Exercise 5F

Type of polygon	Sum of interior	Sum of exterior
Equilateral triangle	180°	360°
Isosceles triangle	180°	360°
Parallelogram	360°	360°
Pentagon	540°	360°

Exercise 5G

1 **(a)** 45° **(b)** 108° **(c)** 144°, 36° **2** 15 sides
3 8°, 45 sides **4** Rectangle, kite, trapezium, rhombus
5 Rhombus, kite and trapezium

Unit 6 Decimals

Exercise 6A

1

	Tens	Units	.	Tenths	Hundredths	Thousandths
(a)	4	1	.	6		
(b)		4	.	1	6	
(c)	3	4	.	6		
(d)		1	.	4	6	3
(e)		0	.	6	4	3
(f)		1	.	0	0	5
(g)		5	.	0	1	
(h)		0	.	0	8	6

2 **(a)** Unit **(b)** Tenth **(c)** Tenth **(d)** Hundredth
   (e) Ten **(f)** Thousandth **(g)** Hundredth
   (h) Ten thousandth **(i)** Hundredth **(j)** Tenth
   (k) Thousandth **(l)** Hundredth

Exercise 6B

1 **(a)** 0.9, 0.76, 0.71, 0.68, 0.62 **(b)** 3.75, 3.4, 3.12, 2.13, 2.09
   (c) 5.2, 5.16, 5.04, 3.6, 3.47 **(d)** 0.42, 0.407, 0.3, 0.09, 0.065
   (e) 6.52, 6.08, 3.7, 3.58, 3.0 **(f)** 0.13, 0.105, 0.06, 0.024, 0.009
   (g) 0.8, 0.525, 0.2, 0.08, 0.05 **(h)** 2.2, 2.09, 1.3, 1.16, 1.08
2 **(a)** 4.09, 4.85, 5.16, 5.23, 5.9 **(b)** 0.021, 0.09, 0.34, 0.37, 0.4
   (c) 5, 5.009, 5.01, 7.07, 7.23 **(d)** 0.07, 0.23, 1.001, 1.08, 1.14
3 £1.09, £1.13, £1.18, £1.20, £1.29, £1.31

4 Latif, Sheila, Ira, Jean, Naomi, Rachel.
5 52.037, 53.027, 53.072, 53.207, 53.702, 57.320
6 0.0014, 0.0015, 0.002, 0.014, 0.02, 1.5

Exercise 6C

1 **(a)** 8 **(b)** 13 **(c)** 14 **(d)** 6 **(e)** 11 **(f)** 20
   (g) 1 **(h)** 20 **(i)** 1 **(j)** 100 **(k)** 20 **(l)** 2
2 **(a)** 3.6 **(b)** 5.3 **(c)** 0.1 **(d)** 9.3 **(e)** 10.7 **(f)** 8.0
   (g) 2.1 **(h)** 0.5 **(i)** 2.5 **(j)** 125.7 **(k)** 0.1 **(l)** 9.9
3 **(a)** 14 mm **(b)** 80 m **(c)** 1 kg **(d)** £204
   (e) 4 lb **(f)** 0 tonne **(g)** 11 g **(h)** 8 min

Exercise 6D

1 6.1 **2** 3.25 **3** 68.9 **4** 126.02 **5** 1.0
6 18.725 **7** 19.8 **8** 11.001 **9** 158.14 **10** 1.914
11 118.17 **12** 5.311

Exercise 6E

1 31.97 **2** 28.71 **3** 19.122 **4** 18.326 **5** 11.064
6 31.006 **7** 15.0976 **8** 178.585

Exercise 6F

1 6.2 **2** 4.14 **3** 14.1 **4** 97.3 **5** 0.11
6 0.35 **7** 6.19 **8** 11.14 **9** 0.9 **10** 4.07
11 14.03 **12** 13.25 **13** 1.225 **14** 11.649
15 2.254 **16** 168.58

Exercise 6G

1 **(a)** £13.50 **(b)** £5.48 **(c)** £5.20 **(d)** £1.20
2 **(a)** 30.4 **(b)** 3.04 **(c)** 0.304 **(d)** 11.25
   (e) 1.125 **(f)** 0.1125 **(g)** 1.125 **(h)** 0.01125
   (i) 0.001125
3 **(a)** 172.2 **(b)** 0.0945 **(c)** 0.0144 **(d)** 0.04
   (e) 0.9 **(f)** 0.0012
4 **(a)** 64.2 **(b)** 642 **(c)** 6.42 **(d)** 562.3
   (e) 56.23 **(f)** 0.5623
   Figures have moved one column left.
5 **(a)** 4.5 **(b)** 45.0 **(c)** 450 **(d)** 2.03
   (e) 20.3 **(f)** 203
   Figures have moved two columns left.

Exercise 6H

1 **(a)** 16.12 **(b)** 0.633 **(c)** 14.84 **(d)** 34.221
   (e) 5.027 **(f)** 0.046
2 **(a)** 3.45 **(b)** 0.345 **(c)** 0.0345 **(d)** 7.8
   (e) 0.78 **(f)** 0.078 **(g)** 6.5 **(h)** 0.65 **(i)** 0.065
   Figures have moved right by number of noughts in the division.
3 15 jugs **4** £15.40

Exercise 6I

1 3.1 **2** 3.6 **3** 1.7 **4** 1.5 **5** 4.3 **6** 2.7
7 10.7 **8** 1.01 **9** 2.5 **10** 52

Exercise 6J

1 **(a)** 0.6 **(b)** 0.5 **(c)** 0.7 **(d)** 0.35 **(e)** 0.16
   (f) 0.06 **(g)** 0.875 **(h)** 0.45 **(i)** 0.76
   (j) 0.3125 **(k)** 0.125 **(l)** 0.54 **(m)** 0.09
   (n) 0.065 **(o)** 0.6666 . . . **(p)** 0.95
2 **(a)** $\frac{3}{10}$ **(b)** $\frac{37}{100}$ **(c)** $\frac{93}{100}$ **(d)** $\frac{137}{1000}$ **(e)** $\frac{293}{1000}$ **(f)** $\frac{7}{100}$
   (g) $\frac{59}{100}$ **(h)** $\frac{3}{1000}$ **(i)** $\frac{3}{100\,000}$ **(j)** $\frac{13}{10\,000}$ **(k)** $\frac{77}{100}$
   (l) $\frac{77}{1000}$ **(m)** $\frac{39}{100}$ **(n)** $\frac{41}{10\,000}$ **(o)** $\frac{19}{1000}$ **(p)** $\frac{31}{1000}$

Unit 7 Measure 1

Exercise 7A

1 (a) 1.8 m   (b) 1.2 m   (c) 2.5 m   (d) 1.5 m
2 (a) 3 m   (b) 10 m   (c) 6 m
3 (a) 7 m   (b) 7 m   (c) 0.9 m
4 (a) 3 m   (b) 2.5 m   (c) 10 m

Exercise 7B

Question	Metric	Imperial
1	570 ml	1 pint
2	300 ml	$\frac{1}{2}$ pint
3	300 ml	$\frac{1}{2}$ pint
4	150 ml	$\frac{1}{4}$ pint
5	1 l	2 pints
6	45 l	10 gallons
7	9 l	2 gallons
8	90 l	20 gallons

Exercise 7C

Question	Metric	Imperial
1	5 kg	11 pounds
2	200 g	$\frac{1}{2}$ pound
3	2 kg	4 pounds
4	100 g	$\frac{1}{4}$ pound
5	1 kg	2 pounds
6	200 g	$\frac{1}{2}$ pound
7	1 kg	2 pounds
8	100 g	$\frac{1}{4}$ pound
9	30 g	1 ounce
10	2 kg	4 pounds

Exercise 7D

1 (a) Not sensible, 1.8 m
   (b) Not sensible, 1.8 m
   (c) Not sensible, 7 m by 6 m
   (d) Not sensible, 2 kg
   (e) Not sensible, 300 ml
   (f) Could be sensible
2 (a) Sensible
   (b) Sensible
   (c) Sensible
   (d) Not sensible, 300 ml
   (e) Not sensible, 300 km
   (f) Not sensible, 10 g
3 (a) Sensible
   (b) Not sensible, 50 mph
   (c) Not sensible, 570 ml
   (d) Not sensible, 50 l
4 (a) Sensible
   (b) Sensible
   (c) Not sensible, 50 l
   (d) Not sensible, 100 g
   (e) Sensible

Exercise 7E

Question	Metric	Imperial
1	metres	feet
2	centimetres	inches
3	kilometres	miles
4	metres	feet
5	kilograms	pounds
6	grams	ounces
7	tonnes	tons
8	litres	gallons
9	millilitres	pints
10	millilitres	fluid ounces/teaspoon
11	millilitres	fluid ounces
12	litres	gallons
13	minutes	minutes
14	seconds	seconds
15	hours	hours
16	days	days
17	centimetres	inches
18	millimetres	inches
19	kilograms	pounds
20	years	years

Exercise 7F

1 (a) 9 o'clock   (b) Quarter past 8   (c) Half past 3
   (d) Quarter to 6   (e) Ten past 3   (f) Twenty to 3
   (g) Five to 2   (h) Twenty-five to 9   (i) Twenty-five past 8
   (j) Ten to 3
2 (a)

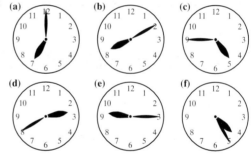

Exercise 7G

1 (a) Half past eight   (b) Ten past ten
   (c) Five past eleven   (d) Quarter to five   (e) Five to four
2 (a) 9:15   (b) 3:30   (c) 4:40   (d) 6:45   (e) 5:00

Exercise 7H

1 (a) 10:00   (b) 22:00   (c) 09:30   (d) 21:30
   (e) 20:20   (f) 08:20   (g) 07:00   (h) 20:00
   (i) 15:30   (j) 04:40   (k) 01:08   (l) 13:08
   (m) 17:50   (n) 05:50   (o) 23:00   (p) 08:00
   (q) 08:15   (r) 20:45   (s) 14:55   (t) 06:40
2 (a) 8 am   (b) 9:20 am   (c) 9:30 pm   (d) 1:10 pm
   (e) 12:10 pm   (f) 12:20 am   (g) 1:40 am   (h) 8 am
   (i) 3:45 pm   (j) 6 pm   (k) 4:30 pm   (l) 9:10 pm
   (m) 11:55 pm   (n) 2.02 pm   (o) 6:25 am   (p) midnight
   (q) midnight   (r) noon   (s) 10:55 am   (t) 8:55 pm

Exercise 7I

1 (a) 6 cm   (b) 40 mph   (c) 40 ml   (d) 100 mph
   (e) 60 ml   (f) 80 mm

Exercise 7J

1 7.7 cm   2 188 cm   3 46 mph
4 22 ml   5 18 mph   6 8.6 cm

Exercise 7K

1 7.8 units   2 5.5 cm   3 37 mph
4 82 mph   5 6.7 units   6 4.4 units

Exercise 7L

1 (a) 47p first class; 36p second class
   (b) 155p first class; 120p second class
   (c) 205p first class; 145p second class

Exercise 7M

1 (a) 2 cm   (b) 3.5 cm   (c) 5.5 cm   (d) 10 cm
   (e) 7.4 cm   (f) 0.8 cm   (g) 1.8 cm   (h) 6.8 cm

Exercise 7N

1 (a) metres   (b) 12 metres
2 (a) 12 seconds   (b) 4 minutes
3 (a) 20 feet   (b) 6 metres

Unit 8 Collecting and recording data

Exercise 8A

1 (a) Y   (b) Y   (c) X   (d) Y   (e) X   (f) Y
   (g) X   (h) Y

Exercise 8B

1 (a) e.g. It suggests Yes is the right answer.
   (b) e.g. Badly for one person may be Well for another.
   (c) e.g. There is no way of answering 'none' or more than 5 hours.
   (d) e.g. Lousy could mean different things to different people.
   (e) e.g. Not enough choices.
   (f) e.g. People may not want to admit to cheating.
   (g) e.g. It suggests Yes is the right answer.

Exercise 8C

1 (a) and (b)   2 (a)   3 (b) and (c)
4 (a) include an option: 4 or more

Exercise 8F

1 (a) Alyssum   (b) Red   (c) Begonia/Dahlia
2 (a) Vauxhall   (b) Volvo   (c) Volvo   (d) 3
   (e) Rover, Vauxhall or VW   (f) 1991   (g) Peugeot GR
   (h) 5   (i) Volvo, Honda, Renault, Ford Escort, Peugeot XT,
   Ford Sierra, Peugeot GR, Vauxhall, Rover, VW.
4 (a) Cherry   (b) Gooseberry and Cherry
   (c) Apple and Pear   (d) Plum or Cherry
   (e) Apple and Cherry   (f) Pear Espalier

Exercise 8G

1 (a) 3.2%   (b) (i) 1987   (ii) 1990   (c) 1991 and 1992
2 (a) (i) Redbridge   (ii) Lambeth   (b) Lambeth
   (c) (i) Redbridge   (ii) Camden   (d) £27
3 (a) Morning   (b) 1992   (c) 1988 – large decrease in traffic flow
4 (a) (i) Hackney   (ii) Bromley   (iii) Lambeth
   (b) (i) Kingston   (ii) Kingston   (iii) Bromley
   (c) (i) £194   (ii) £65

Unit 9 Algebra 2

Exercise 9A

1 (a) Cliffs   (b) Lookout   (c) Vantage Point
   (d) Landing Stage   (e) Tall tree   (f) Beach
2 (a) (2,2)   (b) (3,0)   (c) (8,2)   (d) (1,5)
   (e) (7,5)   (f) (4,4)

Exercise 9B

1
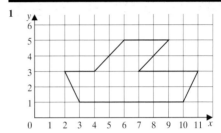

2 (1,1) (8,1) (8,4) (9,4) (7,6) (2,6) (0,4) (1,4) (1,1).
3 (3,0) (8,0) (9,2) (8,2) (8,3) (7,3) (7,4) (7,5) (9,5) (7,6)
   (7,4) (5,4) (4,3) (3,3) (2,2) (1,2) (3,0)

4
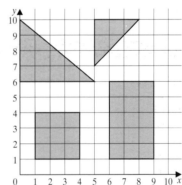

7 (a) $(2\frac{1}{2}, 2)$ (b) $(2, 6\frac{1}{2})$ (c) (8, 5)   (d) $(4\frac{1}{2}, 3\frac{1}{2})$
8 (a) $(2, 2\frac{1}{2})$ (b) $(3\frac{1}{2}, 5)$ (c) $(3, 6\frac{1}{2})$
   (d) (3, 4)   (e) $(6, 8\frac{1}{2})$

Exercise 9C

1 (a)
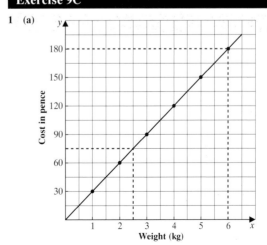

   (b) 75p
   (c) £1.80

2 (a)

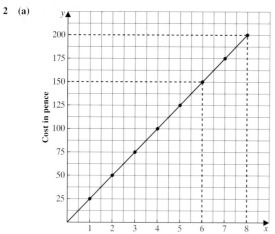

(b) (i) £2   **(ii)** £1.50

3 (a)

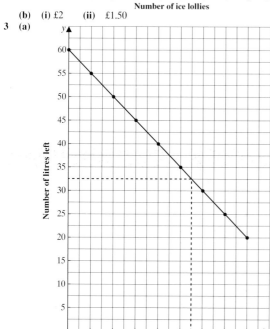

(b) 60 litres   **(c)** 32.5 litres

4 (a)

Distance travelled (km)	0	5	10	15	20	25
Petrol used in l	0	2	4	6	8	10

(b)

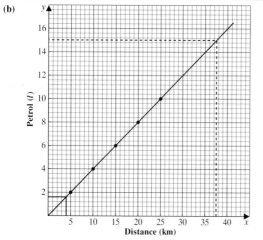

(c) 1.6 litres
(d) 37.5 kilometres

5 (a)

Weeks	0	1	2	3	4	5	6	7	8
Expected depth of water in m	144	142	136	132	128	124	120	116	112

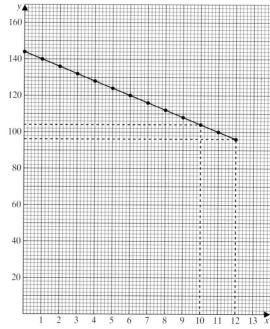

(b) 104 m   **(c)** 12 weeks

Exercise 9D

1

°C	5	20	**27**	28	**10**	**38**	35	80	**93**	40
°F	**41**	**68**	80	**82**	50	100	**95**	**176**	200	**104**

2 (b)

Kilograms	0	**4.5**	**9**	45	30	15	**22.7**	**6.4**	35	50
Pounds	0	10	20	**99**	**66**	**33**	50	14	**77**	110

3

Inches	0	1	2	**4**	**6**	**8**	9	8	**10**	12
Centimetres	0	**2.5**	**5**	10	15	20	**22.5**	**20**	25	30

4

Miles	0	5	**10**	40	**22.5**	30	**45**	**12.5**	24	50
Kilometres	0	**8**	16	**64**	36	**48**	72	20	**38**	80

5

Hectares	0	**8**	**12**	12	15	17	**9.6**	**18**	3	20
Acres	0	20	30	**30**	**37.5**	**42.5**	24	45	**7.5**	50

6 (a) (i) 11 metres   **(ii)** 45 metres
(b) (i) 2.9 seconds   **(ii)** 2.2 seconds

7 (a)

(b) 15 m/s
(c) 17 m

Exercise 9E

1 (a) 10 minutes   (b) 300 metres
   (c) 15 minutes   (d) 15 minutes
   (e) 30 metres per minute or 1.8 km/h
   (f) 20 metres per minute or 1.2 km/h

2 (a) David travelled 20 km in 1 hour, he stopped for $\frac{1}{2}$ hour then travelled 40 km in 1 hour. He then stopped for one hour. He travelled the 60 km return journey in $\frac{1}{2}$ hour.
   (b) First stage 20 km/h, second stage 40 km/h, last stage 120 km/h.

3 (a) Wayne set off from London at 08:00, travelled 75 km in 1 hour and then stopped for 20 minutes. He travelled the remaining 75 km in 40 minutes. He then travelled the 150 km home without stopping in 1 hour.
   (b) Tracey set off at 08:20 and travelled 50 km in 40 minutes, she stopped for 10 minutes and then travelled the remaining 100 km in 40 minutes.
   (c) Wayne and Tracey passed each other at 75 km from London at 09:20.

4 (a) (i) 985 m
     (ii) 800 m
     (iii) 530 m
   (b) (i) 11.02 and 6 seconds
     (ii) 11.02 and 40 seconds
     (iii) 11.03 and 12 seconds

5 Between A and B Imran turns a tap off and the water level rose 10 cm in 5 minutes.
Between B and C Imran had turned both taps off and there was no increase in water level for the 5 minutes it took him to undress and get ready for his bath.
Between C and D Imran gets into the bath and the level rises 30 cm immediately.
Between D and E Imran had his bath and the level stayed the same for 15 minutes.
Between E and F Imran let the water out of the bath and the water level dropped 60 cm in 5 minutes.

6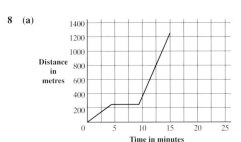

7

8 (a)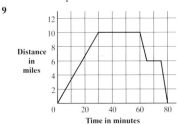

   (b) 1000 metres in 6 minutes   167 metres per minute
     10 km per hour

9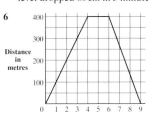

Exercise 9F

1 A (1, 3); B (3, 1); C (4, 0); D (3, –2); E (5, –3); F (1, –3); G (0, –1); H (–1, –3); I (–3, –2); J (–3, 0); K (–4, 2); L (–2, 3)

2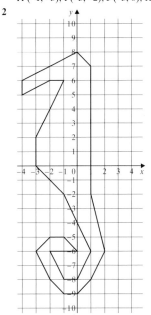

Exercise 9G

1 (a) $y = 3$ (b) $y = 1$ (c) $y = -1$ (d) $y = -3$
2 (a) $x = -4$ (b) $x = -2$ (c) $x = 1$ (d) $x = 4$
3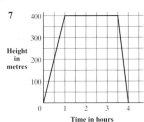

4 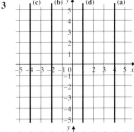   5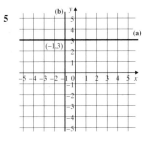

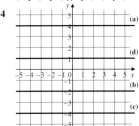

Exercise 9H

1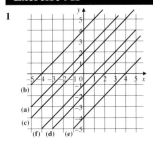

2

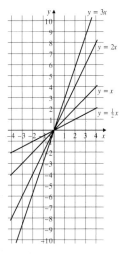

Exercise 9I

1

x	-4	-3	-2	-1	0	1	2	3	4
x^2	16	9	4	1	0	1	4	9	16
-2	-2	-2	-2	-2	-2	-2	-2	-2	-2
y	14	7	2	-1	-2	-1	2	7	14

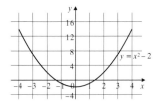

2

x	-4	-3	-2	-1	0	1	2	3	4
$y = x^2 + 1$	17	10	5	2	1	2	5	10	17

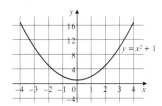

3

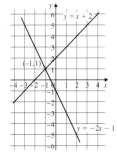

4

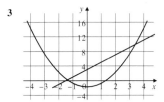

5

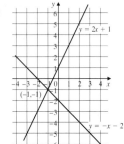

3

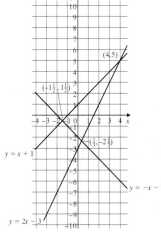

4

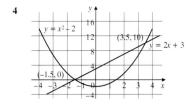

5

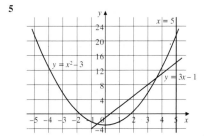

6 (a) $y = 3x - 2$

x	-1	0	1	2
y	-5	-2	1	4

$y = 5 - x$

x	-1	0	1	2
y	6	5	4	3

(b)

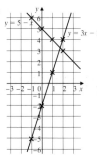

The lines cross at the point $(1\frac{3}{4}, 3\frac{1}{4})$.

Exercise 9J

1 (a) 3-D   (b) 2-D   (c) 1-D   (d) 3-D
2 (a) 2-D   (b) 3-D   (c) 2-D   (d) 3-D
3 (3, 1, 1), (6, 6, 9), (4, 1, 16), (28, 1, 8)

Unit 10 Sorting and presenting data

Exercise 10A

1
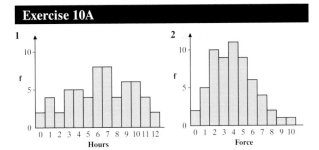

2

3
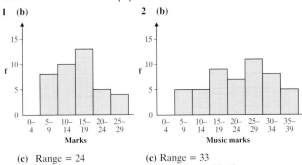

Exercise 10B

1a and 2a teachers to check pupil's work

1 **(b)**
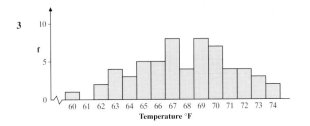

(c) Range = 24

2 **(b)**

(c) Range = 33
(d) Modal class 25–19

Exercise 10C

1 **(a)**

(b) Examples:
October is warmest in London and Majorca.
Majorca is always warmer than London.
January is the coldest month in both places.

2 **(b)** Jan and Feb   **(c)** £70
(d)
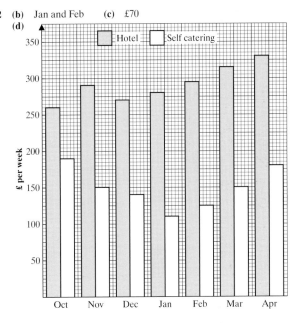

Examples:
Hotel prices are lowest in October.
Self catering prices are highest in October.
Self catering prices are lowest in January.
Difference in prices is lowest in October.

3 **(b)**
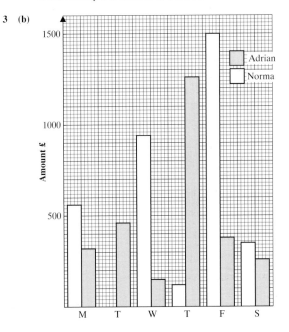

Examples:
Norma's range 1500.
Adrian's range 1110.
Norma spent most on Friday.
Adrian spent most on Thursday.

Exercise 10D

1 **(a)** 50   **(b)** Train   **(c)** 12
3 **(a)** **(i)** Tea   **(ii)** 30   **(iii)** 15
(c) 140   **(d)** 2.19

Exercise 10E

1 (a)

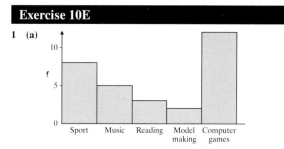

2 (a)

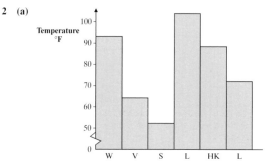

3 (a) Cycleshop   **(b)** 17
   (c) **(i)** 110   **(ii)** 155
   (d) Bikeshop   **(e)** April

4 (a) discrete   **(b)** continuous   **(c)** discrete
   (d) discrete   **(e)** continuous   **(f)** continuous
   (g) discrete   **(h)** continuous   **(i)** continuous
   (j) discrete   **(k)** continuous

5 (a)

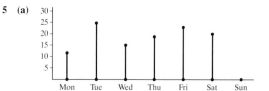

(b)   **(c)**

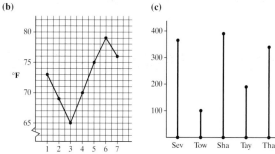

6 (a)

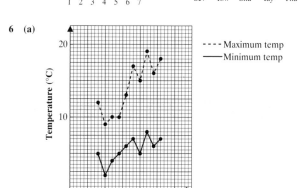

(b) Examples:
Range of max = 10 °C
Range of min = 6 °C
Greatest diff (max – min) = 11 °C

7 (a) 9.00 am
   (b) tank filled up
   (c) slow/steady

8 (a) Morgan in multiples of 10, Rees in multiples of 20

(b)

Petrol sales	Wed	Thur	Fri	Sat	Sun	Mon	Tue
Morgan Cars	60	50	60	70	60	80	50
Rees Motors	60	60	80	100	40	60	40

(c)

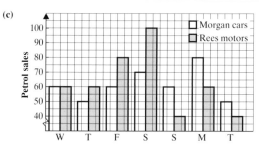

(d) Range for Morgan Cars = 30
Range for Rees Motors = 50
The greatest difference in number of sales between the two garages is on Sunday, and is a difference of 30 sales.

(e) Morgan on Mon, Rees on Sat

Exercise 10F

1 (a)

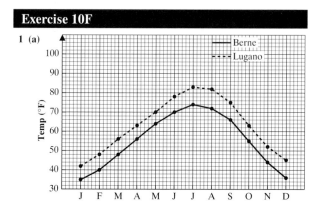

(b) 83 °F, July
(c) Berne

2 (a)

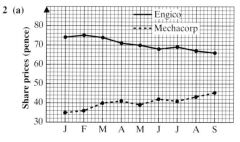

(b) Engico: 64 or 65, steady decrease
Mechacorp: 45 or 46, steady increase

3 (a)

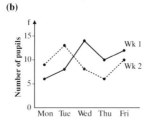

(Note: graph showing Turnover (in millions) vs Year 91–99, rising from 12 at 91, dipping to 9 in 92, up to 17 in 99)

(b) Graph shows general upward trend
Two dips in 1992 and 1997
Lowest value in 1992
Next value could be 18 or 19 million Euros

4 (a)

(Graph showing Sales vs Year & quarter, 1998–2000)

(b) The sales decrease throughout each year but have risen over the last three years.
(c) 87–89. Steady decrease throughout each year.

5 (a)

(Graph showing Anglers vs Year '91–'99, decreasing from ~285 to ~195)

(b) Number of anglers has decreased steadily.
Highest value in 1991.
Lowest value in 1998.
(c) Value in 1999 will probably be 175–185.

Exercise 10G

1 (a) **(b)**

(Two graphs: frequency polygons / bar charts showing Number of pupils vs Mon–Fri; right graph shows Wk 1 and Wk 2)

2 (a)

(Graph with three data series Sept, May, Jan showing scores from 0 to 30)

(c) The scores have gone up between September and May.
There are more people with higher scores and fewer people with lower scores.

Unit 11 Three-dimensional shapes

Exercise 11D

Microwave – cuboid
Kitchen roll – cylinder
Biscuits – cylinder
Various packets – cuboids
Rolling pin – cylinder

Ice cream cone – cone
Food cover – square based pyramid
Cheese grater – triangular prism
Orange – sphere
Stock cubes – cubes

Exercise 11E

1 **(a)** and **(c)**   **2** cylinder   **3** cube, cuboid

Exercise 11F

1 (a) **(b)** **(c)**

2 (a) **(b)**

3 **(a)** and **(c)**   **4** **(b)** and **(c)**

Exercise 11G

1 **(a)** **(i)** cube **(ii)** 12 **(iii)** 6 **(iv)** 8
  (b) **(i)** triangular prism **(ii)** 9 **(iii)** 5 **(iv)** 6
  (c) **(i)** squared-based pyramid **(ii)** 8 **(iii)** 5 **(iv)** 5
2 **(a)** **(i)** 8 **(ii)** 4
  (b) **(i)** 6 **(ii)** 3
  (c) **(i)** 4 **(ii)** 0
3 **A** and **C**

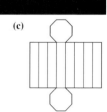

4 Sketches and nets of:

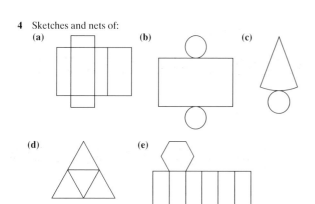

5 Sketches of:
  (a) square-based pyramid base edge 2 cm, slant height 4 cm
  (b) cuboid 2 cm by 4 cm by 9 cm
7 **(a)** cube **(b)** cylinder
8 **(a)** 3 cm by 4 cm **(b)** **(c)** 52 cm²

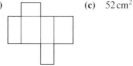

Exercise 11H

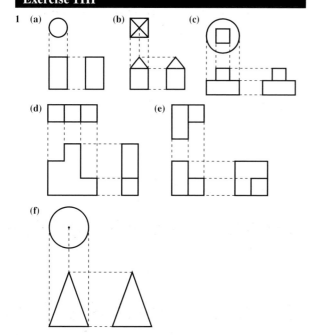

Answers for Units 12A, 12B

Pupils' own investigations.

Unit 13 Measure 2

Exercise 13A

1 **(a)** 300 cm **(b)** 3 cm **(c)** 600 cm **(d)** 1200 cm
  (e) 10 cm
2 **(a)** 20 mm **(b)** 50 mm **(c)** 120 mm **(d)** 200 mm
  (e) 1000 mm
3 **(a)** 5000 m **(b)** 3 m **(c)** 10 000 m **(d)** 20 m
  (e) 60 000 m
4 **(a)** 5000 g **(b)** 40 000 g **(c)** 100 000 g **(d)** 250 000 g
  (e) 1 000 000 g
5 **(a)** 3 *l* **(b)** 8 *l* **(c)** 50 *l* **(d)** 75 *l*
6 **(a)** 6000 ml **(b)** 40 000 ml **(c)** 100 000 ml
  (d) 350 000 ml **(e)** 25 000 ml
7 **(a)** 3 km **(b)** 7 km **(c)** 40 km **(d)** 45 km
8 **(a)** 4 tonnes **(b)** 7 tonnes **(c)** 30 tonnes **(d)** 55 tonnes
9 **(a)** 2 kg **(b)** 3000 kg **(c)** 50 kg **(d)** 12 000 kg
  (e) 100 kg
10 1000 **11** 8 000 000 **12** 1 000 000 **13** 100 000
14 4000 **15** 1 000 000

Exercise 13B

1 B, C, A, F, E, D **2** 250 kg, 3000 g, 2 kg, 250 g, 25 g
3 5 mm, 50 mm, 20 cm, 3 m, 3 km
4 4 cm, 75 cm, 2 m, 3000 mm, 4000 m
5 200 ml, 600 ml, 1 *l*, 2000 ml, 5 *l*

Exercise 13C

1 **(a)** 2.5 m **(b)** 0.5 m **(c)** 3600 m **(d)** 0.75 m
  (e) 5 m **(f)** 0.35 m **(g)** 4.75 m **(h)** 600 m
  (i) 40 m **(j)** 0.005 m
2 **(a)** 4500 g **(b)** 400 g **(c)** 10 300 g **(d)** 30 g **(e)** 5 g
3 **(a)** 35 cm **(b)** 250 cm **(c)** 540 cm **(d)** 0.5 cm
  (e) 8 cm **(f)** 80 cm **(g)** 3.5 cm **(h)** 5 cm
  (i) 8.5 cm **(j)** 27.5 cm
4 **(a)** 3500 ml **(b)** 500 ml **(c)** 15 400 ml **(d)** 50 ml
  (e) 3 ml
5 **(a)** 35 mm **(b)** 7 mm **(c)** 0.8 mm **(d)** 125 mm
  (e) 5 mm
6 **(a)** 0.3 km **(b)** 0.05 km **(c)** 1.25 km **(d)** 0.075 km
  (e) 0.375 km
7 **(a)** 0.25 *l* **(b)** 0.1 *l* **(c)** 0.05 *l* **(d)** 3.5 *l*
  (e) 0.001 *l*
8 **(a)** 0.5 kg **(b)** 300 kg **(c)** 0.05 kg **(d)** 5500 kg
  (e) 6 kg
9 **(a)** 3.5 tonnes **(b)** 0.45 tonnes **(c)** 0.05 tonnes
  (d) 0.003 tonnes **(e)** 0.075 tonnes
10 20 glasses **11** 60 pieces **12** 10 batches

Exercise 13D

1 **(a)** 2.4 cm, 25 mm, 3 cm, 50 mm, 57 mm, 6 cm
  (b) 270 mm, 30 cm, 0.4 m, 45 cm, 500 mm, 1.2 m
  (c) 2 m, 340 cm, 3500 mm, 370 cm, 4 m = 4000 mm
  (d) 0.2 cm, 4 mm, 45 mm, 5 cm, 55 mm, 0.3 m, 36 cm
  (e) 34 cm, 0.4 m, 0.45 m, 50 cm, 560 mm
2 **(a)** 0.05 kg, 250 g, 0.3 kg, 500 g
  (b) 350 g, 0.4 kg, 500 g, 0.52 kg
  (c) 3000 g, 4 kg, 4.5 kg, 5000 g, 400 kg, 0.5 tonnes
3 **(a)** 250 ml, 0.3 *l* = 300 ml, 0.4 *l*, 500 ml
  (b) 45 ml, 0.05 *l*, 360 ml, 0.4 *l*, 450 ml, 500 ml

Exercise 13E

1 640 km **2** 5¼ pints **3** 67.5 litres **4** 220 pounds
5 0.6 litres **6** 6.7 gallons **7** 0.875 pints **8** 93.75 miles
9 1.76 pounds **10** 2.625 pints

Exercise 13F

1 600 g of fat, 1000 g flour, 800 g of dried fruit   **2** $6\frac{2}{3}$ gallons
3 9.1 kg   **4** 10 cm   **5** 60 cm by 37.5 cm   **6** They are the same.
7 11 pounds of bread and 1.1 pounds of spread   **8** 2.86 litres
9 Hazel, her guess was 5.06 pounds   **10** 7.7 pounds

Exercise 13G

1 (a) 120 minutes   **(b)** 300 minutes   **(c)** 150 minutes
   (d) 330 minutes   **(e)** 375 minutes   **(f)** 315 minutes
2 (a) 3 hours   **(b)** 4 hours   **(c)** $1\frac{1}{4}$ hours
   (d) 4 hours 20 minutes   **(e)** 5 hours 25 minutes
   (f) $1\frac{1}{2}$ hours   **(g)** 72 hours   **(h)** 132 hours
   (i) 8 hours 20 minutes
3 3600 seconds   **4** 1440 minutes
5 (a) 31 536 000 seconds   **(b)** 31 622 400 seconds

Exercise 13H

	(a)	(b)	(c)	(d)
1	10:45	10:00	11:55	10:10
2	09:50	11:20	12:30	08:55
3	12:20	13:25	14:30	08:50
4	15:15	20:00	01:35	05:30
5	09:40	11:25	07:55	08:50
6	08:05	10:50	09:40	08:10
7	07:25	09:10	05:40	06:35
8	02:05	04:45	23:15	19:45

Exercise 13I

1 (a) 11th Jan   **(b)** 12th Mar   **(c)** 13th June
   (d) 15th July   **(e)** 20th Sept   **(f)** 30th May
   (g) 5th July   **(h)** 6th Sept
2 (a) 16th Feb   **(b)** 17th Mar   **(c)** 17th Apr
   (d) 21st Dec   **(e)** 29th Nov   **(f)** 2nd Oct
   (g) 6th Apr   **(h)** 9th Dec
3 (a) 5th May   **(b)** 6th July   **(c)** 16th June
   (d) 1st Oct   **(e)** 7th July   **(f)** 19th June
   (g) 30th Dec   **(h)** 4th Jan
4 Wed 10th April   **5** Thurs 2nd May   **6** Mon 13th May
7 Thurs 25th April   **8** Fri 3rd May

Exercise 13J

1 07:55   **2** 08:35   **3** 07:25   **4** 07:40

5
Coate	08:35	09:05	09:35
Piper's Way	08:40	09:10	09:40
Old Town	08:50	09:20	09:50
Drove Rd.	08:55	09:25	09:55
New Town	09:00	09:30	10:00
Bus Station	09:05	09:35	10:05

6
Bristol	08:10	08:25	08:40
Bath	08:30	08:45	09:00
Swindon	08:45	09:00	09:15
Didcot	09:05	09:20	09:35
Reading	09:15	09:30	09:45
London	09:40	09:55	10:10

7 07:45   **8** 08:35   **9** 09:15   **10** 08:10   **11** 30 mins
12 20 mins   **13** $1\frac{1}{2}$ hours   **14** 30 mins   **15** 35 mins
16 (a) 07:10 and 07:45 or 08:00   **(b)** 07:05 and 07:45 or 08:00
    (c) 07:50 and 08:15   **(d)** 08:25 and 08:45 or 09:00
    (e) 08:20 and 08:45 or 09:00   **(f)** 07:55 and 08:15

Exercise 13K

1 (a) 316 km   **(b)** 198 miles
2 (a) 08:20   **(b)** 25 minutes   **3** 76 cm

Unit 14 Percentages

Exercise 14A

1 (a) (i) 11%   **(ii)** 17%   **(iii)** 18%
   (b) 54%   **(c)** $\frac{27}{50}$
2 (a) (i) 11%   **(ii)** 19%   **(iii)** 13%   **(iv)** 7%
   (b) 50%   **(c)** $\frac{1}{2}$

Exercise 14B

1 (a) (i) 1%   **(ii)** $\frac{1}{100}$ **(b)** (i) 9%   **(ii)** $\frac{9}{100}$
   (c) (i) 31%   **(ii)** $\frac{31}{100}$ **(d)** (i) 59%   **(ii)** $\frac{59}{100}$
2 (a) (i) 10%   **(ii)** $\frac{1}{10}$ **(b)** (i) 20%   **(ii)** $\frac{1}{5}$
   (c) (i) 30%   **(ii)** $\frac{3}{10}$ **(d)** (i) 40%   **(ii)** $\frac{2}{5}$
3 (a) $\frac{17}{100}$ **(b)** $\frac{99}{100}$ **(c)** $\frac{41}{100}$ **(d)** $\frac{3}{100}$ **(e)** $\frac{3}{5}$ **(f)** $\frac{4}{5}$
   (g) $\frac{9}{10}$ **(h)** $\frac{3}{10}$ **(i)** $\frac{1}{10}$ **(j)** $\frac{7}{10}$ **(k)** $\frac{11}{50}$ **(l)** $\frac{3}{50}$
   (m) $\frac{16}{25}$ **(n)** $\frac{24}{25}$ **(o)** $\frac{3}{20}$ **(p)** $\frac{13}{20}$
4 (a) 0.37   **(b)** 0.49   **(c)** 0.87   **(d)** 0.07   **(e)** 0.4
   (f) 0.15   **(g)** 0.08   **(h)** 0.28   **(i)** 0.36   **(j)** 0.95
   (k) 0.45   **(l)** 0.03   **(m)** 0.035   **(n)** 0.065   **(o)** 0.125

Exercise 14C

1 (a) 37%   **(b)** 59%   **(c)** 11%   **(d)** 10%
   (e) 36%   **(f)** 70%   **(g)** 3%   **(h)** 77.1%
   (i) 9%   **(j)** 5.5%   **(k)** 83%   **(l)** 56%
   (m) 7.5%   **(n)** 12.5%   **(o)** 67.5%
2 (a) 50%   **(b)** 75%   **(c)** 40%   **(d)** 80%
   (e) 90%   **(f)** 35%   **(g)** 32%   **(h)** 76%
   (i) 15%   **(j)** 37.5%   **(k)** 62.5%   **(l)** 18.75%
   (m) 14%   **(n)** 9%   **(o)** 1.3%
3

Percentage	Decimal	Fraction
40%	0.4	$\frac{2}{5}$
61%	0.61	$\frac{61}{100}$
70%	**0.7**	$\frac{7}{10}$
35%	**0.35**	$\frac{7}{20}$
$8\frac{1}{2}\%$	**0.085**	$\frac{17}{200}$
15%	0.15	$\frac{3}{20}$
12%	**0.12**	$\frac{3}{25}$
7%	0.07	$\frac{7}{100}$
$1\frac{1}{4}\%$	**0.0125**	$\frac{1}{80}$
$\mathbf{66\frac{2}{3}\%}$	**0.666...**	$\frac{2}{3}$

Exercise 14D

1 (a) 52%, 0.53, $\frac{9}{15}$ **(b)** $\frac{7}{10}$, 0.71, 72%
   (c) 0.07, 8%, $\frac{1}{10}$ **(d)** 30%, 0.36, $\frac{3}{8}$
2 (a) maths   **(b)** science and technology

Exercise 14E

1 99   **2** 414 kg   **3** £1190   **4** 72   **5** £25.20
6 £1.68   **7** £28.70   **8** £79.20   **9** £1562
10 0.75 tonnes   **11** £43.20   **12** £30 720
13 9 km   **14** 3.55 km   **15** £669.12   **16** 0.1728 litres

Exercise 14F

1 (a) £0.60   **(b)** £11.40   **2** (a) (i) £12
   (ii) £68   **(b)** (i) £111   **(ii)** £629   **(c)** (i) £6.09
   (ii) £34.51
3
A^*	A	B	C	D	E	F	G
3	9	18	63	30	9	12	6

4 £112.80 **5** (a) £7560 (b) £5040 **6** £29.25
7 (a) £9.17 (b) £4.90 (c) £346.50 (d) £27.79
8 (a) £306 (b) £142.80
9 (a) £65 (b) £234 (c) £522.60
10 £1700 £2300 £2900 **11** £14.19 £7.92
12 £89.60 + £129.60 + £45 + £50.60 = £314.80

Exercise 14G

1 (a) 5% (b) 12% (c) 15% (d) $7\frac{1}{2}$%
   (e) $41\frac{2}{3}$% (f) $8\frac{1}{3}$% (g) 3% (h) 1% (i) 50%
2 (a) $7\frac{1}{2}$% (b) 20% (c) 20% (d) 4% (e) 4%
   (f) 4% (g) $6\frac{1}{4}$% (h) $\frac{1}{2}$% (i) 21.25% (j) $2\frac{1}{2}$%
3 30% **4** 6% **5** 55% **6** 16% **7** $3\frac{1}{8}$%
8 6.4% **9** 8% **10** 42.86%

Exercise 14H

1 £132 £143.40 **2** 15% **3** (a) £3 (b) 5%
4 Nigel £1275
5 (a) £3.36 (b) £86.79 (c) £5.23 (d) £1378.35
6 (a) 21%
   (b) Edwards 2970 votes, Philips 1848 votes, Fortescue 396 votes
7 (a) £6720 (b) £110 430
8 (a) £90 (b) £131

Unit 15 Algebra 3

Exercise 15A

1 $a = 3$ **2** $b = 3$ **3** $c = 7$ **4** $w = 2$ **5** $m = 1$
6 $y = 0$ **7** $x = 5$ **8** $k = 5$ **9** $n = 6$ **10** $h = 8$
11 $g = 4$ **12** $f = 9$ **13** $d = 5$ **14** $e = 3$ **15** $y = 3$
16 $x = 4$ **17** $m = 4$ **18** $d = 3$ **19** $k = 9$ **20** $y = 10$
21 $t = 10$ **22** $z = 4$ **23** $z = 0$ **24** $n = 5$

Exercise 15B

1 $a = 1$ **2** $y = 2$ **3** $h = 7$ **4** $p = 9$ **5** $q = 10$
6 $d = 8$ **7** $x = 0$ **8** $t = 4$ **9** $r = 3$ **10** $k = 1$
11 $n = 1$ **12** $x = 5$ **13** $m = 5$ **14** $y = 16$ **15** $w = 0$
16 $q = 12$ **17** $p = 2$ **18** $t = 0$ **19** $a = 12$ **20** $x = 0$
21 $p = 38$ **22** $a = 4$ **23** $b = 1$ **24** $y = 0$

Exercise 15C

1 $a = 5$ **2** $p = 11$ **3** $q = 8$ **4** $x = 8$ **5** $y = 13$
6 $s = 3$ **7** $x = 22$ **8** $y = 21$ **9** $s = 27$ **10** $a = 1$
11 $p = 11$ **12** $c = 4$ **13** $a = 1$ **14** $p = 5$ **15** $q = 0$
16 $a = 2$ **17** $b = 2$ **18** $c = 15$ **19** $p = 5$ **20** $y = 21$
21 $t = 5$ **22** $p = 0$ **23** $p = 24$ **24** $p = 0$

Exercise 15D

1 $a = 2$ **2** $p = 2$ **3** $p = 3$ **4** $s = 3$ **5** $k = 5$
6 $u = 4$ **7** $g = 7$ **8** $k = 7$ **9** $j = 2$ **10** $f = 4$
11 $r = 9$ **12** $v = 9$ **13** $t = 21$ **14** $d = 12$ **15** $t = 9$

Exercise 15E

1 $a = 10$ **2** $b = 20$ **3** $s = 12$ **4** $c = 30$ **5** $t = 24$
6 $s = 72$ **7** $h = 72$ **8** $f = 28$ **9** $d = 45$ **10** $a = 45$
11 $b = 40$ **12** $r = 52$ **13** $a = 60$ **14** $b = 32$ **15** $k = 48$

Exercise 15F

1 $a = 1$ **2** $b = 3$ **3** $c = 5$ **4** $p = 9$ **5** $q = 4$
6 $d = 8$ **7** $p = 3$ **8** $r = 2$ **9** $t = 4$ **10** $a = 12$
11 $b = 60$ **12** $s = 20$ **13** $r = 3$ **14** $e = 1$ **15** $p = 0$

Exercise 15G

1 $a = 2$ **2** $a = 3$ **3** $a = 2$ **4** $a = 3$ **5** $p = 0$
6 $p = 2$ **7** $q = 3$ **8** $r = 2$ **9** $t = 5$ **10** $f = 3$
11 $r = 13$ **12** $a = 1$ **13** $a = 0$ **14** $d = 3$ **15** $c = 4$
16 $a = 3$ **17** $z = 5$ **18** $r = 18$ **19** $s = -60$ **20** $b = -30$
21 $c = 24$ **22** $f = 27$ **23** $h = 4$ **24** $x = -5$

Exercise 15H

1 $a = 1\frac{1}{2}$ **2** $a = 3\frac{1}{3}$ **3** $a = 2\frac{2}{3}$ **4** $a = 4\frac{1}{3}$ **5** $p = 1\frac{3}{5}$
6 $p = 4\frac{2}{5}$ **7** $e = 0$ **8** $t = 1\frac{1}{2}$ **9** $j = 1\frac{1}{3}$ **10** $c = 1\frac{4}{7}$
11 $k = \frac{1}{4}$ **12** $d = 3\frac{1}{3}$ **13** $u = \frac{2}{9}$ **14** $q = 2\frac{1}{4}$
15 $y = 1\frac{2}{7}$

Exercise 15I

1 $a = -1$ **2** $a = -2$ **3** $a = -4$ **4** $a = -1$ **5** $a = -2$
6 $p = -2$ **7** $s = -5$ **8** $p = -2$ **9** $k = -1$ **10** $h = -1$
11 $y = -5$ **12** $e = -9$ **13** $t = 0$ **14** $w = -1$
15 $c = -2$ **16** $a = 0$

Exercise 15J

1 $s = 3$ **2** $d = 3$ **3** $m = 5$ **4** $h = 4$ **5** $k = 9$
6 $y = 2$ **7** $p = 1\frac{2}{5}$ **8** $f = 3\frac{1}{4}$ **9** $s = 3\frac{2}{3}$ **10** $g = -2\frac{2}{7}$
11 $f = 4\frac{1}{4}$ **12** $k = 3\frac{3}{5}$ **13** $s = -5\frac{2}{3}$ **14** $j = 3\frac{2}{3}$
15 $b = -\frac{5}{9}$ **16** $r = 3\frac{1}{2}$ **17** $t = -5\frac{2}{5}$ **18** $y = -\frac{6}{7}$
19 $e = -\frac{1}{3}$ **20** $f = -1\frac{1}{4}$ **21** $g = -\frac{2}{5}$ **22** $h = -1$
23 $c = -1\frac{2}{3}$ **24** $s = -\frac{5}{8}$ **25** $z = 4$ **26** $x = 25$
27 $p = 4$ **28** $c = -18$ **29** $a = 48$ **30** $e = -24$

Exercise 15K

1 $p = 1$ **2** $d = 5$ **3** $c = 3$ **4** $b = 2\frac{1}{3}$ **5** $g = 3$
6 $g = 5$ **7** $v = -2$ **8** $s = 1$ **9** $d = -\frac{1}{3}$ **10** $t = 5\frac{2}{3}$
11 $h = 3$ **12** $h = 4$ **13** $s = 3$ **14** $y = 3\frac{1}{2}$ **15** $r = 5$

Exercise 15L

1 $a = 1$ **2** $h = 8$ **3** $g = -3$ **4** $f = 6$ **5** $q = 0$
6 $k = 4$ **7** $g = 1$ **8** $h = -2$ **9** $d = -1$ **10** $v = -5\frac{1}{2}$
11 $s = -5\frac{2}{3}$ **12** $n = 3\frac{1}{2}$ **13** $f = -\frac{7}{8}$ **14** $d = 2\frac{3}{7}$
15 $m = -2\frac{1}{5}$

Exercise 15M

1 $k = 5$ **2** $s = 3\frac{1}{2}$ **3** $p = 4$ **4** $g = -1$ **5** $t = 3\frac{2}{3}$
6 $k = 9$ **7** $d = 7$ **8** $c = 3$ **9** $z = -1$ **10** $b = -4\frac{1}{2}$
11 $p = -1\frac{2}{3}$ **12** $g = 8$

Exercise 15N

1 $h = 12$ **2** $t = -8$ **3** $d = -1$ **4** $f = -3$ **5** $s = 7$
6 $d = 6$ **7** $a = 4\frac{1}{2}$ **8** $q = 4$ **9** $y = -\frac{2}{3}$ **10** $e = -1$
11 $s = -1\frac{2}{3}$ **12** $u = -5\frac{1}{2}$ **13** $t = -1\frac{1}{3}$ **14** $s = -1\frac{3}{4}$
15 $q = 2\frac{3}{4}$ **16** $w = -1\frac{1}{5}$ **17** $h = -1\frac{3}{5}$ **18** $s = -7$
19 $r = -1$ **20** $a = 3\frac{2}{3}$

Exercise 15O

1 $p = \frac{2}{3}$ **2** $h = 66$ **3** $r = 3\frac{8}{9}$ **4** $t = 21$ **5** $g = -3$
6 $d = -7\frac{1}{2}$ **7** $k = 13$ **8** $m = 5\frac{1}{2}$ **9** $d = -12\frac{1}{3}$
10 $j = 23$ **11** $y = -1\frac{4}{5}$ **12** $t = -7$

Exercise 15P

1 $t = 1\frac{1}{2}$ **2** $g = 2$ **3** $s = 4$ **4** $q = 6$ **5** $d = 3\frac{1}{3}$
6 $k = 2\frac{1}{2}$ **7** $a = \frac{1}{2}$ **8** $k = 3\frac{1}{2}$ **9** $d = -2\frac{1}{4}$
10 $p = \frac{5}{7}$ **11** $p = -11$ **12** $r = -4\frac{1}{3}$ **13** $t = 7\frac{1}{3}$
14 $g = 6$ **15** $a = -2\frac{3}{7}$

Unit 16 Pie charts

Exercise 16A

1 **IIC** English: 180°; Maths: 60°; Science: 45°; History: 45°; Art: 30°
  IIB Maths: 72°; English: 108°; Science: 60°; History: 50°; Art: 40°; French 30°
  IIA Maths: 72°; English: 120°; French: 18°; Art: 36°; History: 54°; Science: 60°
2 Gwyneth: 45°; Peter: 120°; Wes: 60°; Mario: 30°; Nesta: 105°
3 **(a)** Maths: 108°; French: 36°; Other: 54°; Science: 72°; English: 90°
  (b) Hamster: 50°; Dog: 100°; Cat: 60°; Bird: 40°; Other: 30°; No pet: 80°
  (c) Car: 144°; Van: 36°; Motorcycle: 48°; Lorry: 60°; Bus: 72°
4 Swimming: 120°; Tennis: 48°; Football: 96°; Hockey: 72°; Athletics: 24°

Exercise 16B

1 **(a)** 12 **(b)** 30°
  (c) Fish 150°, Sausages 60°, Salad 90°, Cheese roll 60°
2 Rice 18°, Apple 54°, Ice cream 144°, Trifle 108°, Fruit 36°
3 **(a)** **(ii)**
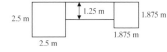

  (iii) Tea 108°, Coffee 99°, Milk 63°, Water 36°, Chocolate 54°
4 **(a)** Chips: 108°; Boiled potatoes: 90°; Carrots: 72°; Peas: 60°; Baked beans: 30° **(b)** 60 portions
5 Rent 90°, Travel 27°, Clothes 42°, Food 120°, Savings 60°, Spare 21°
6 **(a)** Rent £32, Travel £10.22, Clothes £20, Food £49.78, Savings £36, Spare £12 **(b)** None
7 **(a)** All their choices **(b)** chocolate **(c)** orange **(d)** 3
8 **(a)** Bus fares 48°, Going out 100°, Clothes 120°, Records 60°, Other 32°

Unit 17 Ratio and proportion

Exercise 17A

1 **(a)** $\frac{7}{9}$, 7 : 2 **(b)** $\frac{1}{2}$, 16 : 16 **(c)** $\frac{2}{3}$, 8 : 4 **(d)** $\frac{4}{9}$, 12 : 15
2 **(a)** 80 g, 200 g **(b)** 120 g, 300 g **(c)** 200 g, 500 g
  (d) 20 g, 50 g **(e)** 100 g, 250 g
3 **(a)** 400 g, 800 g, 800 g, 8 eggs **(b)** 100 g, 200 g, 200 g, 2 eggs
  (c) 300 g, 600 g, 600 g, 6 eggs
  (d) 500 g, 1000 g, 1000 g, 10 eggs
4 **(a)** 6 **(b)** 14
5 **(a)** 10 kg, 5 kg, 45 kg **(b)** 30 kg, 15 kg, 135 kg
  (c) 2 kg, 1 kg, 9 kg **6** 680 g
7 **(a)** 12 kg, 38 kg **(b)** 60 kg, 190 kg **(c)** 1.2 kg, 3.8 kg

Exercise 17B

1 **(a)** 1 : 2 **(b)** 1 : 3 **(c)** 1 : 6 **(d)** 1 : 4 **(e)** 1 : 3
  (f) 2 : 3 **(g)** 5 : 1 **(h)** 7 : 1 **(i)** 2 : 3 **(j)** 5 : 3
2 **(a)** 1 : 5 **(b)** 1 : 5 **(c)** 1 : 8 **(d)** 5 : 1
3 **(a)** 7 : 9 **(b)** 10 : 11 **(c)** 22 : 25 **(d)** 1 : 2
  (e) 9 : 7 **(f)** 11 : 3 : 2
4 £200, £160 **5** £160, £200, £120 **6** 10 : 25 : 5
7 **(a)** 40 min **(b)** 1 hour 20 min
8 4 lb   3 lb   2 lb
  16 km   12 km   8 km
  28 miles   21 miles   14 miles
  36 t   27 t   18 t
  £64   £48   £32
9 5 : 3
10 7000 kg

Exercise 17C

1 **(a)** 20 **(b)** 28 **(c)** 2.5 **(d)** 28 **(e)** 40 **(f)** 9
2 **(a)** 18 ft **(b)** 8 ft **(c)** 5 ft **3** **(a)** 135 **(b)** 84
4 455 **5** **(a)** 1400 **(b)** 6 hours **6** 4.8 m
7 **(a)** 280 miles **(b)** 77.14 litres
8 **(a)** £28.80 **(b)** 15 hours

Exercise 17D

1 £120 **2** £4.50 **3** £10.80 **4** £64 **5** £3.24
6 21
7 **(a)** 1 hr 45 mins **(b)** $5\frac{1}{2}$ hours **(c)** 8 hours
8 **(a)** 64 km **(b)** 10 hours
9 **(a)** 2.4 hours **(b)** 2
10 **(a)** 12 days **(b)** 28.8 days **(c)** 48

Exercise 17E

1 **(a)** 1.25 km **(b)** 1.8 km **(c)** 2.6 km **(d)** 3.1 km
2 **(a)** 12 cm **(b)** 10.4 cm **(c)** 16.8 cm **(d)** 51.2 cm
3 **(a)** 25 cm **(b)** 276 m **(c)** 2.19 km
4
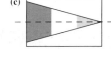

5 **(a)** 1 : 180 **(b)** 23.33 cm

Unit 18 Symmetry

Exercise 18A

1 **(a)** A, D **(b)** B, C **(c)** F **(d)** E **(e)** G, H
2
(a)   **(b)**   **(c)**
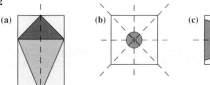
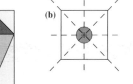

3 (a)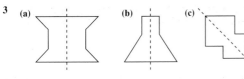

(b)

(c)

4 (a)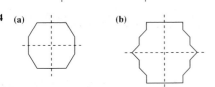

(b)

8 (a)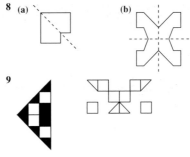

(b)

9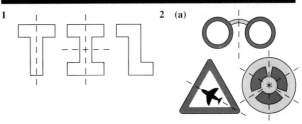

5 (a) A, B, C, D, E, K, M, T, U, V, W, Y
(b) H, I, O, X
(c) F, G, J, L, N, P, Q, R, S, Z

Rotational symmetry: (a) 1 (b) 6 (c) 2 (d) 2 (e) 2
10 (a) Infinite (b) 5 (c) 5
11 (a) 3 planes (b) 5 planes

Exercise 18B

1 (a) A, C (b) D, E (c) B, F

2 (a)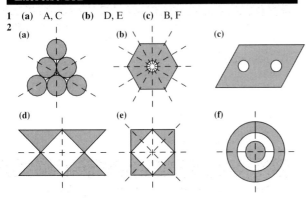

(b)

(c)

(d)

(e)

(f)

Exercise 18G

1

2 (a)

(b) Use a mirror or tracing paper

3 (a) 5 (b) 1 **4** (a), (b), (d)
5 Plane that bisects right angle or plane parallel to and equidistant from end faces.

Exercise 18C

1 (a) No (b) Yes (c) Yes (d) Yes (e) No
2 (a) 3 (b) 2 (c) 4 (d) 5 (e) 5

Unit 19 Measure 3

Exercise 19A

1 (a) 10 cm (b) 20 cm (c) 9 cm (d) 16 cm
(e) 16 cm
2 (a) 20 cm (b) 12 cm (c) 18 cm (d) 16 cm

Exercise 18D

1 (a) (i) 3 (ii) 3 (b) (i) 5 (ii) 5

Exercise 19B

1 20 cm **2** 22 cm **3** 21 m **4** 22 cm **5** 25 cm
6 22 m

Exercise 18E

1 (a) 2 (b) 3 (c) 4 (d) 1
2 (a) 2 planes (b) 5 planes (c) 6 planes (d) 3 planes
3 (a) isosceles triangular prism (b) cone

Exercise 19C

1 16.9 cm **2** 13.4 cm **3** 14.6 cm **4** 14.4 cm **5** 20.7 cm

Exercise 18F

1 1 **2** **3** pentagon **4**

5 8
6 equilateral triangle
7 (a) 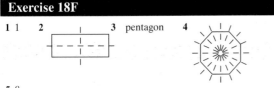 (b) (c) (d) (e)

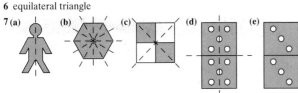

Exercise 19D

1 (a) 6.28 cm (b) 9.42 cm (c) 12.6 cm (d) 31.4 cm
(e) 25.1 cm (f) 37.7 cm
2 (a) 62.8 cm (b) 15.7 m (c) 157 cm (d) 7.85 cm
(e) 11.3 cm (f) 25.9 cm

Exercise 19E

1 12.6 cm **2** 18.8 cm **3** 31.4 cm **4** 62.8 cm
5 50.3 cm **6** 75.4 cm **7** 126 cm **8** 31.4 m
9 314 cm **10** 15.7 cm **11** 22.6 cm **12** 51.8 cm

Exercise 19F

1	56.5 cm	**2**	28.3 cm	**3**	283 cm	**4**	377 cm	
5	94.2 cm	**6**	7.85 cm	**7**	4.71 cm	**8**	18.8 inches	
9	1.88 cm	**10**	3.14 cm					

Exercise 19G

1 3 cm, 1.5 cm   **2** 4 cm, 2 cm   **3** 1 m, 0.5 m
4 3.18 m, 1.59 m   **5** 28.7 cm, 14.3 cm   **6** 2.39 cm, 1.19 cm
7 1.59 m, 0.796 m   **8** 1.91 in, 0.955 in   **9** 0.318 cm, 0.159 cm
10 0.796 cm, 0.398 cm

Exercise 19H

1 8 cm^2 **2** 7 cm^2 **3** 12 cm^2 **4** 16 cm^2 **5** 7 cm^2
6 6 cm^2

Exercise 19I

1 $11\frac{1}{2} \text{ cm}^2$ **2** $7\frac{1}{2} \text{ cm}^2$ **3** 10 cm^2 **4** $7\frac{1}{2} \text{ cm}^2$ **5** 15 cm^2
6 $17\frac{1}{2} \text{ cm}^2$

Exercise 19J

1 16 cm^3 **2** 3 cm^3 **3** 24 cm^3 **4** 72 cm^3 **5** 44 cm^3

Exercise 19K

1 (a) 8 cm^2 (b) 15 cm^2 (c) 24 cm^2 (d) 30 cm^2
2 (a) 16 cm^2 (b) 64 cm^2 (c) 92.16 cm^2 (d) 146.41 cm^2
3 (a) 4 cm^2 (b) 7.5 cm^2 (c) 23.4 cm^2 (d) 42.5 cm^2
4 (a) 24 cm^2 (b) 40.8 cm^2 (c) 9.84 cm^2

Exercise 19L

1 (a) 20 cm^2 (b) 48 cm^2 (c) 47.7 cm^2 (d) 23.4 cm^2
2 (a) 7.2 cm^2 (b) 16 cm^2 (c) 22.4 cm^2

3

Shape	Length	Width	Area
(a)	5 cm	6 cm	**30 cm²**
(b)	4 cm	**5 cm**	20 cm²
(c)	2 cm	**15 cm**	30 cm²
(d)	**8 cm**	5 cm	40 cm²
(e)	**5 cm**	12 cm	60 cm²

4

Shape	Base	Vertical height	Area
(a)	10 cm	5 cm	**25 cm²**
(b)	**10 cm**	12 cm	60 cm²
(c)	**16 cm**	5 cm	40 cm²
(d)	4 cm	**16 cm**	32 cm²
(e)	16 cm	**8 cm**	64 cm²

5 96 cm^2
6 (a) 90 cm^2 (b) 102 cm^2 (c) 300 cm^2 (d) 60 cm^2

Exercise 19M

1 (a) 12.6 cm^2 (b) 28.3 cm^2 (c) 50.2 cm^2
    (d) 314 cm^2 (e) 201 cm^2 (f) 452 cm^2
2 (a) 1256 cm^2 (b) 78.5 m^2 (c) 7850 cm^2
    (d) 19.6 cm^2 (e) 40.69 cm^2 (f) 213.7 cm^2
3 (a) 3.14 cm^2 (b) 7.07 cm^2 (c) 19.63 cm^2
    (d) 78.5 cm^2 (e) 50.2 cm^2 (f) 113 cm^2
4 (a) 314 cm^2 (b) 19.6 m^2 (c) 1962.5 cm^2
    (d) 4.91 cm^2 (e) 10.2 cm^2 (f) 53.4 cm^2

Exercise 19N

1 254 cm^2 **2** 63.6 cm^2 **3** 6360 cm^2 **4** $11\,300 \text{ cm}^2$
5 707 cm^2 **6** 4.91 cm^2 **7** 1.77 cm^2 **8** 28.3 inches^2
9 0.283 cm^2 **10** 0.785 cm^2

Exercise 19O

1 6 cm   **2** 2 cm   **3** 2 m   **4** 3.57 m   **5** 5.35 cm
6 3.09 cm   **7** 1.26 cm   **8** 2.76 inches   **9** 1.13 cm
10 0.892 cm

Exercise 19P

1 24 cm^3 **2** 36 cm^3 **3** 108 cm^3 **4** 64 cm^3 **5** 125 cm^3
6 512 cm^3 **7** 160 cm^3 **8** 250 cm^3 **9** 288 cm^3
10 192 m^3 **11** 150 m^3

Exercise 19Q

1 £95.40   **2** £125
3 (a) 10 tiles   (b) 20 tiles   (c) 800 tiles   (d) £520
4 7 rolls   **5** 2 tins   **6** $3750 l$ **7** $11.7 l$ **8** 2667 tiles
9 400 turves   **10** (a) 28.8 m^3 (b) 8 skips

Exercise 19R

1 (a) 20 000   (b) 130 000   (c) 24 000   (d) 152 000
    (e) 12   (f) 2.3   (g) 16.43   (h) 4200   (i) 300
    (j) 0.3   (k) 0.01
2 (a) 7 000 000   (b) 15 000 000   (c) 3 500 000
    (d) 4 780 000   (e) 4   (f) 3,78   (g) 14.789
    (h) 800 000   (i) 2000   (j) 24   (k) 0.0378
    (l) 0.142   (m) 0.003
3 1.2 m^3
4 (a) 1.6 m^2 (b) $16\,000 \text{ cm}^2$
5 $2\,300\,000 \text{ cm}^3$
6 (a) $324\,000 \text{ cm}^3$ (b) 0.324 m^3

Exercise 19S

1 12 miles   **2** 25 mph   **3** 5 hours   **4** 3.75 mph
5 7 mph   **6** 8 mph   **7** $1\frac{1}{3}$ mph   **8** $66\frac{2}{3}$ mph
9 20 mph   **10** 15 km/h

Exercise 19T

1 26.4 cm   **2** (a) (i) 5 m   (ii) 2 m   (b) 24 m   (c) 22 m^2
3 (a) 24 m^3 (b) 68 m^2
4 (a) 500 cm^3 (b) (i) 400 cm^3 (ii) 80%   **5** 80 m
6 (a) 283 m   (b) 6360 m^2
7 (a) 30 cm   (b) 27 cm^2 **8** 18.9 cm

Unit 20 Averages

Exercise 20A

1 (a) 28   (b) 34, 41   (c) 27, 35   (d) 28, 31
    (e) 28, 30   (f) 28
2 (a) 6   (b) 5, 5   (c) 4, 4, 4   (d) 11   (e) 7, 7, 7
    (f) 5, 5, 5, 6, 6
3 (a) Maths 8, 17, 8, 5, 20   English 14, 6, 15, 14, 9
    (b) Maths 8   English 14   **4** (a) 1

Exercise 20B

1 (a) 6   (b) 11   (c) 5   (d) $9\frac{1}{2}$ (e) 6   (f) 12
3 (a) any $\geqslant 5$ (b) any $\leqslant 4$ (c) any $\geqslant 7$ (d) 5
4 (a) $7\frac{1}{2}$ **5** (a) 26   (b) 32

Exercise 20C

1 (a) 13   (b) 24.25   (c) 22.5   (d) 23.75   (e) 19.5
    (f) 25.25
2 (a) £25   (b) £36   **3** (a) 162 cm   (b) 161.5 cm
4 (a) 560 g   (b) 1310 g
5 (a) 13   (b) 16   (c) 6   (d) 25

Exercise 20D

1 (a) History 28, Maths 21, English 14, Art 58, Science 33
   (b) (i) Maths   (ii) English   (iii) Art

2

Make	Standard	De Luxe	Price range
Ford 1.1	£7145	£7545	£400
Peugeot 1.5	£7460	£9190	**£1730**
Rover 200	£9995	£13895	**£3900**
Citroen 1.9	£10220	**£14205**	£3985
Volvo 1.6	**£11570**	£15480	£3910
Audi 1.9	£17693	**£19514**	£1821
Porsche 3.6	£58995	£93950	**£34955**
Lada 1.3	**£4995**	£6495	£1500
Toyota 1.3	£8040	**£12767**	£4727

3 (a) London 22, Paris 22, Moscow 29, New York 13, Luxor 13
   (b) Moscow, London and Paris (equal),
        Luxor and New York (equal)

Exercise 20E

1

	Mode	Median	Mean
(a)	6	10	10
(b)	24, 32	28	27.4
(c)	£3.60	£3.60	£4.74

2 (a) £1715
   (b) (i) £140   (ii) £155   (iii) £171.50   (iv) £300

Exercise 20F

1 (a) 15   (b) 100   (c) 3.66
2 10
3 (a) 10.86 seconds   (b) 10.5 seconds
4 (i) (a) 8   (b) None   (c) 105   (d) 3.1   (e) 4, 8
   (ii) (a) 5   (b) 6   (c) 106   (d) 3.1   (e) 5
5 (a) £3.29   (b) 3 h 8 min   (c) 4.32 m
6 5
7 (a) 14   (b) any number except 1, 3 or 5   (c) 2
   (d) 4   (e) 3.7   (f) 5
8 £13.83
9 (a) 22   (b) 20   (c) 22
10 (a) (i) 24.5   (ii) 31   (iii) 28
    (b) Mean influenced by low figures 4 and 11
11 (a) 10, 14, 18   (b) 14   (c) 14
12 Mean 12, Range 4. Brand B is more consistent.
13 (a)

	Mode	Median	Mean	Range
11A	20	20	18.3	11
11B	17, 18	17	15.9	13
11C	16	16	14.7	12
10A	20, 22, 24	21	19.6	17
10B	18, 19	18	17.3	10
10C	17	13.5	13.2	11
9A	16, 20, 21	19.5	18.5	9
9B	18	17	16.1	6
9C	14, 18, 20	18	18.2	9
9D	10, 11, 16	11.5	12.3	8

    (b) (i) 16, 22 and 17, 18, 21   (ii) 19.5, 19
         (iii) 19.9, 19.4
    (c) (i) 18   (ii) 16   (iii) 14.8
14 (a) See table above for question **13** (a)   (b) 20
15 (a) (i) 3.35   (ii) 4   (iii) 3
    (b) Mode – gives highest number of certificates.

Exercise 20G

1 (a)

Last two digits of phone no.	Tally	Frequency
0–9	||	2
10–19	||||	4
20–29	||||| |||||	10
30–39	|||||		6
40–49	|||||		6
50–59	|||||	5
60–69	|||	3
70–79			1
80–89			1
90–99	||	2

   (b)

Stem	Leaf
0	8, 8
10	4, 5, 7, 7
20	2, 3, 4, 4, 5, 6, 6, 6, 7, 9
30	0, 6, 6, 6, 6, 7
40	3, 3, 4, 5, 5, 9
50	2, 3, 5, 5, 7
60	7, 8, 9
70	2
80	1
90	0, 8

   (c) (i) 30–39
        (ii) 40–49
   (d) 40.7

2 (a)

Stem	Leaf
0	4, 6, 7, 8, 9, 9
10	1, 2, 4, 5, 7, 7, 8
20	0, 2, 3, 4, 5, 6, 7, 9
30	2, 3, 4, 5, 5, 6, 7, 8, 8, 9
40	1, 3, 4, 7, 8, 9
50	0, 2, 7, 8

   (b) 29 seconds

3 (a)

Stem	Leaf
0	5, 6, 8, 8, 9
10	0, 2, 4, 4, 5, 7, 8
20	0, 3, 5, 5, 7, 9
30	2, 4, 5, 6, 7, 8

   (b) 18 minutes
   (c) 20.4 minutes

4 (a)

Marks	Tally	Frequency
0–9			1
10–19	||||	4
20–29	||||| |||	8
30–39	||||| ||||| ||||	14
40–49	||||| ||||| ||	12
50–59	||||| |||||		11

   (b)

Stem	Leaf
0	9
10	4, 5, 8, 9
20	0, 4, 5, 6, 7, 7, 8, 9
30	2, 4, 4, 4, 4, 5, 5, 5, 5, 5, 6, 7, 7, 8
40	0, 2, 3, 3, 3, 4, 4, 5, 5, 6, 6, 7
50	0, 2, 2, 3, 3, 4, 4, 5, 5, 6, 8, 9

   (c) (i) 30–39   (ii) 30–39
   (d) 38.0 marks   (e) 37 marks   (f) 46%

5 (a)

Stem	Leaf	Frequency
50	2, 4, 5, 7, 8, 8	6
60	0, 0, 3, 4, 5, 5, 5, 6, 6, 8, 8, 8, 9	13
70	0, 0, 0, 1, 2, 4, 4, 4, 5, 6, 7, 7	12
80	0, 0, 1, 1, 1, 2, 2, 2, 3, 4, 5, 6, 6	13
90	2, 2, 3, 5	4

   (b) 73 kg   (c) 17

Unit 21 Algebra 4

Exercise 21A

1 £130   **2** £137.50   **3** 180 cm   **4** 150 cm   **5** £4.25

Exercise 21B

1 £160   **2** £65   **3** £172.50   **4** £14.40   **5** £12
6 £3.36   **7** £3.75   **8** £3.80   **9** £20   **10** £41.40
11 12 years   **12** 5 coins   **13** 25 bars   **14** 12 people
15 8 sweets   **16** £12.60

Exercise 21C

1 8   **2** 7   **3** 10   **4** 6   **5** 15   **6** 0   **7** 15
8 0   **9** 16   **10** 16   **11** 19   **12** 32   **13** 2
14 5   **15** 1   **16** 8   **17** 11   **18** 0   **19** 9   **20** 30

Exercise 21D

1 6   **2** 16   **3** 15   **4** 0   **5** 6   **6** 7   **7** 3
8 9   **9** 20   **10** 6   **11** 7   **12** 0   **13** 8
14 12   **15** 0   **16** 24   **17** 4   **18** 0   **19** 8   **20** 6

Exercise 21E

1 (a) 15   (b) 21   (c) 12   (d) 37.2
2 (a) 24   (b) 15   (c) 56   (d) 25.5
3 (a) 160   (b) 125   (c) 210   (d) 79
4 (a) 13   (b) 35   (c) 60   (d) 7
5 (a) 53   (b) 210   (c) 7   (d) 17.5

Exercise 21F

1 −3   **2** 2   **3** −6   **4** −4   **5** 2   **6** −8   **7** 5
8 11   **9** 0   **10** −9   **11** 8   **12** −2

Exercise 21G

1 1   **2** −2   **3** −7   **4** −5   **5** 5   **6** −8   **7** 3
8 −5   **9** −5   **10** 8   **11** −3   **12** 5   **13** −4
14 6   **15** 1   **16** −3   **17** 0   **18** 10   **19** −5   **20** 10

Exercise 21H

1 −12   **2** −12   **3** −10   **4** 4   **5** −21   **6** −32
7 30   **8** 27

Exercise 21I

1 −8   **2** −3   **3** −6   **4** −6   **5** −15   **6** 0   **7** 15
8 0   **9** 4   **10** −24   **11** −11   **12** −20   **13** −2
14 −5   **15** −1   **16** 8   **17** −19   **18** 0   **19** 21
20 30
Exercise 21J
1 (a) −7   (b) 25   (c) −60   (d) −7
2 (a) 0   (b) 12   (c) −15   (d) 3
3 (a) −11   (b) −11   (c) −1   (d) 5   (e) 6   (f) −29

Exercise 21K

1 (a) 5   (b) 20   (c) 125   (d) 500   (e) 61.25
2

x	−3	−2	−1	0	1	2	3
$y = 3x^2 + 4$	31	16	7	4	7	16	31

3 (a) 11.0   (b) 11.2   (c) 5   (d) 6.7   (e) 0   (f) 14.1
4 (a) 12   (b) 8   (c) 508   (d) 48   (e) 8
    (f) 2   (g) 200   (h) −4000   (i) 104   (j) −400

Exercise 21L

1 (a) 4 cm   (b) 8 cm   (c) 4 cm   (d) 20 cm
2 9 cm   **3** 2   **4** 7   **5** 22   **6** 10   **7** 4 chocolates
8 4   **9** 24

Exercise 21M

1 (a) 4 < 6   (b) 5 > 2   (c) 12 > 8   (d) 6 = 6
    (e) 15 > 8   (f) 3 < 24   (g) 10 > 3   (h) 0 < 0.1
    (i) 6 > 0.7   (j) 4.5 = 4.5   (k) 0.2 < 0.5   (l) 4.8 > 4.79
2 (a) True   (b) False, 2 < 6   (c) False, 6 = 6
    (d) False, 6 < 8   (e) False, 6 > 4   (f) False, 8 < 14
    (g) False, 7 > 6.99   (h) False, 6 < 6.01   (i) False, 7 > 0
    (j) False, 4 = 4   (k) False, 6 > 4   (l) True
3 (a) 5   (b) 4, 5, 6, 7   (c) 1, 2, 3   (d) 4, 5
    (e) 2, 3   (f) 3, 4, 5   (g) 2, 3   (h) 4, 5, 6
    (i) 1, 2, 3, 4   (j) 3, 4, 5, 6, 7   (k) 6, 7, 8, 9   (l) 7

Exercise 21N

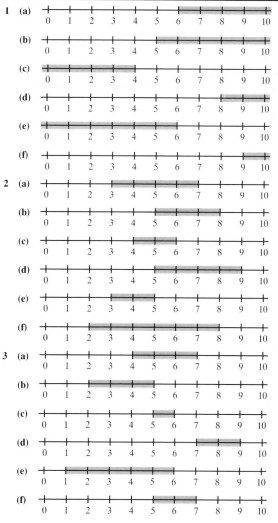

4 (a) 2 < x < 6   (b) 7 < x < 8

Unit 22 Transformations

Exercise 22A

4 **(a)** 2 sq right, 2 sq down
   (b) 1 sq left, 4 sq up
   (c) 2 sq left, 3 sq up

Exercise 22B

1 **(a)**

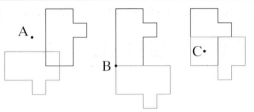

   (b)

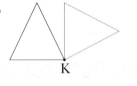

   (c)

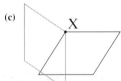

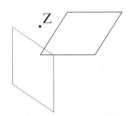

3 **(a)** $\frac{1}{2}$ turn clockwise or anticlockwise
   (b) $\frac{1}{4}$ turn anticlockwise
   (c) $\frac{1}{4}$ turn clockwise
   (d) $\frac{1}{2}$ turn clockwise or anticlockwise

Exercise 22C

2

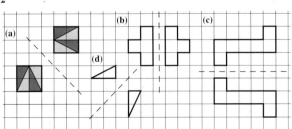

Exercise 22D

Check answers with your teacher.

Exercise 22E

3 **(b)** 2   **4** **(b)** rotation $\frac{1}{4}$ turn clockwise

Unit 23 Probability

Exercise 23B

1 **(a)** $\frac{1}{6}$ **(b)** $\frac{2}{6}$ or $\frac{1}{3}$ **(c)** $\frac{1}{2}$ **(d)** $\frac{1}{2}$ **(e)** $\frac{1}{3}$
   (f) $\frac{2}{3}$ **(g)** $\frac{2}{3}$ **(h)** $\frac{1}{2}$ **(i)** 0   **(j)** 0
2 **(a)** $\frac{1}{13}$ **(b)** $\frac{1}{52}$ **(c)** $\frac{1}{2}$ **(d)** $\frac{2}{13}$
   (e) $\frac{2}{13}$, or $\frac{3}{13}$ if Ace is low   **(f)** $\frac{3}{13}$ **(g)** $\frac{3}{13}$ **(h)** $\frac{1}{4}$
   (i) $\frac{1}{2}$ **(j)** $\frac{3}{4}$
3 **(a)** $\frac{5}{11}$ **(b)** $\frac{6}{11}$ **(c)** 0   **(d)** 1

Exercise 23C

1 **(a)** 0   **(b)** 1   **(c)** $\frac{1}{2}$ **(d)** 0   **(e)** 1   **(f)** $\frac{1}{2}$
3

	(i) Dice A	(ii) Dice B
(a)	$\frac{1}{6}$	$\frac{1}{6}$
(b)	$\frac{1}{6}$	$\frac{2}{6}$
(c)	$\frac{3}{6}$	$\frac{4}{6}$
(d)	$\frac{1}{6}$	0
(e)	$\frac{3}{6}$	$\frac{2}{6}$
(f)	$\frac{3}{6}$	$\frac{4}{6}$
(g)	$\frac{3}{6}$	$\frac{3}{6}$
(h)	$\frac{2}{6}$	$\frac{2}{6}$

4 **(a)** **(i)** $\frac{3}{10}$ **(ii)** 0.3   **(iii)** 30%
   (b) **(i)** $\frac{4}{10}$ **(ii)** 0.4   **(iii)** 40%
   (c) **(i)** $\frac{2}{10}$ **(ii)** 0.2   **(iii)** 20%
   (d) **(i)** $\frac{1}{10}$ **(ii)** 0.1   **(iii)** 10%
5 **(a)** $\frac{25}{100}$, 0.25, 25%   **(b)** $\frac{8}{100}$, 0.08, 8%   **(c)** $\frac{5}{100}$, 0.05, 5%
   (d) $\frac{88}{100}$, 0.88, 88%
6 $\frac{5}{12}$

Exercise 23D

1 **(a)** $\frac{2}{3}$ **(b)** 0.8   **(c)** 70%
2 **(a)** $\frac{1}{6}$ **(b)** $\frac{5}{6}$ **(c)** $\frac{2}{3}$ **(d)** $\frac{1}{3}$ **(e)** $\frac{1}{2}$
3 **(a)** 0.14   **(b)** **(i)** 0.3844   **(ii)** 0.0576

Exercise 23F

1 **(a)** **(i)** $\frac{4}{36}$ **(ii)** $\frac{4}{36}$ **(iii)** $\frac{1}{36}$ **(b)** **(i)** $\frac{6}{36}$ **(ii)** $\frac{6}{36}$
   (c) $\frac{12}{36}$ **(d)** $\frac{6}{36}$ **(e)** $\frac{1}{36}$
2 **(a)** $\frac{1}{25}$ **(b)** $\frac{4}{25}$ **(c)** $\frac{5}{25}$ **(d)** $\frac{9}{25}$ **(e)** $\frac{16}{25}$ **(f)** $\frac{9}{25}$
   (g) 1   **(h)** 0
3 **(a)** HH   HT   TH   TT
   (b) EE   EO   OE   OO
   (c)

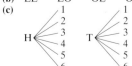

4 (a) 36 (b) 8

5

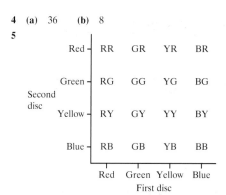

Red	RR	GR	YR	BR
Green	RG	GG	YG	BG
Yellow	RY	GY	YY	BY
Blue	RB	GB	YB	BB
	Red	Green	Yellow	Blue

Second disc (vertical), First disc (horizontal)

6 (a)

Total score	Ordered pairs	Theoretical probability
2	(1,1)	$\frac{1}{36}$
3	(1,2) (2,1)	$\frac{2}{36}$
4	(1,3) (2,2) (3,1)	$\frac{3}{36}$
5	(1,4) (2,3) (3,2) (4,1)	$\frac{4}{36}$
6	(1,5) (2,4) (3,3) (4,2) (5,1)	$\frac{5}{36}$
7	(1,6) (2,5) (3,4) (4,3) (5,2) (6,1)	$\frac{6}{36}$
8	(2,6) (3,5) (4,4) (5,3) (6,2)	$\frac{5}{36}$
9	(3,6) (4,5) (5,4) (6,3)	$\frac{4}{36}$
10	(4,6) (5,5) (6,4)	$\frac{3}{36}$
11	(5,6) (6,5)	$\frac{2}{36}$
12	(6,6)	$\frac{1}{36}$

(b) 7 (c) 1, one of these outcomes must occur

7 RR – Long Melford; RL – Lavenham; LR – Bury St. Edmunds; LL – Stoke by Clare

Exercise 23G

1 (a) 60 (b) 30 (c) 45

(d)

	Mon	Tues	Total
Jackie	25	35	**60**
Sam	**30**	**10**	40
Total	55	**45**	100

(e) (i) $\frac{25}{100}$ (ii) $\frac{10}{100}$ (iii) $\frac{60}{100}$ (iv) $\frac{65}{100}$

2 (a)

	Air	Le Shuttle	Boat	Total
July	8	**2**	10	20
Aug	15	5	**30**	50
Sept	**7**	8	**15**	**30**
Total	30	**15**	**55**	100

(b) (i) $\frac{20}{100}$ (ii) $\frac{80}{100}$ (iii) $\frac{30}{100}$ (iv) $\frac{55}{100}$ (v) $\frac{5}{100}$
(vi) $\frac{30}{100}$ (vii) $\frac{70}{100}$ (viii) $\frac{90}{100}$

3 (a)

	France	Italy	Elsewhere	Total
July	**6**	10	4	20
Aug	18	**24**	8	**50**
Sept	**4**	16	**10**	30
Total	28	50	**22**	**100**

(b) (i) $\frac{50}{100}$ (ii) $\frac{28}{100}$ (iii) $\frac{20}{100}$ (iv) $\frac{50}{100}$ (v) $\frac{8}{100}$
(vi) $\frac{4}{100}$ (vii) $\frac{16}{100}$

4 (a)

	Hotel	Caravan	Camping	Other	Total
July	11	4	3	**2**	**20**
Aug	**22**	14	**8**	6	**50**
Sept	**16**	7	4	3	30
Total	49	**25**	15	11	100

(b) (i) $\frac{22}{100}$ (ii) $\frac{4}{100}$ (iii) $\frac{49}{100}$ (iv) $\frac{51}{100}$ (v) $\frac{25}{100}$
(vi) $\frac{11}{100}$ (vii) $\frac{8}{100}$

Unit 24 Calculators and computers

Exercise 24A

1 [1] [a^{b}/$_{c}$] [2] [a^{b}/$_{c}$] [3] [−] [4] [a^{b}/$_{c}$] [5] [=]

2 (a) $1\frac{40}{161}$ (b) $3\frac{5}{8}$ (c) $\frac{173}{240}$ (d) $43\frac{58}{77}$

3 (a) 30.9 (b) $428\frac{4}{7}$ (c) 8.74
(d) 3318.48 (e) 1530.9 (f) 112.94 (2 d.p.)

4 £132.30

5 (a) 3.14 (b) 5.58 (c) 1.56 (d) 96 (e) 252
(f) 1131 (g) 33.33 (h) 560.11 (i) 19.5

Exercise 24B

1 [7] [3] [→] [L] [EXE]
[4] [9] [→] [W] [EXE]
[L] [W] [→] [A] [EXE]
[2] [L] [+] [2] [W] [→] [P] [EXE]

$A = 3577\,\text{cm}^2$, $P = 244\,\text{cm}$

2 $A = 452.39\,\text{cm}^2$, $C = 75.398\,\text{cm}$

3 $A = 67.5\,\text{cm}^2$

4 $V = 144\,000\,\text{cm}^3$

Exercise 24C

1 [L] [×] [L] [×] [L] [→] [V]
or [L] [x^y] [3] [→] [V]

$V = 125\,\text{cm}^3,\ 216\,\text{cm}^3,\ 343\,\text{cm}^3$

2 [(] [F] [−] [3] [2] [)] [×] [5] [÷] [9] [→] [C]

$C = 30\,°\text{C},\ 10\,°\text{C},\ 20\,°\text{C}$

Exercise 24D

1 [?] [→] [L] [:] [L] [x^y] [3] [EXE]

$L = 4.64\,\text{cm}$

2 $F = 71.6\,°\text{F}$

Exercise 24E

1 [1] [EXE] [Ans] [+] [2] [EXE]

2 [5] [EXE] [Ans] [+] [5] [EXE]

3 [2] [EXE] [Ans] [×] [2] [EXE]

4 [3] [EXE] [Ans] [×] [3] [EXE]

5 [1] [0] [EXE] [Ans] [−] [1] [EXE]

6 [1] [6] [EXE] [Ans] [÷] [2] [EXE]

7 [2] [0] [0] [EXE] [Ans] [÷] [1] [0] [EXE]

8 [(−)] [5] [EXE] [Ans] [−] [2] [EXE]

Unit 25 Scatter diagrams

Exercise 25A

1 (a)

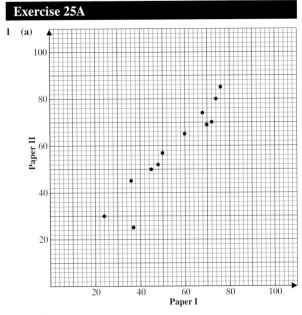

(b) Positive

2 (a) Positive   (b) Negative   (c) None
  (d) None (positive)   (e) Positive
  Answer to (d) would be affected by where the gardens are.

4 (a) (i)

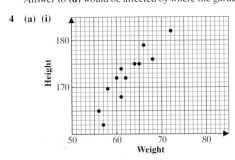

(ii)

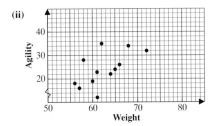

(iii)   (iv)

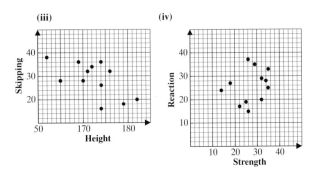

(v)   (vi)

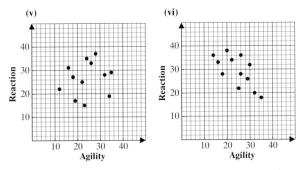

(b) (i) Positive   (ii) Positive (none)   (iii) None
    (iv) None   (v) None   (vi) None

5 (a) C, D, F   (b) A, E   (c) B
6 (a) Positive   (b) 54%
7 (a) Negative correlation. When it is raining you do not go out if you do not have to.
   (b) No correlation. There is no connection between eating apples and test results.

8 (a)

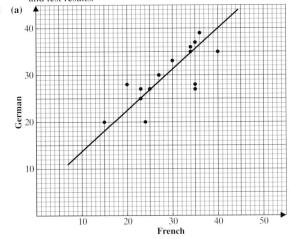

(b) Positive   (d) 30–32

Examination practice paper:
Non-calculator

1 (a) Three thousand, six hundred and four   (b) 3600
2 (a) 17   (b) 14
3 (a) (i) Isosceles triangle   (ii) Right-angled triangle
      (iii) Rectangle   (iv) Kite

(b)

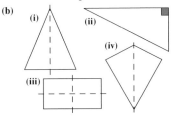

(c) (iii) has order 2, the others have order 1

4 (a)

No.	Tally	Freq.
1					3
2					3
3									7
4									7
5						4
6								6

(b)

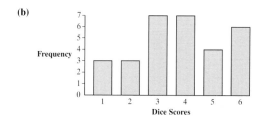

(c) **(i)** 3 and 4   **(ii)** 3.8   **(iii)** 5

5 5643, 4904, 3667, 3641
6 **(a)** 17   **(b)** 17
7 **(a)** 18, 23
   (b) Add 5 to the previous number
   (c) 98   **(d)** $5n - 2$
8 **(b)** 83°   **(c)** **(i)** 25 cm   **(d)** 27.3 cm²
9 **(a)** $\frac{5}{12}$ **(b)** $\frac{2}{3}$ **(c)** 0
10 37 cm²
11 H1, H2, H3, H4, H5, H6, T1, T2, T3, T4, T5, T6
12 **(a)** 3456   **(b)** 16 packets
13 **(a)** $a = 3$ **(b)** $s = 3\frac{3}{8}$ **(c)** $c = 10\frac{3}{4}$
14 **(a)** **(i)** 162   **(ii)** $1\frac{3}{10}$ or 1.3   **(b)** $\frac{1}{6}$
15 **(a)**

x	–3	–2	–1	0	1	2	3
y	–7	**–5**	–3	**–1**	1	**3**	5

x	–3	–2	–1	0	1	2	3
y	6	**1**	–2	–3	**–2**	1	6

(b)

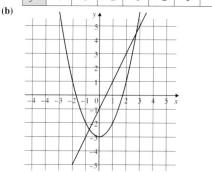

16 **(a)** 4 hours   **(b)** 44.4 mph   **(c)** 640 km
17

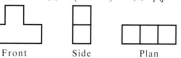

18 **(a)** Graph A, Good – Negative   Graph B, Good – Positive
      Graph C, No correlation
   (b) Graph A, Miles per gallon and size of engine
      Graph B, Miles per gallon and speed up to 50 mph
      Graph C, Miles per gallon and time of day
19 £50 : £200   **20** 6000 m or 6 km

Examination practice paper: Calculator

1 **(a)** 3.2 m   **(b)** 280 g   **(c)** 52 kph
2 **(a)** 15.96   **(b)** £3.94
       3.60
       1.50
      _____
      £21.06
3 **(a)** Manchester   **(b)** Exeter
4 **(a)** Trapezium   **(b)** Acute
5 **(a)** 36, 9   **(b)** 3, 6   **(c)** 8, 36, 48   **(d)** 3, 19
6 **(a)** 247.9   **(b)** 36.7
7 **(a)** $5a - 2b$ **(b)** $6a + 9b$ **(c)** $2x^2 + xy$
8 **(a)** 207°   **(b)** 027°   **9** 0.6
10 Walk: 252; Car: 140°; Bus: 88°; Bicycle: 12, 6°
11 105°. Angles in a triangle and on a straight line are 180°.
12 550 g   **13** **(a)** 29   **(b)** 25   **(c)** 27
14 £22.50, £23.50, £22.67. Ben's shoes are the cheapest.
15 **(a)** 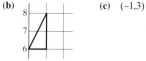   **(b)** 1 2 3 4 5 6   **(c)** 305
                                    5 8 11 14 17 20
16 **(a)** 272 km   **(b)** 182.6 lb or 183 lb
17 **(a)** e.g. Rotation, 90° anti-clockwise, centre (0,0)
   (b)   **(c)** (–1,3)
18 $10x - 8$ **19** **(a)** $2(2a + 5b)$ **(b)** $p(p + 2q)$
20

Front   Side   Plan

21 **(a)** (2, 3)   **(b)** (7, 3) and (5, 7) or (1, –3) and (–1, 1)
22 **(a)** 138.47 Lira   **(b)** £24.59
23 Student's own graph and line of best fit
   (c) £23 000–£25 000 read from their graph.
24 **(a)** **(i)** 40°   **(ii)** corresponding (F) angles   **(b)** **(i)** 80°
      (ii) isosceles triangle so third angle in triangle = 100°
         angles on a straight line = 180°
25 141.4 cm

Index